AF581638

Cell Death, Inflammation and Oxidative Stress in Neurodegenerative Diseases: Mechanisms and Cytoprotective Molecules

Cell Death, Inflammation and Oxidative Stress in Neurodegenerative Diseases: Mechanisms and Cytoprotective Molecules

Editor

Anne Vejux

MDPI • Basel • Beijing • Wuhan • Barcelona • Belgrade • Manchester • Tokyo • Cluj • Tianjin

Editor
Anne Vejux
University of Bourgogne
Franche-Comté
France

Editorial Office
MDPI
St. Alban-Anlage 66
4052 Basel, Switzerland

This is a reprint of articles from the Special Issue published online in the open access journal *International Journal of Molecular Sciences* (ISSN 1422-0067) (available at: https://www.mdpi.com/journal/ijms/special_issues/Cell_Death_ND_2).

For citation purposes, cite each article independently as indicated on the article page online and as indicated below:

LastName, A.A.; LastName, B.B.; LastName, C.C. Article Title. *Journal Name* **Year**, *Volume Number*, Page Range.

ISBN 978-3-0365-2910-3 (Hbk)
ISBN 978-3-0365-2911-0 (PDF)

Cover image courtesy of Anne Vejux

Contents

About the Editor

Anne Vejux, has obtained his Ph.D. in Biochemistry at the University of Burgundy (France) (cytotoxic oxysterols, apoptosis and lipid metabolism in cardiovascular diseases). After, she was appointed post-doctoral researcher at TIRO Laboratory then assistant teacher-researcher at the University of Nice/CEA (France) (effects of iodine on cell functions of different cell lines, expressing or not Natrium iodide symporter). Then, she was also assistant teacher-researcher at the Faculty of Sciences of the University of Franche-Comté and in the "Estrogen, Gene Expression and Central Nervous System Diseases" laboratory (France) (regulation of qsox1 gene expression, roles of QSOX1 protein in cell functions of cancer cells). She was researcher in clinical research at University Hospital of Besançon/Centre d'Investigation Clinique/Inserm (France) (development of medical devices in virology and oncology). In 2011, she becomes assistant professor in Biochemistry at the University of Burgundy and joined the Team 'Biochemistry of the Peroxisome, Inflammation and Lipid Metabolism' (study of the relationships between peroxisomal metabolism, inflammation, oxidation, and lipid status and the research of biomarkers of X-Adrenoleukodystrophy, multiple sclerosis and Parkinson's disease).

International Journal of
Molecular Sciences

Editorial

Cell Death, Inflammation and Oxidative Stress in Neurodegenerative Diseases: Mechanisms and Cytoprotective Molecules

Anne Vejux

Team Bio-PeroxIL, "Biochemistry of the Peroxisome, Inflammation and Lipid Metabolism" (EA7270), Université de Bourgogne Franche-Comté, INSERM, UFR Sciences Vie Terre et Environnement, 21000 Dijon, France; anne.vejux@u-bourgogne.fr

Citation: Vejux, A. Cell Death, Inflammation and Oxidative Stress in Neurodegenerative Diseases: Mechanisms and Cytoprotective Molecules. *Int. J. Mol. Sci.* **2021**, *22*, 13657. https://doi.org/10.3390/ijms222413657

Received: 10 December 2021
Accepted: 13 December 2021
Published: 20 December 2021

Neurodegenerative diseases are the most common chronic neurological pathologies associated with age, with a major impact on the patient's quality of life. These pathologies are a heavy medical, social and economic burden, yet there is no causal treatment available. Among the factors contributing to neurodegeneration are cell death caused by different mechanisms (apoptosis, autophagy, reticulum stress, necrosis, necroptosis), high oxidative stress resulting from a disturbed balance between prooxidant and endogenous antioxidant systems, and also inflammation. As treatments for the causes are not available, research has been focused on evaluating molecules of natural or synthetic origin to expand the therapeutic arsenal that could be made available.

The study of the different processes in pathologies that are not neurodegenerative but that are associated with neurodegeneration also advances our knowledge.

In the context of Alzheimer's disease, it has been shown that energy metabolism can have an impact on the development of the disease. Kubis-Kubiak et al. studied the impact of hyperglycemia or insulinemia on the secretion of the neuronal protein S100B in relation to oxidative stress, nitrosative stress and DNA damage in neurons [1]. This team was able to show that the S100B protein could play a key role in the local toxicity induced by high glucose or insulin concentrations in the early stages of the disease [1]. It would then be interesting to evaluate, as proposed by the authors, the protective mechanisms linked to this protein, in particular by looking at the relationships between S100B and the glucose transporters GLUT1 and GLUT3 in the brain or within the insulin receptor (IR). We might use the S100B protein as a diagnostic marker for the early stages of neuropathological disorders [1]. As the abovementioned authors have suggested, identifying the markers capable of helping in the diagnosis or in following the evolution of the disease is important. Laura Gomez-Virgilio and her collaborators, through a review of the literature, have evaluated the monitoring of oxidative stress by measuring NADH using the FLIM technique on olfactory neuron precursors, which had been isolated from patients in a non-invasive way [2]. This approach could allow researchers to find oxidative therapies in a more efficient way and then to personalize the follow-up of this disease.

Concerning Parkinson's disease, the second most common neurodegenerative disorder, some authors have focused on a particular type of death, necroptosis, a cell death that is independent of the caspases but which involves receptor-interacting protein 3 (RIP3) and the pseudokinase mixed lineage domain-like protein (MLKL) or RIP1 kinase. Oliveira et al. identified a compound, Oxa12, that acts as an inhibitor of necroptosis not only in a BV-2 cell model when treated with the pan-caspase inhibitor zVAD-fmk, but also in vivo in a sub-acute 1-methyl-1-4-phenyl-1,2,3,6-tetrahydropyridine hydrochloride (MPTP) PD-related mouse model [3]. Another team targeted microglia-mediated neuroinflammation. They tested Epigallocatechin-3-gallate (EGCG), a natural antioxidant in green tea, by loading it into liposomes (phosphatidylcholine (PC) or phosphatidylserine (PS) coated with or without vitamin E) in an in vitro lipopolysaccharide (LPS)-induced BV-2 microglial cell activation model and in an in vivo model by targeting inflammation in the substantia nigra of Sprague-Dawley rats treated with LPS. The team of Cheng and collaborators was able to show that EGCG could inhibit inflammation and promote neuroprotection, making it a candidate for anti-Parkinson's therapy [4].

In an effort to identify the molecules capable of reinforcing the available therapeutic arsenal, a team focused on multiple sclerosis and the evaluation of Apolipoprotein D protection in a cuprizone-induced cell model. The experiments that they carried out showed that the increase in Apolipoprotein D levels, whether exogenous or endogenous, moderately prevents the cytotoxic effects of cuprizone [5].

A study of repetitive mild traumatic brain injury (mTBI) showed that the TDP-43 proteinopathy identified in most cases of amyotrophic lateral sclerosis (ALS) did not impact the response in terms of chronic damage and inflammation in the optic tract [6].

More generally, different teams have highlighted the interrelationships between oxidative stress, cell death and inflammation in the context of neurodegenerative diseases [7] and have also compared these mechanisms with other pathologies, such as cancer or type 2 diabetes [8]. Others have targeted a specific mechanism, such as the team of Pluta et al., who chose to present the mechanisms involved in neuroinflammation in brain tissue after ischemia [9], or the team of Cadiele Oliana Reichert et al., who described the mechanisms of ferroptosis present in neurodegenerative diseases: lipid peroxidation, glutathione peroxidase 4 enzyme activity and iron metabolism [10].

We have seen that identifying a molecule as being able to inhibit the mechanisms responsible for neurodegeneration was not the only important step, as in the case of EGCG, but that the delivery method was also important. Karine Charrière and her collaborators from the Bio-PeroxIL Laboratory offered a summary of the known effects of docosahexaenoic acid (DHA), based on in vitro and in vivo targeting microglia and also based on clinical trials, while describing the nanomedicine techniques that have already been tested at the level of the microglia that could facilitate the action of DHA [11].

The editorial board would like to thank all the authors who participated in the success of this Special Issue by submitting quality articles. We hope that the articles published will help advance research on neurodegenerative diseases and that we will see the continuation of this work in the next special issue.

Funding: This research was funding by University of Burgundy.

Conflicts of Interest: The author declare no conflict of interest.

References

1. Kubis-Kubiak, A.; Wiatrak, B.; Piwowar, A. The Impact of High Glucose or Insulin Exposure on S100B Protein Levels, Oxidative and Nitrosative Stress and DNA Damage in Neuron-Like Cells. *Int. J. Mol. Sci.* **2021**, *22*, 5526. [CrossRef] [PubMed]
2. Gómez-Virgilio, L.; Luarte, A.; Ponce, D.P.; Bruna, B.A.; Behrens, M.I. Analyzing Olfactory Neuron Precursors Non-Invasively Isolated through NADH FLIM as a Potential Tool to Study Oxidative Stress in Alzheimer's Disease. *Int. J. Mol. Sci.* **2021**, *22*, 6311. [CrossRef] [PubMed]
3. Oliveira, S.R.; Dionísio, P.A.; Gaspar, M.M.; Ferreira, M.B.T.; Rodrigues, C.A.B.; Pereira, R.G.; Estevão, M.S.; Perry, M.J.; Moreira, R.; Afonso, C.A.M.; et al. Discovery of a Necroptosis Inhibitor Improving Dopaminergic Neuronal Loss after MPTP Exposure in Mice. *Int. J. Mol. Sci.* **2021**, *22*, 5289. [CrossRef] [PubMed]
4. Cheng, C.-Y.; Barro, L.; Tsai, S.-T.; Feng, T.-W.; Wu, X.-Y.; Chao, C.-W.; Yu, R.-S.; Chin, T.-Y.; Hsieh, M.F. Epigallocatechin-3-Gallate-Loaded Liposomes Favor Anti-Inflammation of Microglia Cells and Promote Neuroprotection. *Int. J. Mol. Sci.* **2021**, *22*, 3037. [CrossRef] [PubMed]
5. Martínez-Pinilla, E.; Rubio-Sardón, N.; Peláez, R.; García-Álvarez, E.; del Valle, E.; Tolivia, J.; Larrayoz, I.M.; Navarro, A. Neuroprotective Effect of Apolipoprotein D in Cuprizone-Induced Cell Line Models: A Potential Therapeutic Approach for Multiple Sclerosis and Demyelinating Diseases. *Int. J. Mol. Sci.* **2021**, *22*, 1260. [CrossRef] [PubMed]
6. Pilipović, K.; Rajič Bumber, J.; Dolenec, P.; Gržeta, N.; Janković, T.; Križ, J.; Župan, G. Long-Term Effects of Repetitive Mild Traumatic Injury on the Visual System in Wild-Type and TDP-43 Transgenic Mice. *Int. J. Mol. Sci.* **2021**, *22*, 6584. [CrossRef] [PubMed]
7. Behl, T.; Makkar, R.; Sehgal, A.; Singh, S.; Sharma, N.; Zengin, G.; Bungau, S.; Andronie-Cioara, F.L.; Munteanu, M.A.; Brisc, M.C.; et al. Current Trends in Neurodegeneration: Cross Talks between Oxidative Stress, Cell Death, and Inflammation. *Int. J. Mol. Sci.* **2021**, *22*, 7432. [CrossRef] [PubMed]

8. Asslih, S.; Damri, O.; Agam, G. Neuroinflammation as a Common Denominator of Complex Diseases (Cancer, Diabetes Type 2, and Neuropsychiatric Disorders). *Int. J. Mol. Sci.* **2021**, *22*, 6138. [CrossRef] [PubMed]
9. Pluta, R.; Januszewski, S.; Czuczwar, S.J. Neuroinflammation in Post-Ischemic Neurodegeneration of the Brain: Friend, Foe, or Both? *Int. J. Mol. Sci.* **2021**, *22*, 4405. [CrossRef] [PubMed]
10. Reichert, C.O.; de Freitas, F.A.; Sampaio-Silva, J.; Rokita-Rosa, L.; Barros, P.; de Lima Barros, P.; Levy, D.; Bydlowski, S.P. Ferroptosis Mechanisms Involved in Neurodegenerative Diseases. *Int. J. Mol. Sci.* **2020**, *21*, 8765. [CrossRef] [PubMed]
11. Charrière, K.; Ghzaiel, I.; Lizard, G.; Vejux, A. Involvement of Microglia in Neurodegenerative Diseases: Beneficial Effects of Docosahexahenoic Acid (DHA) Supplied by Food or Combined with Nanoparticles. *Int. J. Mol. Sci.* **2021**, *22*, 10639. [CrossRef] [PubMed]

International Journal of
Molecular Sciences

MDPI

Article

The Impact of High Glucose or Insulin Exposure on S100B Protein Levels, Oxidative and Nitrosative Stress and DNA Damage in Neuron-Like Cells

Adriana Kubis-Kubiak [1,*], Benita Wiatrak [2] and Agnieszka Piwowar [1]

1 Department of Toxicology, Faculty of Pharmacy, Wroclaw Medical University, Borowska 211, 50-556 Wroclaw, Poland; agnieszka.piwowar@umed.wroc.pl
2 Department of Pharmacology, Faculty of Medicine, Wroclaw Medical University, Mikulicza-Radeckiego 2, 50-345 Wroclaw, Poland; benita.wiatrak@umed.wroc.pl
* Correspondence: adriana.kubis-kubiak@umed.wroc.pl

Abstract: Alzheimer's disease (AD) is attracting considerable interest due to its increasing number of cases as a consequence of the aging of the global population. The mainstream concept of AD neuropathology based on pathological changes of amyloid β metabolism and the formation of neurofibrillary tangles is under criticism due to the failure of Aβ-targeting drug trials. Recent findings have shown that AD is a highly complex disease involving a broad range of clinical manifestations as well as cellular and biochemical disturbances. The past decade has seen a renewed importance of metabolic disturbances in disease-relevant early pathology with challenging areas in establishing the role of local micro-fluctuations in glucose concentrations and the impact of insulin on neuronal function. The role of the S100 protein family in this interplay remains unclear and is the aim of this research. Intracellularly the S100B protein has a protective effect on neurons against the toxic effects of glutamate and stimulates neurites outgrowth and neuronal survival. At high concentrations, it can induce apoptosis. The aim of our study was to extend current knowledge of the possible impact of hyper-glycemia and -insulinemia directly on neuronal S100B secretion and comparison to oxidative stress markers such as ROS, NO and DBSs levels. In this paper, we have shown that S100B secretion decreases in neurons cultured in a high-glucose or high-insulin medium, while levels in cell lysates are increased with statistical significance. Our findings demonstrate the strong toxic impact of energetic disturbances on neuronal metabolism and the potential neuroprotective role of S100B protein.

Keywords: Alzheimer's disease; hyperglycemia; hyperinsulinemia; S100B protein; oxidative stress

Citation: Kubis-Kubiak, A.; Wiatrak, B.; Piwowar, A. The Impact of High Glucose or Insulin Exposure on S100B Protein Levels, Oxidative and Nitrosative Stress and DNA Damage in Neuron-Like Cells. *Int. J. Mol. Sci.* **2021**, *22*, 5526. https://doi.org/10.3390/ijms22115526

Academic Editor: Anne Vejux

Received: 10 April 2021
Accepted: 21 May 2021
Published: 24 May 2021

Publisher's Note: MDPI stays neutral with regard to jurisdictional claims in published maps and institutional affiliations.

1. Introduction

Alzheimer's disease (AD), first described in 1907, is a neurodegenerative disorder characterized essentially by β-amyloid plaques and tangles of hyperphosphorylated tau proteins alongside cholinergic dysfunction [1–3]. The pathological heterogeneity characteristic of AD creates difficulties in establishing a single theory as to its cause, characterizations, and possible treatments [4]. Important risk factors include age-related biochemical and metabolic changes, vascular disease, traumatic brain injury, epigenetic factors, diet, mitochondrial malfunction, metal exposure, infections, hypertension, obesity, dyslipidemia and diabetes mellitus type 2 (T2DM) [5,6]. In diabetes mellitus, the feedback loops between insulin action and insulin secretion do not function properly. In T2DM, the pancreatic islet β-cell's dysfunction, together with insulin resistance in insulin-sensitive tissues, leads to increased glucose production in the liver and decreased glucose uptake in muscle, which results in an excessive amount of glucose circulating in the blood [7]. The growing occurrence of both diabetes and dementia leading to AD is becoming a social and economic challenge worldwide [8,9].

Current research studies point out that AD and T2DM might share some underlying mechanisms and putative biochemical pathways, i.e., the desensitization of insulin signaling, most likely driven by chronic inflammation, and mitochondrial dysfunction, as a consequence of increased oxidative stress or vasculopathy [10,11]. Moreover, T2DM is a known risk factor of AD since hyperinsulinemia and insulin resistance, hallmarks of this metabolic disturbance, can lead to memory impairment [12,13]. Almost 70% of T2DM cases demonstrate diverse forms of nervous system impairment such as diabetic neuropathy, slowed digestion of food in the stomach, carpal tunnel syndrome, erectile dysfunction and other peripheral nerve problems or even central nervous system complications, including strokes and possibly cognitive impairment [14,15]. Given that numerous data points to coincidence as well as co-morbidity between those two pathological states, in 2008 prof. De la Monte [16] proposed that AD might be called "type 3 diabetes mellitus," but this term is under criticism by others [17,18]. Nowadays, despite the vast number of papers published on AD and diabetes, the underlying link between these two disorders is still unclear [19,20]. Observations from clinical trials are highly ambiguous and have been futile in deciphering clear paths responsible for the augmented risk of dementia in T2DM patients [21–23]. The origin of brain neurodegeneration connected to continuous hyperglycemia is heterogeneous and comprised of modifications in neurotransmitters' metabolism, neuronal inflammation, mitochondrial disturbances and micro- and macrovascular dysfunction [24]. In T2DM, chronic hyperglycemia, among others, is associated with an increased risk of developing memory deficits or decreased psychomotor speed [25]. The molecular and cellular pathophysiology underlying this complication is not yet well understood. Therefore, it is important to conduct research aimed at explaining the complexity of this relationship.

S100B is an extracellular alarmin, a small helix-loop-helix protein that binds up to four Ca^{2+} per dimer in EF-hands motifs, counteracts amyloid-β accumulation, and is upregulated in AD [26]. Despite its most common localization in a subtype of mature astrocytes that ensheath blood vessels and in neural/glial antigen 2-expressing cells, S100B protein is also expressed in different cell types, ranging from arterial smooth muscle to melanoma cells [27]. Uniquely for the S100 protein family, S100B location is also observed in distinct subpopulations of neurons and in adipocytes, suggesting its potential new role in the regulation of energetic metabolism [28]. The levels of S100B mRNA expressed in the cerebral cortex and adipose tissue have been shown to be very similar [29]. S100B levels are raised in the adult organism as a consequence of nervous system damage, which makes it a potential clinical marker. Moreover, it was found that before any detectable changes in intracerebral pressure, neuroimaging, and neurological examination, S100B concentrations are elevated in serum or cerebrospinal fluids, enabling fast and crucial medical treatment before permanent damage occurs [30,31]. Depending on the concentration, this Janus-faced molecule can represent both beneficial and toxic effects on neuropathological changes in cellular metabolism characteristic for AD [32]. The extracellular action of the S100B protein is primarily based on its interaction with RAGE receptors, which are highly activated during T2DM or in hyperglycemic states. S100B binds to RAGE in the extracellular space and activates a number of intracellular biochemical pathways in microglia and neurons. RAGE expression is potentiated by increased extracellular concentrations of ligands, including the S100B protein. These receptors mediate both the trophic and the toxic effects of the S100B protein on cells [33,34]. At lower concentrations, the binding of S100B protein to RAGE causes activation of the Ras/ERK pathway and stimulation of MEK/MAP kinase regulated by extracellular signals, ERK/NF-κB/Bcl-2 and Ras/Cdc42-Rac1. At higher concentrations, S100B protein leads to hyperactivation of the Ras/MEK/ERK pathway, leading to overproduction of ROS [27]. Wartchow et al. [35] exposed the existence of insulin-S100B regulation of glucose utilization in the brain tissue. Moreover, it regulates glial fibrillary acidic protein and two glycolytic enzymes: fructose-1,6-bisphosphate aldolase and phosphoglucomutase in the brain, which is why it is so crucial for the potential role of S100B protein to be revealed [36,37].

The purpose of this article is to elucidate the potential role of S100B protein in correlation with oxidative and nitrosative stress in disturbed glucose/insulin homeostasis in neuron-like cells differentiated with nerve growth factor (NGF) pheochromocytoma (PC12) cells. This cellular model is widely used in both neurobiological and neurotoxicological studies to study the mechanisms of action of neurotoxicants as well as the potential for chemicals to alter neuronal differentiation [38,39]. To our knowledge, there is no information available regarding the role of S100B protein in the pathological mechanism underlying the "dance macabre" between T2DM and AD. We hypothesized that S100B levels are increased in neurons with simultaneous conditions of metabolic disturbance and neuropathological changes, and this fact could probably be used for the detection of the first cellular disruptions connected to dementia and T2DM.

2. Results

The MTT assay evaluated mitochondrial activity, which can be treated as a measurement of metabolic activity, and thus viability, of cells. Cell dysfunction and cytotoxicity after exposure to high concentrations of glucose or insulin were assessed by MTT assay and presented in Figure 1.

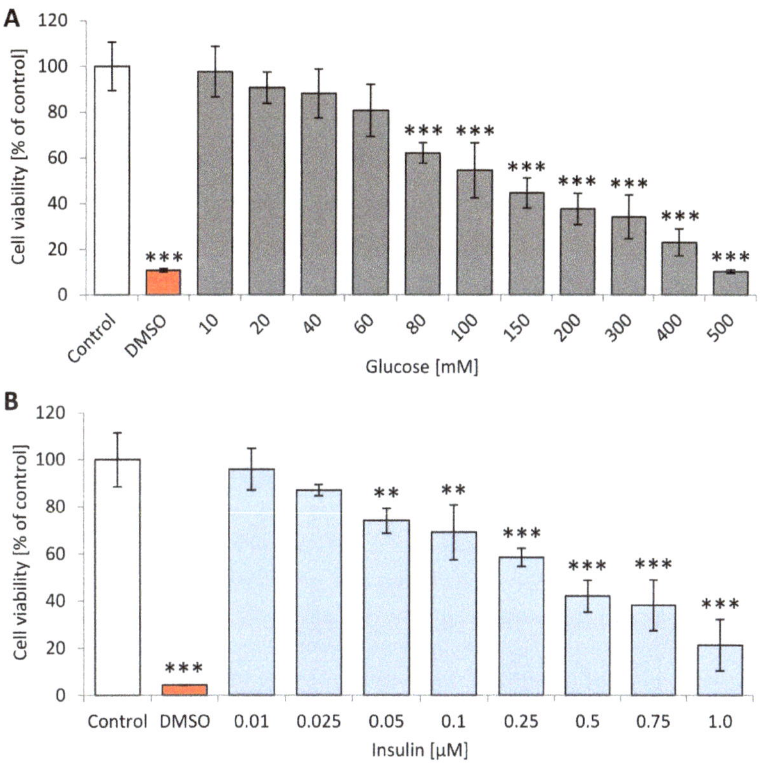

Figure 1. Metabolic activity measured in MTT assay of neuron-like cells after 24 h incubation with different concentrations of (**A**) glucose and (**B**) insulin. Control—untreated neuron-like cells; DMSO—PC12 cells treated with DMSO. Statistically significant differences compared to the untreated neuron-like cells: ** $p < 0.01$, *** $p < 0.001$.

Analysis of neuron-like cells' viability after 24 h incubation with glucose (5–500 mM) and insulin (10–750 µM) will aid the selection of optimal concentrations (40–60% viability) for further studies. The metabolic activity of neuron-like cells was decreased when they were cultured for 24 h with solutions of both glucose and insulin as calculated with reference to untreated neuron-like cells. DMSO treatment resulted in a substantial 95.6% decrease in mitochondrial activity, and a comparable reduction (89.8%) was obtained after incubation with 500 mM glucose. Administration of 10 mM, 20 mM glucose and 0.01 µM insulin did not affect the viability of neuron-like cells, as the results are similar (90.5%, 97.5%, 99.7% and 95.9%, respectively) to those obtained with untreated cells (100%). The results obtained after incubation with glucose or insulin were clearly dependent on concentration. The optimal concentrations—50–150 mM for glucose and 50–250 µM for insulin—were chosen for planned experiments based on the obtained MTT results for tested substances.

2.1. Nitrite Levels

The levels of cellular nitrogen free radicals were measured after 1 h and 24 h incubation with glucose or insulin. The obtained results are presented in Figure 2.

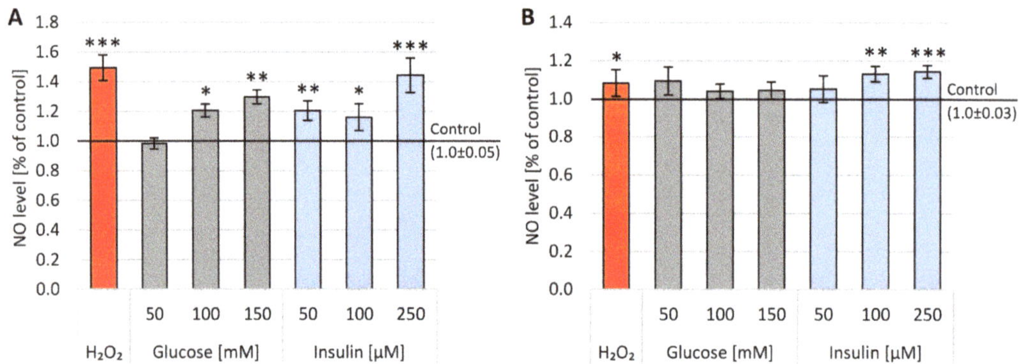

Figure 2. NO levels in neuron-like cells after: (**A**) 1 h incubation with glucose or insulin and (**B**) 24 h incubation with glucose (50, 100 or 250 mM) or insulin (50, 100 or 250 µM). Control—untreated neuron-like cells; H_2O_2—cells incubated with 50 µM H_2O_2 (positive control). Statistically significant differences compared to the untreated neuron-like cells: * $p < 0.05$, ** $p < 0.01$, *** $p < 0.001$.

After 1 h incubation, a statistically significant increase in the level of nitric oxide was observed both in the presence of glucose and insulin. The results obtained after 1 h incubation with glucose were clearly dependent on concentration, with a statistically substantial rise after 100 mM (x-fold—1.21) and 150 mM (x-fold—1.3) glucose. In the case of insulin, all three concentrations instigated a statistically significant upsurge in nitric oxide levels after 1 h treatment compared to untreated cells. It is interesting that the addition of 250 µM insulin for 1 h produced a similar amount (x-fold—1.44) of nitric oxide to H_2O_2 (x-fold—1.49).

Significantly, the impact of glucose and insulin on nitric oxide levels after 24 h incubation was less expressed; only after treatment with 100 µM and 250 µM insulin were the levels of nitric oxide considerably higher (x-fold—1.13 and 1.14, respectively) in comparison to untreated cells.

2.2. Reactive Oxygen Species Concentration

The DCF-DA test was used to predict extracellular ROS accumulation after 1 h or 24 h incubation with glucose (50–150 mM) and insulin (50–250 µM) (Figure 3).

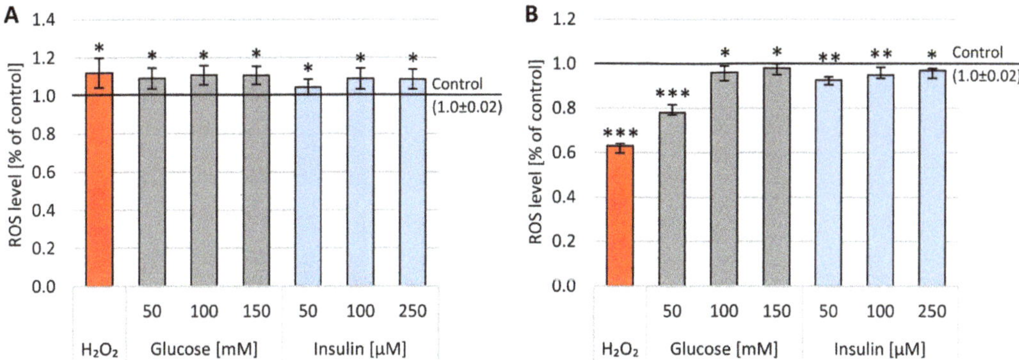

Figure 3. ROS levels in neuron-like cells after: (**A**) 1 h incubation with glucose or insulin and (**B**) 24 h incubation with glucose (50, 100 or 150 mM) or insulin (50, 100 or 150 µM). Control—untreated neuron-like cells; H_2O_2—cells incubated with 50 µM H_2O_2 (positive control). Statistically significant differences compared to the untreated neuron-like cells: * $p < 0.05$, ** $p < 0.01$, *** $p < 0.001$.

After 1 h incubation, a statistically significant increase in the level of ROS was observed both in the presence of glucose and insulin. A statistically significant increase in ROS levels was observed after 1 h incubation with glucose or insulin in all analyzed concentrations, with the highest concentration obtained for 150 mM glucose (x-fold—1.2). After insulin administration, the highest level of ROS was obtained after 100 µM (x-fold—0.8). No concentration dependence was observed 1 h or 24 h after glucose or insulin administration.

It is worth noting that the impact of glucose and insulin on ROS levels was less expressed after 24 h incubation, as there was an ample decrease in the level of free oxygen radicals. Furthermore, these data were supported by the results obtained with positive assay control (H_2O_2), confirming the observation that reactive oxygen species are the first messengers of stress in the cells, and their levels are diminished after 24 h of treatment.

2.3. Double-Stranded DNA Breaks

High levels of oxygen and nitrogen free radicals can lead to DNA damage, including DBSs. The number of double-stranded DNA breaks (DSBs) was assessed in the FHA by measuring the size of the nuclear halo (chromatin dispersion). The results obtained after 1 h incubation are presented on sample photos with the nuclear halo and analysis of relative NDF in Figure 4.

In neuron-like cells treated with glucose for 1 h, the increase in DBSs was generally lower than after incubation with insulin in all used concentrations. The dependence of disruption on concentration has been demonstrated for all concentrations and substances tested—the higher the concentration, the stronger the chromatin dispersion halo. The highest DBSs were observed after administration of 250 M insulin (x-fold—3.59), while the lowest DBSs levels were measured after treatment with 50 mM glucose (x-fold—2.0). The observed effect was statistically significant ($p < 0.001$) in all experimental settings.

Results obtained after 24 h incubation are presented as sample photos with the chromatin dispersion and analysis of relative NDF in Figure 5.

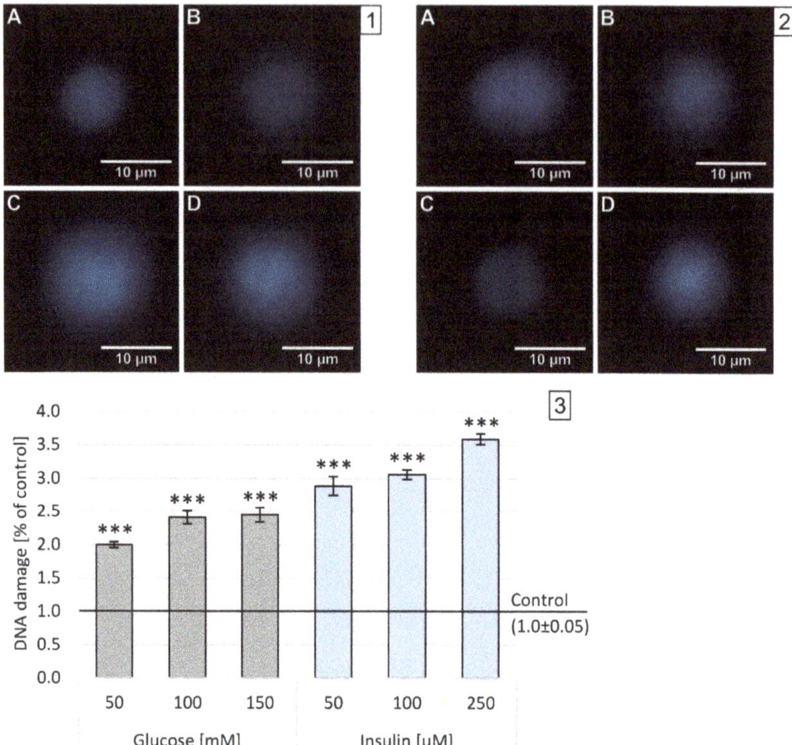

Figure 4. Analysis of double-stranded DNA breaks in neuron-like cells after 1 h incubation with glucose or insulin. (**1**) Sample microphotographs of a nuclear diffusion halo: A—control (untreated cells), B—glucose 50 mM, C—glucose 100 mM and D—glucose 150 mM. (**2**) Sample microphotographs of nuclear diffusion halo: A—control (untreated cells), B—insulin 50 µM, C—insulin 100 µM and D—insulin 150 µM. (**3**) A comparison of relative NDF for neuronal-like cells incubated for 1 h with glucose or insulin. Statistically significant differences compared to the untreated neuron-like cells: *** $p < 0.001$.

Similar to the results obtained in the DCF-DA, the incubation of neuron-like cells with glucose or insulin for 24 h resulted in less DNA strand damage than incubation for 1 h. After 24 h incubation with all three concentrations of insulin, the damaged DNA strand was observed to regenerate. There was a strong concentration dependence in both glucose and insulin. Glucose caused weak DNA damage with maximum levels after 24 h incubation with 150 mM glucose (x-fold—1.43). Insulin treatment had the opposite effect, leading to the regeneration of DNA double strands, with minimal values after 24 h incubation with 50 µM (x-fold—1.43).

2.4. S100B Protein Concentration

The study assessed the effects of glucose and insulin on the intracellular and extracellular S100B protein levels in neuron-like cells. The obtained concentrations of S100B protein after 24 h incubation with a broad range of glucose concentrations are shown in Figure 6. The extracellular concentration is presented per the amount of cells seeded in one well (1×10^4), while the intracellular concentration was adjusted per 5 µg of total protein concentration in 100 µL of cell lysates measured by BCA assay.

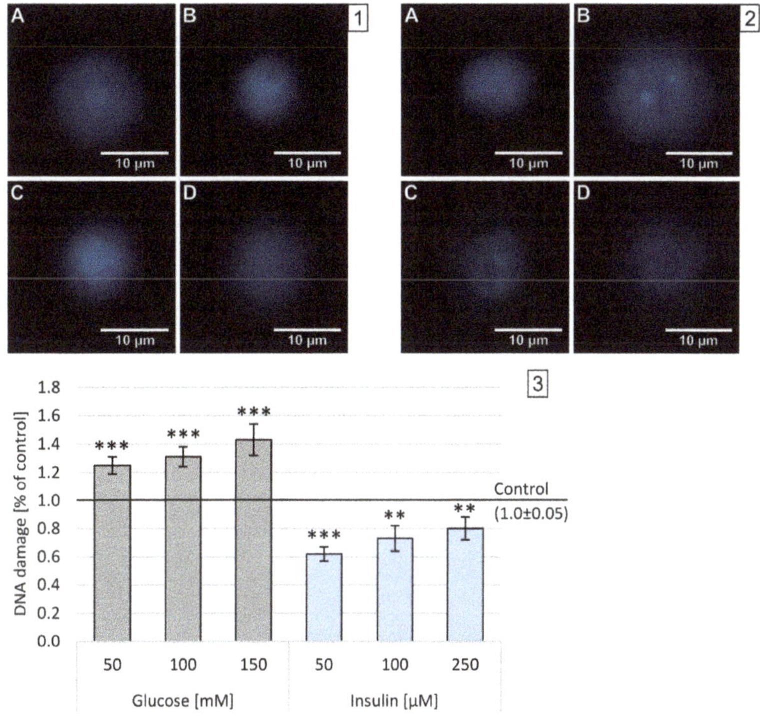

Figure 5. Analysis of double-stranded DNA breaks in neuron-like cells after 24 h incubation with glucose or insulin. (**1**) Sample microphotographs of a nuclear diffusion halo: A—control (untreated cells), B—glucose 50 mM, C—glucose 100 mM and D—glucose 150 mM. (**2**) Sample microphotographs of nuclear diffusion halo: A—control (untreated cells), B—insulin 50 μM, C—insulin 100 μM and D—insulin 150 μM. (**3**) A comparison of relative NDF for neuronal-like cells incubated for 24 h with glucose or insulin. Statistically significant differences compared to the untreated neuron-like cells: ** $p < 0.01$, *** $p < 0.001$.

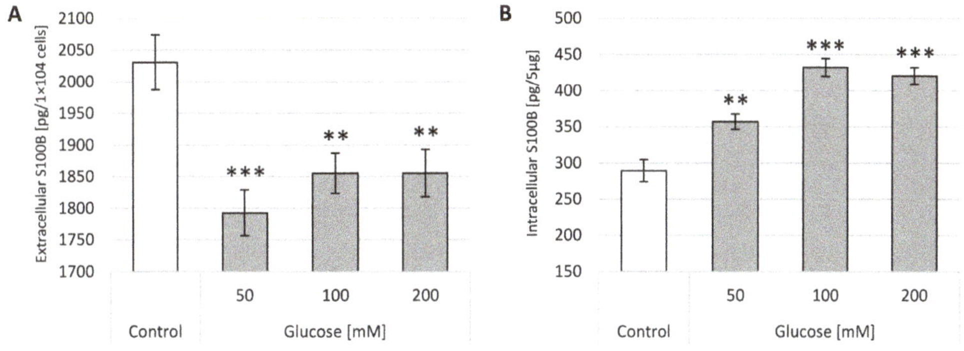

Figure 6. S100B protein concentration in neuron-like cells after incubation with different glucose concentrations (50, 100 and 200 mM). Control—untreated cells. (**A**) extracellular S100B levels per 1×10^4 cells; (**B**) intracellular S100B levels per 5 μg of total cellular protein concentration. Statistically significant differences compared to the untreated neuron-like cells: ** $p < 0.01$, *** $p < 0.001$.

At 50–200 mM glucose concentrations, significantly lower S100B protein levels (for around 10%) in the supernatant were observed compared to the control. In the case of extracellular S100B protein levels, there was no a dose-response relation observed. All obtained outcomes for extracellular S100B protein concentrations were statistically significant with $p < 0.01$ or $p < 0.001$. Cellular levels of S100B protein after incubation with glucose show an opposite trend compared to data obtained for supernatants, which stays in line with the results observed. As shown in Figure 6B, all treatments caused a rise in S100B intracellular concentrations, which reached their maximum after the administration of 100 mM glucose (432 pg/5 µg total protein). The increase after 50 mM glucose treatment is 23%, the next 100 mM of glucose caused a rise of 49% and, after 200 mM glucose, the upsurge of 45%. Figure 6B depicts the normal distribution of values of S100B concentrations after glucose administration confirmed by a bell-shaped dose-response dependence. Significant differences in intracellular S100B values were obtained in the case of 200 mM glucose treatment ($p < 0.05$).

Figure 7 presents the changes in S100B protein levels identified after 24 h incubation with a range of insulin concentrations. The extracellular concentration is presented per the amount of cells seeded in one well (1×10^4), while the intracellular concentration was adjusted per 5 µg of total protein concentration in 100 µL of cell lysates measured by BCA assay.

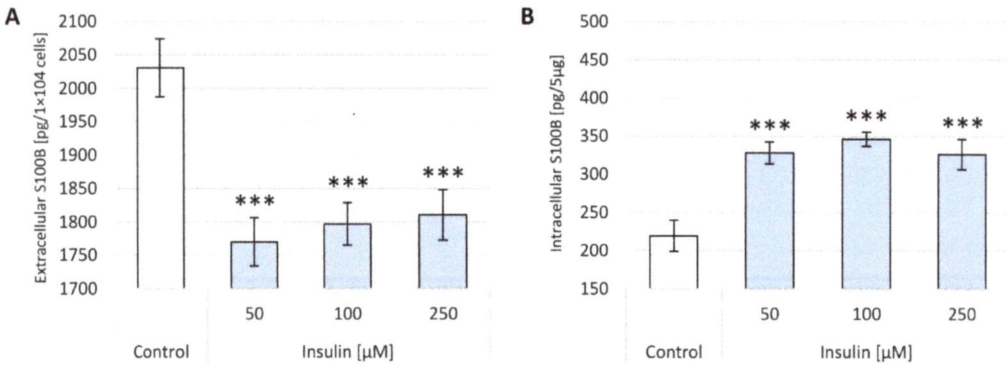

Figure 7. S100B protein concentration in neuron-like cells after incubation with different insulin concentrations (50, 100 and 250 µM). Control—untreated cells. (**A**) Extracellular S100B levels per 1×10^4 cells; (**B**) intracellular S100B levels per 5 µg of total cellular protein concentration. Statistically significant differences compared to the untreated neuron-like cells: *** $p < 0.001$.

The incubation with each selected insulin concentration (50–250 µM) significantly diminished the S100B protein level in the supernatants, and the strongest effect was observed after the administration of 50 µM (1769 pg/1×10^4 cells). There is a weak direct proportional correlation between increasing insulin concentration and S100B protein extracellular levels with maximum concentration after 250 µM insulin treatment (1810 pg/1×10^4 cells). All obtained data for extracellular S100B protein concentrations were statistically significant with $p < 0.001$. As illustrated in Figure 7B, the highest value of S100B protein was detected after administration of 100 µM insulin (341 pg/5 µg total protein), while after the administration of 250 µM insulin, the increase was the lowest compared to the control (326 pg/5 µg total protein and 219 pg/5 µg total protein, respectively). Significant differences ($p < 0.05$) in intracellular S100B values were obtained in cases of all insulin treatments. A bell-shaped graph can also be observed in the case of incubation with insulin in Figure 7B, and these values correlate favorably with data obtained with glucose. It is interesting to note that treatment incubation with insulin caused similar effects to glucose—a decrease in extracellular and increase in intracellular S100 levels.

2.5. Statistical Correlation between ROS, NO and DBSs Results

Correlation coefficients between DNA damage and ROS or NO levels were determined and shown in Table 1.

Table 1. Correlations between measured parameters ROS, NO and DBSs in neuron-like cells after glucose or insulin treatment for 1 and 24 h.

	1 h			24 h		
	ROS vs. DBSs	**NO vs. DBSs**	**ROS vs. NO**	**ROS vs. DBSs**	**NO vs. DBSs**	**ROS vs. NO**
Glucose	0.98	0.99	0.94	−0.70	0.44	−0.95
Insulin	0.93	0.66	0.34	0.96	−0.65	−0.42

Based on the calculated Pearson correlation coefficients in all tested concentrations, correlations between DCF-A, FHA and Griess results for glucose after 1 h incubation were strongly positive and statistically significant, whereas correlations between the results obtained after 24 h incubation with all tested glucose concentrations were negative. After insulin treatment, a strong correlation was observed only between ROS and DNA damage after 1 h incubation as well as after 24 h treatment. In the case of insulin, there was a positive correlation only between ROS level and DNA damage after 1 h and 24 h incubation.

2.6. Statistical Correlation between ROS, NO, DBSs and S100B Results

Correlations between extracellular and cellular S100B protein levels and DNA damage, ROS or NO levels, analyzed by using Pearson correlation coefficients, are shown in Table 2.

Table 2. Correlations between 100B protein levels and measured parameters: ROS, NO and DBSs in neuron-like cells after glucose or insulin treatment for 24 h.

	Intracellular S100B			Extracellular S100B		
	vs. DBSs	**vs. NO**	**vs. ROS**	**vs. DBSs**	**vs. NO**	**vs. ROS**
Glucose	−0.62	−0.62	0.33	−0.99	−0.99	0.92
Insulin	0.90	0.90	−0.77	0.98	0.98	−0.60

Based on the calculated Pearson correlation coefficients in almost all glucose concentration tested after 24 h incubation, correlations between the results of intracellular and extracellular levels of S100B protein were negative compared to outcomes from ROS, NO and DBSs levels. The obtained correlation was strong only in the case of extracellular S100B levels vs. ROS levels, whereas after insulin administration, a strong correlation was observed only between intracellular S100B protein levels and DBSs or NO as well as among extracellular S100B protein levels and DBSs or NO. To reiterate, the correlation between extracellular and intracellular S100B levels and DNA damage or NO levels after insulin administration was strong. Such a consequent effect was not observed in a hyperglycemic environment.

3. Discussion

Regardless of the fact that "type 3 diabetes mellitus" has received much attention in the past decade, it is still unclear whether an overall failure of brain glucose and insulin metabolism regulations is associated with AD pathogenesis. It is generally accepted that altered cerebral glucose uptake and insulin resistance are some of the hallmarks of the progression of neurodegenerative processes [40]. Investigators from Thiambisetty laboratory were one of the first to find, in 2017, that lower rates of glycolysis and higher brain glucose levels were correlated to more severe amyloid plaques and neurofibrillary tangles found in the brains of people with AD [41]. Current evidence from the AD-T2DM animal model, where scientists combined the APP/PS1 mouse model with the genetic db/db model of type 2 diabetes, highlights neuroinflammatory processes manifested by a significant increase of microglia burden in the cortex and general brain atrophy in animals of 14 weeks of age when T2DM was started, but no AD pathology was observed [42]. Moreover, recent studies have found that AD-T2DM mice are characterized by the upregulation of a broad profile of pro-inflammatory cytokines, such as IL-1α, IFN-γ and IL-3 in the brain [43].

In our study, we demonstrated that environments with high levels of glucose or insulin have a negative impact on neuronal metabolism by elevating levels of oxidative stress markers. Our present data support the view that hyperglycemia and/or insulinemia causes activation of ROS and NO levels in neurons, which also affects double-stranded DNA breaks. Furthermore, there was a significant positive correlation between those parameters after 1 h from glucose administration. The statistical analysis did not confirm any significant differences between oxidative stress parameters DBSs after insulin treatment with two exceptions for ROS vs. DBSs after 1 and 24 h of incubation. It is suggested that, over the years, these small changes in the neuronal tissue may lead to the deactivation or shift of crucial cell cycle pathways such as the MEK-ERK1/2-NF-κB pathway as well as to the upregulation of pro-apoptotic factors, which leads to amyloid β accumulation. Although caution must be exerted in extrapolating in vitro findings to the in vivo situation, our data are in line with the suggestion that the fast-shifting concentrations of glucose and/or insulin in neuronal tissue may be one of the first pathological changes connected to AD and begin several years prior to the onset of clinical symptoms. Some clinical studies have suggested a significant role for S100B in neurodegeneration processes, pointing to its increased levels in the body fluids of patients with AD [44–46]. Christl et al. [47] estimated that S100B protein levels in cerebrospinal fluid might have a diagnostic value, particularly at the early stages of the disease, as it declines to normal levels in more advanced stages. Animal studies provide additional data showing the importance of S100B in neurodegenerative processes. The performance of S100B-overexpressing Tg2576 mice was inferior in the spatial learning study, dependent on the functioning of the hippocampus. Animals showed higher levels of brain parenchymal β-amyloid depositions and cerebral amyloid angiopathy, enhanced amy-loidogenic APP metabolism, augmented reactive astrocytosis and microgliosis and increased levels of pro-inflammatory cytokines (TNF-α, IL-1β and IL-6) as early as at 7–9 months of age. Mice whose gene encoding S100B was inactivated showed increased spatial memory and memorization under the influence of anxiety and increased long-term synaptic enhancement in the CA1 sector of the hippocampus. These results indicate that despite playing an important role in encouraging brain inflammatory responses, S100B has a role in directly promoting amyloidogenic APP processing [48].

Regarding the role of S100B protein in T2DM, Kheirouri et al. [49] measured its concentration in the blood serum of patients with metabolic syndromes characterized by intermittent fasting, central obesity, dyslipidemia and arterial hypertension. Moreover, the participants of the study had elevated insulin levels, showed high values on the HOMA-IR index of insulin resistance, and the serum level of S100B protein was significantly elevated compared to healthy volunteers. The extracellular action of the S100B protein is primarily based on its interaction with RAGE—the best-known class of advanced glycation end receptors [50]. S100B binds to RAGE in the extracellular space and activates a number of intracellular biochemical pathways such as MAPKs in microglia and neurons. These receptors mediate both the trophic and the toxic effects of the S100B protein on microglia and endothelial cells. RAGE is also one of the known receptors for Aβ peptide [51,52]. In the physiological state, RAGE expression in cells is kept at a low level. Increasing the amount of RAGE ligands observed in inflammation, oxidative stress, diabetes and AD results in the induction of the expression of this receptor [53]. Literature data show a relationship between glucose concentration and S100B protein secretion. Cultures of astrocytes obtained from rat brains under conditions of metabolic stress: glucose, oxygen and serum (FBS) deprivation increased the secretion of S100B protein. However, after 12 and 24 h of exposure to metabolic stress, mRNA expression for the S100B protein decreased considerably, and a significant reduction in S100B protein secretion was observed after 48 h of incubation. Based on these observations, it is hypothesized that the S100B protein may be actively secreted in the microglia in the early stages of metabolic stress [54].

The aim of our work was to broaden current knowledge on the role of S100B protein in oxidative stress instigated by local variations in glucose or insulin concentrations in neurons. We analyzed extracellular and intracellular S100B protein levels in order to evaluate the potential role of S100 protein in metabolic disturbances occurring in neuronal cells. The single marked observation to emerge from the data comparison was that neither hyperglycemic nor insulinemic conditions provoked S100B protein efflux from the neurons, and moreover, they even decreased secretion. Interestingly, a small but still statistically significant rise in S100B concentrations was observed in cell lysates. No substantial statistical correlation between intracellular and extracellular S100B levels and analyzed markers of oxidative stress and DNA damage was found after glucose administration, excluding levels in supernatants vs. ROS, where significance was observed. In contradiction, hyperinsulinemic conditions provoked a strong correlation between S100B concentration vs. NO or double-stranded DNA break levels. Collectively, these data support the notion that cellular S100B protein could behave as a neuroprotective factor against extracellular glucose/insulin fluctuations. Our observations are consistent with the results carried out on the primary cortical astrocytes of rats by Nardin et al. [55]. After 24 h, the extracellular S100B levels were reduced by around 45% in astrocytes cultured in a high-glucose medium. A decrease in glutathione content but not glutamate uptake activity was also observed. It is important to emphasize the fact that S100 protein family members exert a dual effect (neurotrophic or neuroprotective) on neurons and astrocytes, depending on the concentration attained in the brain's extracellular space. Our findings are consistent with the previous results presented by Alhemeyer et al. [56] who showed that neurotoxicity caused by glutamate and staurosporine can be counteracted by the S100B protein. Another study performed on PC12 cells found that the S100B protein could inhibit NGF-induced cell differentiation, but increased expression of S100B did not reverse the effects of differentiation in PC12 already differentiated with NGF cells. This may suggest an effect of the S100B protein on the inhibition of cell differentiation in the early stages of cell development [57].

In conclusion, our data support the view that there is a putative relationship between S100B and metabolic disturbances in AD-like pathology and that the S100B protein acts probably as a cytoprotective factor and may protect neurons against the toxicity of local high levels of glucose or insulin during the initial phases of the pathophysiology of AD. This study is the first step towards enhancing our understanding of the role of S100 proteins in neuronal metabolism in the specific condition of "type 3 diabetes mellitus". Further research is necessary to clarify the mechanism of action of S100B protein as well as to explain the reasons for the significant differences in their effect on neuronal properties. As the brain's energetic regulation is entirely insulin-dependent, it would be especially interesting to evaluate the interrelation between S100 protein family members and the glucose transporters: GLUT1 and GLUT3 in the human brain as well as their relation to the receptor for insulin (IR), and these results may represent an excellent initial step toward using S100B levels as diagnostic markers in the early stages of neuropathological disorders.

4. Materials and Methods

4.1. Materials

Glucose and human insulin solutions (Sigma-Aldrich, St. Louis, MO, USA) were freshly prepared in RPMI-1640 (low glucose—5.5 mM, Biological Industries, Cromwell, CT, USA) supplemented with 2% fetal bovine serum (FBS) and 100 µg/mL penicillin-streptavidin (both from Sigma-Aldrich, St. Louis, MO, USA). Concentration ranges from 5–500 mM for glucose and 10–750 µM for insulin were chosen for the evaluation of the cytotoxicity effect. PC12 cells—pheochromocytoma cells derived from the adrenal gland of *Rattus Norvegicus*—were purchased from ATCC (Manassas, VA, USA). The differentiation was performed using human recombinant β nerve growth factor, NGF-β (Sigma-Aldrich,

St. Louis, MO, USA) dissolved in PBS to a stock concentration 1 μg/mL and stored at −20 °C for up to 2 weeks.

4.2. Modification of the Surface of Culture Plates

The surface of the well on plates was modified using type I collagen (Sigma-Aldrich, St. Louis, MO, USA) and dissolved in 0.1 M acetic acid to a stock concentration of 0.1% (*w/v*). Before using type I collagen, a solution was dissolved in water to a final concentration of 0.01% (*w/v*), which was added to cover the surface of the wells. The 96-well plates were left at 4 °C overnight. After removing the solution, the surface was washed 3 times for 5 min with PBS (300 μL). The plates so prepared were stored at 4 °C for up to 1 month. The plates in the biological assays were irradiated with UV for 30 min before use.

4.3. Cell Culture Conditions and Differentiation

The differentiation process of PC12 cells was performed according to the protocol established by Greene and Tischler with some modifications [58]. PC12 cells were grown in 25 or 75-cm^2 culture flasks in a CO_2-incubator (37 °C, 5% CO_2 and 95% humidity) in RPMI-1690 medium supplemented with 10% FBS and 100 μg/mL penicillin-streptomycin. The cells were used at logarithmic growth between passage 4 and 20. The PC12 cells were dissociated with TrypLE (Gibco, Thermo Fisher Scientific, Waltham, MA, USA), seeded on type I collagen-coated 96-well plates in a concentration of 5×10^3 cells per well and incubated for 24 h prior to differentiation in order to let the cells adhere to the plates' surface. For differentiation, cells were treated with 100 ng/mL of NGF-β freshly dissolved in RPMI 1640 media supplemented with 2% FBS and 100 μg/mL penicillin-streptavidin. For the bioassays, cells were treated with NGF-β for 5 days. Medium and NGF were replenished every 48 h. The differentiation process was analyzed with a holo-tomographic microscope (3D Cell Explorer, Ecublens, Switzerland). The morphological changes of NGF-treated PC12 cells versus untreated PC12 cells are shown in Figure 8. Further, the described experiments were performed after 5 days of human NGF-β-induced differentiation.

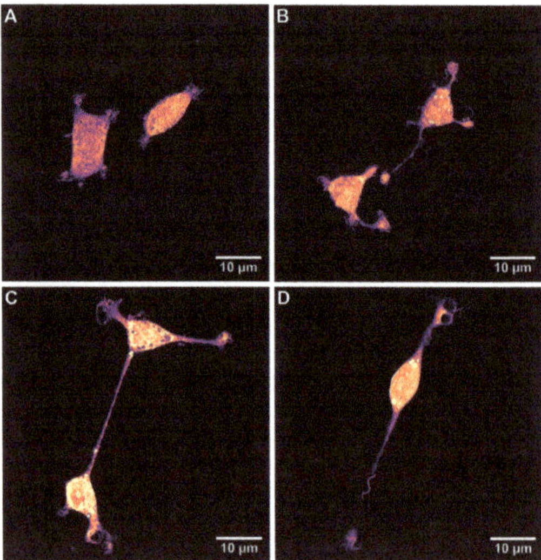

Figure 8. The representative microphotograph of a non-differentiated PC12 cells (**A**); the representative microphotographs of PC12 cells differentiated into neuron-like cells: after 2 days (**B**), after 4 days (**C**), and after 5 days (**D**) of incubation with human NGF-β. The morphological changes during the differentiation process were analyzed with the 3D Cell Explorer free-label microscope.

4.4. MTT Test

To determine the cytotoxicity of glucose and insulin treatment on neuron-like cells, MTT (3-(4,5-dimethylthiazol-2-yl)-2,5-diphenyltetrazolium bromide) reduction assay was performed according to the protocol of Liu et al. [59] with some modifications. This colorimetric assay is based on the ability of succinate dehydrogenase to reduce yellow MTT tetrazolium salt into blue MTT formazan crystals in living cells. The level of conversion provides an indication of mitochondrial metabolic function. Then, 5×10^3 neuron-like cells were incubated with diverse concentrations of glucose or insulin for 24 h in a CO_2-incubator (37 °C, 5% CO_2 and 95% humidity). Next, the supernatants were removed, and the cells were washed 3 times with 100 µL of PBS to remove phenol red residue, and 100 µL of 0.5 mg/mL MTT solution in PBS was added. The cells were incubated for 4 h at 37 °C in 5% CO_2 and 95% humidity in the darkness. Subsequently, the MTT solution was carefully discarded, the formazan crystals were dissolved in 100 µL DMSO acidified with 1N HCl, and plates were shacked for 30 min at room temperature. Cell viability was determined by measuring the absorbance at 570 nm on the Synergie multiwell scanning spectrophotometer (BIOKOM, Janki, Poland). Cell viability for each glucose or insulin concentration was calculated as the percentage of the untreated neuron-like cells. All experiments were performed on 5 wells per concentration and repeated at least 3 times.

4.5. Nitrite Levels Quantitation Assay

Griess reagent assay was performed to evaluate the influence of glucose or insulin on nitric oxide (NO) levels in neuron-like cells [60]. This spectrophotometric assay is based on the formation of an azo dye by the reaction of NO_2^- present in the sample with the Griess reagent. For each od experimental setting, the 1×10^4 cells were seeded in a 96-well plate. After 1 h and 24 h incubation with a series of glucose or insulin concentrations, 50 µL of supernatant was transferred into new 96-well plates. The 1:1 mixture (*v/v*) reagent A (1% sulfanilamide in 5% phosphoric acid) and reagent B (0.1% N-(1-Naphthyl) ethylenediamine dihydrochloride) was added to supernatants and left for 20 min in the dark at room temperature. After incubation, the absorbance at 548 nm was measured using a Varioskan LUX microplate reader (Thermo Scientific, Waltham, MA, USA).

4.6. Reactive Oxygen Species Concentration Measurement

To evaluate the influence of glucose or insulin on reactive oxygen species (ROS) in neuron-like cells, we used an assay with DCF-DA [61], a fluorogenic dye that measures hydroxyl, peroxyl and other ROS activity within the cell. It is deacetylated by esterases to a non-fluorescent compound, which is later oxidized by ROS into highly fluorescent 2′, 7′–dichlorofluorescein. The rest of the supernatant left on the plates used for the Griess assay was removed, and the cell pellet was washed 3 times with PBS. Next, 25 µM of DCF-DA solution in MEM medium, without supplementation or phenol red, was added to the treated PC12 cells and left for 1 h in a CO_2-incubator (37 °C, 5% CO_2, 95% humidity). After incubation, the DCF-DA solution was removed, cells were washed 3 time with MEM, and fresh MEM was added. The fluorescence was immediately measured with excitation at 485 nm and emission at 535 nm using a Varioskan LUX microplate reader (Thermo Scientific). All experiments were performed with 5 wells per concentration and repeated at least 3 times.

4.7. DNA Double-Strand Breaks Assessment

A fast halo assay (FHA) [62] was performed to assess DSBs in the DNA of differentiated PC12 cells treated with glucose or insulin. This test enables the rapid assessment of the extent of DNA breakage caused by different types of DNA lesions. After 1 h and 24 h incubation with glucose or insulin, the supernatants were collected into tubes. The trypsinization process was performed with a TrypLE solution (Gibco, Thermo Fisher Scientific, Waltham, MA, USA) for 3–5 min in a CO_2-incubator (37 °C, 5% CO_2, 95% humidity). After detaching from surfaces, suspended cells were collected into tubes and centrifuged for 5 min at 1000× *g* (Eppendorf, Hamburg, Germany) to get rid of cellular debris. Next, the supernatant was removed, and the cell pellet was washed with PBS and centrifuged again under the same conditions. The cells were then re-suspended at the density of 1000 cells/μL in sterile PBS and put in bathwater (37 °C). Then, 120 μl of 1.25% agarose (low melting point) in sterile PBS was added to cells and immediately sandwiched between an agarose-coated (high melting point) slide and a coverslip. After complete gelling (cooling block for 10 min), the coverslips were removed, and the slides were placed into the lysis buffer overnight at 4 °C in the dark. The next day, the slides were transferred into a Tris-HCl buffer (pH = 13) for 30 min in the dark and then twice into a neutralization buffer for 5 min. Finally, the slides were stained using 5 μM of 4′,6-diamidino-2-phenylindole (DAPI) for 20 min and analyzed under a fluorescence microscope. The DAPI-labelled DNA was visualized using a fluorescence microscope (Leica Microsystems, Wetzlar, Germany), and the subsequent images were digitally recorded on a PC and analyzed with image-analysis software developed by one of co-authors. The slides were numerically coded before reading to reduce operator bias. The extent of strand scission was quantified by calculating the nuclear diffusion factor (NDF), which represents the ratio between the total area of the halo plus nucleus and that of the nucleus. Data are expressed as relative NDF, calculated by subtracting the NDF of control cells from that of treated cells. All experiments were performed at least 3 times.

4.8. Human S100B ELISA Test

The S100B protein concentration was determined in the incubation medium and in the cell lysates after 24 h of incubation with glucose or insulin and measured by adapting the enzyme-linked immunosorbent assay (ELISA) kit following the manufacturer's protocols (human S100B ELISA Kit, Genorise, England). Briefly, the standards or samples (100 μL of lysates or supernatants) were added per well, and S100B was bound by the immobilized antibody during a 1 h incubation at room temperature. After triplicate washing with Assay Buffer (300 μL each) using an auto-washer (DiaWasher ELX50, Dialab GmbH, Austria), a detection antibody specific for human S100B (100 μL) was added to the wells and incubated for 1 h in RT. Following 3 washes with Assay Buffer (300 μL each), an HRP Conjugate (100 μL) was added for 1 h in RT. After a triplicate wash with Assay Buffer (300 μL each), a substrate solution (100 μL) was added to the wells, and color developed in proportion to the amount of S100B bounded in the initial step. The color development was stopped after 20 min by adding a stop solution. The intensity of the color was measured immediately using a microplate reader set to 450 nm with subtraction of readings at 540 nm or 570 nm to correct for optical imperfections in the plate. For each od ELISA experimental setting, the 1×10^4 cells were seeded in a 96-well plate. For intracellular concentration measurements, cells after incubation with glucose or insulin were centrifuged (3 min, 1200× *g*, RT), the supernatant was collected to Eppendorf vials. Cellular pellet was lysed according to freeze-thaw protocol and standardized for cellular protein concentration by BCA assay. Briefly, protein concentration was measured in cell lysates and adjusted to 50 μg protein per 100 μL. Such prepared samples were added per well and processed for ELISA assay.

4.9. Statistical Analysis

All experiments were carried out 3 times in triplicate. Statistical significance was calculated compared to the control. All results are presented as mean ± SEM (standard error of the mean) relative to the control—differentiated and untreated PC12 cells. Positive assay control (DMSO) was the reference value in the MTT assay. Positive assay control (H_2O_2) was the reference value in the Griess and DCF-DA assays. As data have normal distribution confirmed by Shapiro–Wilk test, a parametric test was used (one-way ANOVA with Tukey post hoc tests), and the Pearson correlation was performed between Griess, DCF-DA and DBS results.

Author Contributions: Conceptualization, A.K.-K. and B.W.; methodology, A.K.-K. and B.W.; writing—original draft preparation, A.K.-K., B.W.; writing—review and editing, A.K.-K. and B.W.; visualization, A.K.-K. and B.W.; revision and funding acquisition, A.P.; final approval of manuscript—A.P. All authors have read and agreed to the published version of the manuscript.

Funding: This research received no external funding.

Institutional Review Board Statement: Not applicable.

Informed Consent Statement: Not applicable.

Data Availability Statement: The data generated and analyzed during the current study are available from the corresponding author upon reasonable request.

Conflicts of Interest: The authors declare no conflict of interest.

References

1. Stelzma, R.A.; Schnitzlein, H.N.; Murllagh, F.R. An English translation of Alzheimer's 1907 paper, "Uber eine eigenartige Erkankung der Hirnrinde". *Clin. Anat.* **1995**, *8*, 429–431. [CrossRef] [PubMed]
2. Bloom, G.S. Amyloid-β and Tau. *JAMA Neurol.* **2014**, *71*, 505–508. [CrossRef] [PubMed]
3. Hampel, H.; Mesulam, M.-M.; Cuello, A.C.; Farlow, M.R.; Giacobini, E.; Grossberg, G.T.; Khachaturian, A.S.; Vergallo, A.; Cavedo, E.; Snyder, P.J.; et al. The cholinergic system in the pathophysiology and treatment of Alzheimer's disease. *Brain* **2018**, *141*, 1917–1933. [CrossRef]
4. Bondi, M.W.; Edmonds, E.C.; Salmon, D.P. Alzheimer's Disease: Past, Present, and Future. *J. Int. Neuropsychol. Soc.* **2017**, *23*, 818–831. [CrossRef]
5. Armstrong, R.A. Risk factors for Alzheimer's disease. *Folia Neuropathol.* **2019**, *57*, 87–105. [CrossRef]
6. Silva, M.V.F.; Loures, C.D.M.G.; Alves, L.C.V.; De Souza, L.C.; Borges, K.B.G.; Carvalho, M.D.G. Alzheimer's disease: Risk factors and potentially protective measures. *J. Biomed. Sci.* **2019**, *26*, 1–11. [CrossRef] [PubMed]
7. Zheng, Y.; Ley, S.H.; Hu, F.B. Global aetiology and epidemiology of type 2 diabetes mellitus and its complications. *Nat. Rev. Endocrinol.* **2018**, *14*, 88–98. [CrossRef]
8. Seuring, T.; Archangelidi, O.; Suhrcke, M. The Economic Costs of Type 2 Diabetes: A Global Systematic Review. *PharmacoEconomics* **2015**, *33*, 811–831. [CrossRef]
9. El-Hayek, Y.H.; Wiley, R.E.; Khoury, C.P.; Daya, R.P.; Ballard, C.; Evans, A.R.; Karran, M.; Molinuevo, J.L.; Norton, M.; Atri, A. Tip of the Iceberg: Assessing the Global Socioeconomic Costs of Alzheimer's Disease and Related Dementias and Strategic Implications for Stakeholders. *J. Alzheimer's Dis.* **2019**, *70*, 323–341. [CrossRef]
10. Hölscher, C. Evidence for pathophysiological commonalities between metabolic and neurodegenerative diseases. *Int. Rev. Neurobiol.* **2020**, 65–89. [CrossRef]
11. De La Monte, S.M.; Tong, M.; Wands, J.R. The 20-Year Voyage Aboard the Journal of Alzheimer's Disease: Docking at 'Type 3 Diabetes', Environmental/Exposure Factors, Pathogenic Mechanisms, and Potential Treatments. *J. Alzheimer's Dis.* **2018**, *62*, 1381–1390. [CrossRef] [PubMed]
12. Hayden, M.R. Type 2 Diabetes Mellitus Increases The Risk of Late-Onset Alzheimer's Disease: Ultrastructural Remodeling of the Neurovascular Unit and Diabetic Gliopathy. *Brain Sci.* **2019**, *9*, 262. [CrossRef] [PubMed]
13. Kubis-Kubiak, A.; Rorbach-Dolata, A.; Piwowar, A. Crucial players in Alzheimer's disease and diabetes mellitus: Friends or foes? *Mech. Ageing Dev.* **2019**, *181*, 7–21. [CrossRef] [PubMed]
14. Thakur, A.K.; Tyagi, S.; Shekhar, N. Comorbid brain disorders associated with diabetes: Therapeutic potentials of prebiotics, probiotics and herbal drugs. *Transl. Med. Commun.* **2019**, *4*, 1–13. [CrossRef]
15. Lundqvist, M.H.; Almby, K.; Abrahamsson, N.; Eriksson, J.W. Is the Brain a Key Player in Glucose Regulation and Development of Type 2 Diabetes? *Front. Physiol.* **2019**, *10*. [CrossRef]
16. De La Monte, S.M.; Wands, J.R. Alzheimer's Disease is Type 3 Diabetes—Evidence Reviewed. *J. Diabetes Sci. Technol.* **2008**, *2*, 1101–1113. [CrossRef]
17. Kandimalla, R.; Thirumala, V.; Reddy, P.H. Is Alzheimer's disease a Type 3 Diabetes? A critical appraisal. *Biochim. Biophys. Acta (BBA) Mol. Basis Dis.* **2017**, *1863*, 1078–1089. [CrossRef]
18. Chatterjee, S.; Mudher, A. Alzheimer's Disease and Type 2 Diabetes: A Critical Assessment of the Shared Pathological Traits. *Front. Neurosci.* **2018**, *12*, 383. [CrossRef]
19. Salas, I.H.; De Strooper, B.; De Strooper, B. Diabetes and Alzheimer's Disease: A Link not as Simple as it Seems. *Neurochem. Res.* **2018**, *44*, 1271–1278. [CrossRef]
20. De Sousa, R.A.L.; Harmer, A.R.; Freitas, D.A., Mendonça, V.A.; Lacerda, A.C.R.; Leite, H.R. An update on potential links between type 2 diabetes mellitus and Alzheimer's disease. *Mol. Biol. Rep.* **2020**, *47*, 6347–6356. [CrossRef]
21. Cheng, G.; Huang, C.; Deng, H.; Wang, H. Diabetes as a risk factor for dementia and mild cognitive impairment: A meta-analysis of longitudinal studies. *Intern. Med. J.* **2012**, *42*, 484–491. [CrossRef]

22. Li, X.; Leng, S.; Song, D. Link between type 2 diabetes and Alzheimer's disease: From epidemiology to mechanism and treatment. *Clin. Interv. Aging* **2015**, *10*, 549–560. [CrossRef] [PubMed]
23. Antidiabetic therapies and Alzheimer disease. *Dialog. Clin. Neurosci.* **2019**, *21*, 83–91. [CrossRef]
24. Koekkoek, P.S.; Kappelle, L.J.; Berg, E.V.D.; Rutten, G.E.H.M.; Biessels, G.J. Cognitive function in patients with diabetes mellitus: Guidance for daily care. *Lancet Neurol.* **2015**, *14*, 329–340. [CrossRef]
25. Moheet, A.; Mangia, S.; Seaquist, E.R. Impact of diabetes on cognitive function and brain structure. *Ann. N. Y. Acad. Sci.* **2015**, *1353*, 60–71. [CrossRef] [PubMed]
26. Langeh, U. Targeting S100B Protein as a Surrogate Biomarker and its Role in Various Neurological Disorders. *Curr. Neuropharmacol.* **2020**, *19*, 265–277. [CrossRef]
27. Sorci, G.; Bianchi, R.; Riuzzi, F.; Tubaro, C.; Arcuri, C.; Giambanco, I.; Donato, R. S100B Protein, a Damage-Associated Molecular Pattern Protein in the Brain and Heart, and Beyond. *Cardiovasc. Psychiatry Neurol.* **2010**, *2010*, 1–13. [CrossRef]
28. Rickmann, M.; Wolff, J. S100 protein expression in subpopulations of neurons of rat brain. *Neuroscience* **1995**, *67*, 977–991. [CrossRef]
29. Sjöstedt, E.; Fagerberg, L.; Hallström, B.M.; Häggmark, A.; Mitsios, N.; Nilsson, P.; Pontén, F.; Hökfelt, T.; Uhlen, M.; Mulder, J. Defining the Human Brain Proteome Using Transcriptomics and Antibody-Based Profiling with a Focus on the Cerebral Cortex. *PLoS ONE* **2015**, *10*, e0130028. [CrossRef]
30. Hajduková, L.; Sobek, O.; Prchalová, D.; Bílková, Z.; Koudelková, M.; Lukášková, J.; Matuchová, I. Biomarkers of Brain Damage: S100B and NSE Concentrations in Cerebrospinal Fluid—A Normative Study. *BioMed Res. Int.* **2015**, *2015*, 1–7. [CrossRef]
31. Lindblad, C.; Nelson, D.W.; Zeiler, F.A.; Ercole, A.; Ghatan, P.H.; Von Horn, H.; Risling, M.; Svensson, M.; Agoston, D.V.; Bellander, B.-M.; et al. Influence of Blood–Brain Barrier Integrity on Brain Protein Biomarker Clearance in Severe Traumatic Brain Injury: A Longitudinal Prospective Study. *J. Neurotrauma* **2020**, *37*, 1381–1391. [CrossRef] [PubMed]
32. Donato, R.; Sorci, G.; Riuzzi, F.; Arcuri, C.; Bianchi, R.; Brozzi, F.; Tubaro, C.; Giambanco, I. S100B's double life: Intracellular regulator and extracellular signal. *Biochim. Biophys. Acta (BBA) Bioenerg.* **2009**, *1793*, 1008–1022. [CrossRef] [PubMed]
33. Ostendorp, T.; Leclerc, E.; Galichet, A.; Koch, M.; Demling, N.; Weigle, B.; Heizmann, C.W.; Kroneck, P.M.H.; Fritz, G. Structural and functional insights into RAGE activation by multimeric S100B. *EMBO J.* **2007**, *26*, 3868–3878. [CrossRef]
34. Hofmann, M.A.; Drury, S.; Fu, C.; Qu, W.; Taguchi, A.; Lu, Y.; Avila, C.; Kambham, N.; Bierhaus, A.; Nawroth, P.; et al. RAGE Mediates a Novel Proinflammatory Axis. *Cell* **1999**, *97*, 889–901. [CrossRef]
35. Wartchow, K.M.; Tramontina, A.C.; De Souza, D.F.; Biasibetti, R.; Bobermin, L.D.; Gonçalves, C.-A. Insulin Stimulates S100B Secretion and These Proteins Antagonistically Modulate Brain Glucose Metabolism. *Neurochem. Res.* **2016**, *41*, 1420–1429. [CrossRef]
36. Landar, A.; Caddell, G.; Chessher, J.; Zimmer, D.B. Identification of an S100A1/S100B target protein: Phosphoglucomutase. *Cell Calcium* **1996**, *20*, 279–285. [CrossRef]
37. Frizzo, J.K.; Tramontina, F.; Bortoli, E.; Gottfried, C.; Leal, R.B.; Lengyel, I.; Donato, R.; Dunkley, P.R.; Gonçalves, C.-A. S100B-mediated inhibition of the phosphorylation of GFAP is prevented by TRTK-12. *Neurochem. Res.* **2004**, *29*, 735–740. [CrossRef] [PubMed]
38. Shafer, T.J.; Meacham, C.A.; Barone, S., Jr. Effects of prolonged exposure to nanomolar concentrations of methylmercury on voltage-sensitive sodium and calcium currents in PC12 cells. *Dev. Brain Res.* **2002**, *136*, 151–164. [CrossRef]
39. Parran, D.K.; Barone, S.; Mundy, W.R. Methylmercury decreases NGF-induced TrkA autophosphorylation and neurite outgrowth in PC12 cells. *Dev. Brain Res.* **2003**, *141*, 71–81. [CrossRef]
40. Bosco, D.; Fava, A.; Plastino, M.; Montalcini, T.; Pujia, A. Possible implications of insulin resistance and glucose metabolism in Alzheimer's disease pathogenesis. *J. Cell. Mol. Med.* **2011**, *15*, 1807–1821. [CrossRef]
41. An, Y.; Varma, V.R.; Varma, S.; Casanova, R.; Dammer, E.; Pletnikova, O.; Chia, C.W.; Egan, J.M.; Ferrucci, L.; Troncoso, J.; et al. Evidence for brain glucose dysregulation in Alzheimer's disease. *Alzheimer's Dement.* **2018**, *14*, 318–329. [CrossRef] [PubMed]
42. Ramos-Rodriguez, J.J.; Jimenez-Palomares, M.; Murillo-Carretero, M.I.; García, C.I.; Berrocoso, E.; Hernandez-Pacho, F.; Lechuga-Sancho, A.M.; Cozar-Castellano, I.; Garcia-Alloza, M. Central vascular disease and exacerbated pathology in a mixed model of type 2 diabetes and Alzheimer's disease. *Psychoneuroendocrinology* **2015**, *62*, 69–79. [CrossRef] [PubMed]
43. Sankar, S.B.; García, C.I.; Weinstock, L.D.; Ramos-Rodriguez, J.J.; Hierro-Bujalance, C.; Fernandez-Ponce, C.; Wood, L.B.; Garcia-Alloza, M. Amyloid beta and diabetic pathology cooperatively stimulate cytokine expression in an Alzheimer's mouse model. *J. Neuroinflammation* **2020**, *17*, 1–15. [CrossRef] [PubMed]
44. Chaudhuri, J.R.; Mridula, K.R.; Rathnakishore, C.; Anamika, A.; Samala, N.R.; Balaraju, B.; Bandaru, V.S. Association Serum S100B Protein in Alzheimer's Disease: A Case Control Study from South India. *Curr. Alzheimer Res.* **2021**, *17*, 1095–1101. [CrossRef]
45. Chaves, M.L.; Camozzato, A.L.; Ferreira, E.D.; Piazenski, I.; Kochhann, R.; Dall'Igna, O.; Mazzini, G.S.; Souza, D.O.; Portela, L.V. Serum levels of S100B and NSE proteins in Alzheimer's disease patients. *J. Neuroinflammation* **2010**, *7*, 6. [CrossRef]
46. Peskind, E.R.; Griffin, W.T.; Akama, K.T.; Raskind, M.A.; Van Eldik, L.J. Cerebrospinal fluid S100B is elevated in the earlier stages of Alzheimer's disease. *Neurochem. Int.* **2001**, *39*, 409–413. [CrossRef]
47. Christl, J.; Verhülsdonk, S.; Pessanha, F.; Menge, T.; Seitz, R.J.; Kujovic, M.; Höft, B.; Supprian, T.; Lange-Asschenfeldt, C. Association of Cerebrospinal Fluid S100B Protein with Core Biomarkers and Cognitive Deficits in Prodromal and Mild Alzheimer's Disease. *J. Alzheimer's Dis.* **2019**, *72*, 1119–1127. [CrossRef]

48. Mori, T.; Koyama, N.; Arendash, G.W.; Horikoshi-Sakuraba, Y.; Tan, J.; Town, T. Overexpression of human S100B exacerbates cerebral amyloidosis and gliosis in the Tg2576 mouse model of Alzheimer's disease. *Glia* **2009**, *58*, 300–314. [CrossRef]
49. Kheirouri, S.; Ebrahimi, E.; Alizadeh, M. Association of S100B Serum Levels with Metabolic Syndrome and its Components. *Acta Médica Port.* **2018**, *31*, 201–206. [CrossRef]
50. Michetti, F.; Corvino, V.; Geloso, M.C.; Lattanzi, W.; Bernardini, C.; Serpero, L.; Gazzolo, D. The S100B protein in biological fluids: More than a lifelong biomarker of brain distress. *J. Neurochem.* **2012**, *120*, 644–659. [CrossRef]
51. Leclerc, E.; Sturchler, E.; Vetter, S.W. The S100B/RAGE Axis in Alzheimer's Disease. *Cardiovasc. Psychiatry Neurol.* **2010**, *2010*, 1–11. [CrossRef] [PubMed]
52. Prasad, K. AGE–RAGE stress: A changing landscape in pathology and treatment of Alzheimer's disease. *Mol. Cell. Biochem.* **2019**, *459*, 95–112. [CrossRef] [PubMed]
53. Palanissami, G.; Paul, S.F.D. RAGE and Its Ligands: Molecular Interplay Between Glycation, Inflammation, and Hallmarks of Cancer—A Review. *Horm. Cancer* **2018**, *9*, 295–325. [CrossRef] [PubMed]
54. Gerlach, R.; Demel, G.; König, H.-G.; Gross, U.; Prehn, J.; Raabe, A.; Seifert, V.; Kögel, D. Active secretion of S100B from astrocytes during metabolic stress. *Neuroscience* **2006**, *141*, 1697–1701. [CrossRef]
55. Nardin, P.; Tramontina, F.; Leite, M.C.; Tramontina, A.C.; Quincozes-Santos, A.; De Almeida, L.M.V.; Battastini, A.M.; Gottfried, C.; Gonçalves, C.-A. S100B content and secretion decrease in astrocytes cultured in high-glucose medium. *Neurochem. Int.* **2007**, *50*, 774–782. [CrossRef]
56. Ahlemeyer, B.; Beier, H.; Semkova, I.; Schaper, C.; Krieglstein, J. S-100β protects cultured neurons against glutamate- and staurosporine-induced damage and is involved in the antiapoptotic action of the 5 HT(1A)-receptor agonist, Bay x 3702. *Brain Res.* **2000**, *858*, 121–128. [CrossRef]
57. Arcuri, C.; Bianchi, R.; Brozzi, F.; Donato, R. S100B Increases Proliferation in PC12 Neuronal Cells and Reduces Their Responsiveness to Nerve Growth Factor via Akt Activation. *J. Biol. Chem.* **2005**, *280*, 4402–4414. [CrossRef]
58. Greene, L.A.; Tischler, A.S. Establishment of a noradrenergic clonal line of rat adrenal pheochromocytoma cells which respond to nerve growth factor. *Proc. Natl. Acad. Sci. USA* **1976**, *73*, 2424–2428. [CrossRef]
59. Liu, Y.; Peterson, D.A.; Kimura, H.; Schubert, D. Mechanism of Cellular 3-(4,5-Dimethylthiazol-2-yl)-2,5-Diphenyltetrazolium Bromide (MTT) Reduction. *J. Neurochem.* **2002**, *69*, 581–593. [CrossRef]
60. Fox, J.B. Kinetics and mechanisms of the Griess reaction. *Anal. Chem.* **1979**, *51*, 1493–1502. [CrossRef]
61. Wang, H.; Joseph, J.A. Quantifying cellular oxidative stress by dichlorofluorescein assay using microplate reader11Mention of a trade name, proprietary product, or specific equipment does not constitute a guarantee by the United States Department of Agriculture and does not imply its approval to the exclusion of other products that may be suitable. *Free. Radic. Biol. Med.* **1999**, *27*, 612–616. [CrossRef]
62. Sestili, P. The Fast-Halo Assay for the Assessment of DNA Damage at the Single-Cell Level. *Methods Mol. Biol.* **2009**, *521*, 517–533. [CrossRef]

International Journal of *Molecular Sciences*

MDPI

Review

Analyzing Olfactory Neuron Precursors Non-Invasively Isolated through NADH FLIM as a Potential Tool to Study Oxidative Stress in Alzheimer's Disease

Laura Gómez-Virgilio [1], Alejandro Luarte [1,2,3], Daniela P. Ponce [1], Bárbara A. Bruna [1] and María I. Behrens [1,2,4,5,*]

1 Centro de Investigación Clínica Avanzada, Facultad de Medicina and Hospital Clínico Universidad de Chile, Santiago 8380453, CP, Chile; jalim166@gmail.com (L.G.-V.); aluarte@bni.cl (A.L.); dponcedelavega@gmail.com (D.P.P.); bbruna@ug.uchile.cl (B.A.B.)
2 Departamento de Neurociencia, Facultad de Medicina, Universidad de Chile, Santiago 8380453, CP, Chile
3 Faculty of Medicine, Biomedical Neuroscience Institute, Universidad de Chile, Santiago 8380453, CP, Chile
4 Departamento de Neurología y Psiquiatría, Clínica Alemana de Santiago y Universidad del Desarrollo, Santiago 7650729, CP, Chile
5 Departamento de Neurología y Neurocirugía, Hospital Clínico Universidad de Chile, Santiago 8380453, CP, Chile
* Correspondence: behrensl@uchile.cl; Tel.: +562-22101061

Abstract: Among all the proposed pathogenic mechanisms to understand the etiology of Alzheimer's disease (AD), increased oxidative stress seems to be a robust and early disease feature where many of those hypotheses converge. However, despite the significant lines of evidence accumulated, an effective diagnosis and treatment of AD are not yet available. This limitation might be partially explained by the use of cellular and animal models that recapitulate partial aspects of the disease and do not account for the particular biology of patients. As such, cultures of patient-derived cells of peripheral origin may provide a convenient solution for this problem. Peripheral cells of neuronal lineage such as olfactory neuronal precursors (ONPs) can be easily cultured through non-invasive isolation, reproducing AD-related oxidative stress. Interestingly, the autofluorescence of key metabolic cofactors such as reduced nicotinamide adenine dinucleotide (NADH) can be highly correlated with the oxidative state and antioxidant capacity of cells in a non-destructive and label-free manner. In particular, imaging NADH through fluorescence lifetime imaging microscopy (FLIM) has greatly improved the sensitivity in detecting oxidative shifts with minimal intervention to cell physiology. Here, we discuss the translational potential of analyzing patient-derived ONPs non-invasively isolated through NADH FLIM to reveal AD-related oxidative stress. We believe this approach may potentially accelerate the discovery of effective antioxidant therapies and contribute to early diagnosis and personalized monitoring of this devastating disease.

Keywords: oxidative stress; FLIM; Alzheimer's disease

Citation: Gómez-Virgilio, L.; Luarte, A.; Ponce, D.P.; Bruna, B.A.; Behrens, M.I. Analyzing Olfactory Neuron Precursors Non-Invasively Isolated through NADH FLIM as a Potential Tool to Study Oxidative Stress in Alzheimer's Disease. *Int. J. Mol. Sci.* **2021**, 22, 6311. https://doi.org/10.3390/ijms22126311

Academic Editor: Anne Vejux

Received: 22 March 2021
Accepted: 29 April 2021
Published: 12 June 2021

1. Introduction

Alzheimer's disease (AD) is the most common cause of dementia and the sixth cause of death in the world, constituting a major health problem for aging societies [1]. This disease is a neurodegenerative continuum with well-established pathology hallmarks, namely the deposition of amyloid-β (Aβ) peptides in extracellular plaques and intracellular hyperphosphorylated forms of the microtubule associated protein tau forming neurofibrillary tangles (NFTs), accompanied by neuronal and synaptic loss [2]. Interestingly, patients who will eventually develop AD manifest brain pathology decades before clinical symptoms appear [3,4]. Nevertheless, AD is still frequently diagnosed when symptoms are highly disabling and yet there is no satisfactory treatment.

Although the manifestations of AD are preponderantly cerebral, cumulative evidence shows that AD is a systemic disorder [5]. Accordingly, molecular changes associated with AD are not exclusively manifested in the brain but include cells from different parts of the body, ranging from the blood and skin to peripheral olfactory cells. More recently, neurons derived from induced pluripotent stem cells (iPSCs) from AD patients have contributed to glean a more realistic insight of brain pathogenic mechanisms [6]. Alternatively, the culture of olfactory neuronal precursors (ONPs) has emerged as a relatively simpler tool to study different brain disorders, taking advantage of their neuronal lineage and their readily non-invasive isolation [7,8]. For instance, patient-derived ONPs manifest abnormal amyloid components together with tau hyperphosphorylation, which have recently led to the proposal of these cells as a novel diagnostic tool for AD [9–11].

Different hypotheses have attempted to explain AD pathogenesis. Some of them include Aβ cascade, tau hyperphosphorylation, mitochondrial damage, endoplasmic reticulum (ER) stress, and oxidative stress. Interestingly, although it has been difficult to establish a prevailing causative mechanism, increased levels of oxidative stress seem to be a common feature for many of these models. Furthermore, oxidative stress due to increased levels of reactive oxygen species (ROS) has been broadly recognized as a very early signature during the course of AD [12–14]. Interestingly, AD-related oxidative stress is by no means restricted to neuronal cells but is also related to astrocytes' oxidative damage and antioxidant capacity [15]. Indeed, since the acknowledgment of the tripartite synapse, it has become increasingly clear that different antioxidant mechanisms of astrocytes can be harnessed by synaptically active neurons and surrounding cells [16–18]. In the tripartite synapse, the astrocyte's endfeet are close to synapses and can be activated by the spillover of synaptic glutamate to provide a timely antioxidant response [19,20]. Moreover, it is not entirely understood how other glial cells such as pericytes may contribute to the damage induced by AD-related oxidative stress. For instance, oxidative damage may compromise the integrity of pericytes, which in turn could alter the blood-brain barrier's integrity, favoring the infiltration of cytotoxic cells and the emergence of brain edema [21,22]. In coherence with a broader systemic manifestation of this disease, the peripheral olfactory system shows AD-associated oxidative stress, which has been measured both in the olfactory neuroepithelium and in cultured ONPs [23–25]. However, while the intriguing relationship between oxidative stress and AD has been long known, their translational impact has remained limited.

Interestingly, the oxidative status of cells is highly correlated with the content of autofluorescent metabolic co-factors such as NADH and its phosphorylated version NADPH [26–29]. In addition, NADH is required to synthesize NADPH, which is at the core of the antioxidant response of different cells by sustaining the synthesis of antioxidants such as glutathione (GSH) and thioredoxin [30]. Furthermore, it has been shown in AD animal models that the provision of NADH is upstream the levels of GSH in order to counterbalance increased ROS levels and neuronal death [27]. Interestingly, external manipulation of oxidative or reducing conditions of cultured neurons are directly manifested as changes in mitochondrial and cytosolic NADH content [28]. As such, by imaging NADH autofluorescence, it might be possible to obtain a real-time monitoring of redox imbalance without the need to use exogenous staining or recombinant sensors. Complementary to methodologies purely based on fluorescence intensity, Fluorescence Lifetime Imaging Microscopy (FLIM) has received increasing attention [31,32]. Fluorescence lifetime is the average time in which a fluorophore remains excited to emit photons before descending to the ground state, providing unique information about its biochemical environment. Importantly, NADH FLIM can be harnessed to increase the sensitivity to its autofluorescence and to discriminate its binding to enzymes from different signaling pathways. In this review, we explore the idea of using ONPs non-invasively isolated coupled to NADH FLIM to reveal AD-associated oxidative stress. This approach may have a broad impact for early AD diagnosis and treatment.

2. Olfactory Neuroepithelium and the Non-Invasive Isolation of ONPs

The olfactory neuroepithelium is a key structure for odor sensing. It consists of a pseudostratified columnar epithelium located on the outer domain of the olfactory mucosa settled on the basement membrane (BM) and the lamina propria (LP) [33]. The cellular composition of these layers has been widely documented based on morphological analysis and the use of characteristic markers for each cell type [34–37]. Figure 1 schematizes the location, cellular components, and molecular markers of the human olfactory mucosa.

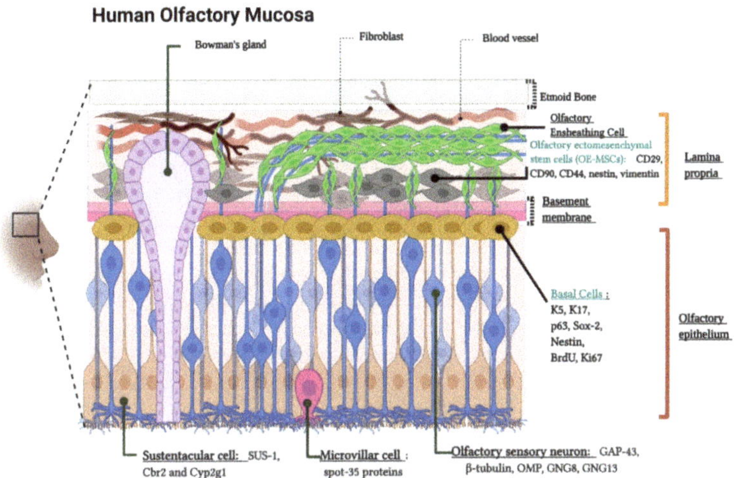

Figure 1. Cytoarchitecture and cellular components of the human olfactory mucosa. Lamina propria components. Olfactory Ensheathing Cells, Bowman's gland and Olfactory Ectomesenchymal Stem Cells (OE-MSCs). The image indicates the OE-MSCs markers: CD29, CD90, CD44, Nestin, and Vimentin. Olfactory epithelium components. Basal Cells, Olfactory sensory neurons (OSNs) or Olfactory receptor neurons (ORNs), Sustentacular cells, and Microvillar cells. The figure shows basal cell markers: K5 (Keratin 5), K17 (Keratin 17), p63, Sox-2 (SRY-Box Transcription Factor 2), Nestin, BrdU (Bromodeoxyuridine), and Ki-67; ORNs markers: GAP-43 (Growth Associated Protein 43), β-tubulin, OMP (Olfactory Marker Protein), GNG8 (Guanine Nucleotide-binding protein subunit Gamma), and GNG13 (Guanine Nucleotide-binding protein G(I)/G(S)/G(O) subunit Gamma-13)); sustentacular cell markers (SUS-1, Cbr2 (Carbonyl Reductase 2) and Cyp2g1 (Cytochrome P450, family 2, subfamily G, polypeptide 1)) and, microvillar cell marker: (spot-35 proteins). Created with BioRender.com.

The olfactory neuroepithelium is also a source of stem cells, which are capable of self-renewal and can generate neuronal precursors throughout the entire human lifetime. These precursors include neural stem cells known as basal cells. As expected for neural stem cells, basal cells are multipotent and allow the continuous replacement of neuronal and non-neuronal cells such as olfactory receptor neurons (ORNs) and sustentacular cells (of astrocytic lineage), respectively [38–40]. In addition, the LP contains another less accessible population of stem cells, whose features meet most of the minimum criteria of the mesenchymal and Tissue Stem Cell Committee of the International Society for Cellular Therapy [41]. As such, they are named as olfactory ectomesenchymal stem cells (OE-MSCs) [42–44].

Isolation of cells of the olfactory neuroepithelium from patients provides a source of cultured neural stem cells, which has been used to model different brain disorders such as schizophrenia, Parkinson's disease, autism, ataxia-telangiectasia, hereditary spastic paraplegia (HSP), and AD [7,45–49]. These neural stem cells can be frozen and stored for subsequent use and tolerate several passages without significantly losing their main properties. Furthermore, purified cultures obtained by cloning selection through limiting dilution significantly increases cell viability at least until passage 60 [50]. In this work, we will refer to neural stem cells isolated from the olfactory neuroepithelium as olfactory neuronal precursors (ONPs), similar to [8,9,50,51].

Different strategies have been used to isolate and culture patient-derived ONPs, ranging from biopsies to non-invasive exfoliation of the nasal turbinate. Human ONPs were first isolated by Wolozin et al. from the olfactory neuroepithelium of cadavers or from adult biopsied samples [10,52]. Another similar isolation approach demonstrated that a significant subpopulation of these cells express markers of mature olfactory neurons such as OMP, Golf, NCAM, and NST and look small and bright to the microscope, in contrast to the remaining "dark phase" cells that do not express OMP, but glial markers [53]. However, a systematic characterization of these cultures has shown that after a few days in vitro, both dark and bright phase cells show an intracellular calcium increase in response to odorants, highlighting the neuronal features of these cells [54]. In addition, cells with features of ONPs have also been obtained from dissociated neurospheres, which have been denominated "olfactory neurosphere-derived" (ONS) cells [43]. Alternatively, ONPs can be non-invasively isolated by an exfoliation of the nasal cavity [51]. These exfoliated cells can be cultured in a modified media to propitiate neural lineage maintenance and proliferation. Notably, these neuronal precursors conserve their capability to differentiate into ORNs in the presence of dibutyryl adenosine 3′,5′-cyclic monophosphate (Db-cAMP) and, strikingly, maintain their electrical response to odorants [51]. Thus, non-invasively isolated ONPs retain neuronal features similar to those obtained by biopsy. A simplified extraction protocol and the molecular characterization of non-invasively isolated ONPs is shown in Figure 2.

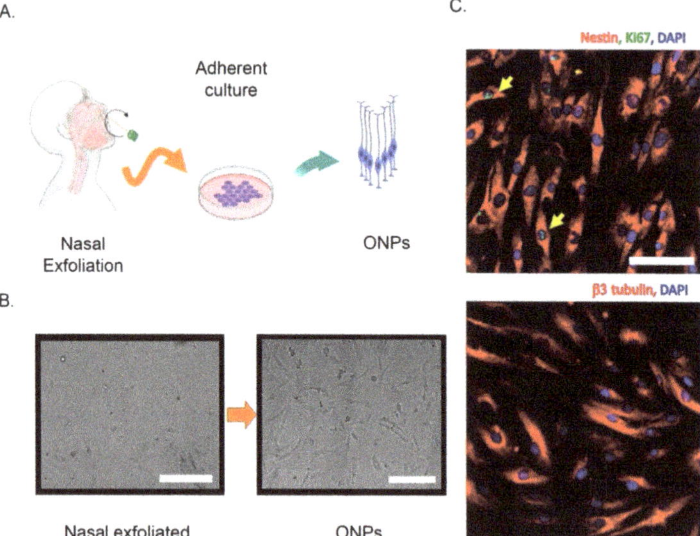

Figure 2. Non-invasive isolation of olfactory neuronal precursors (ONPs). (**A**) Schematic cartoon of the isolation protocol based on the extraction of nasal exfoliate with the subsequent adherent culture and enrichment of ONPs. (**B**) Left, the nasal exfoliate is directly seeded on adherent plates, showing a mixture of cell morphologies. Right, after 1–2 weeks ONPs dividing colonies are easily observed with their characteristic morphologies. (**C**) Upper panel, immunofluorescence of cultured ONPs, depicting the stem cell marker Nestin and Ki67 (yellow arrows) to show active cell proliferation. Lower panel, cultured ONPs express neuronal markers such as β3 tubulin. Cell nuclei are shown by DAPI staining. All scale bars = 100 μm. All images were generated in our lab. Created with BioRender.com.

3. Alzheimer's Disease-Related Oxidative Stress in the Olfactory Epithelium and ONPs

Oxidative stress is the result of an imbalance between oxidant and antioxidant cellular pathways. One of the most studied oxidant compounds are ROS, which are highly reactive molecules, including peroxide (H_2O_2), superoxide anion radical ($O_2 \bullet -$), and hydroxyl radical ($\bullet$ OH), among others. These molecules may covalently interact with lipids, proteins, and carbohydrates, generating molecular adducts and cumulative damage that, when sensed by cells, may actively trigger different death programs [55].

It was well established almost three decades ago that oxidative stress damage is linked to AD [14]. Furthermore, it has been proposed that oxidative stress at different brain neuronal and non-neuronal cells might be the earliest event of a pathogenic cascade [13]. Whether oxidative stress is a causative agent or just a consequence in neurodegenerative disorders has been thoroughly debated for several years, but still remains an open question [56–58]. The most parsimonious interpretation of this evidence is that oxidative stress as well as other potential AD causative agents (such as Aβ accumulation) are part of a highly interconnected vicious cycle rather than a linear chain of events with a unique origin. The molecular mechanisms and implications of oxidative stress on the nervous system and, potentially, during AD pathogenesis have been thoroughly reviewed elsewhere [12,59]. Here, we focus on evidence showing AD-associated oxidative stress in the peripheral olfactory system rather than reviewing mechanistic explanations.

Oxidative stress associated with AD is manifested in the olfactory neuroepithelium. Accordingly, increased immunoreactivity of the antioxidant enzyme manganese and Copper-Zinc superoxide dismutases have been detected in ORNs and basal and sustentacular cells of the olfactory neuroepithelium of AD patients compared with age-matched controls [60]. Analogously, AD patients harbor a higher immunoreactivity against the antioxidant protein Metallothionein both in the olfactory neuroepithelium and the Bowman's Glands and the LP [61]. Both results suggest that cells from olfactory neuroepithelium elicit an increased antioxidant defense, due to increased oxidative stress during AD. With respect to the direct measurement of oxidation products, post-mortem staining showed an increase in 3-nitrotyrosine (3-NT) in the brain and olfactory neuroepithelium of AD patients [23]. Figure 3 schematizes the antioxidant response and oxidative damage reported in ONPs and OE from AD patients. It would be of interest to uncover whether some AD genetic factors such as the *ApoE* ε4 allele (*ApoE4*) (the single most important genetic risk factor for AD) also manifests oxidative stress signatures in the olfactory epithelium. It is plausible that this is the case because deficits in odor fluency, identification, recognition memory, and odor threshold sensitivity have been associated with the inheritance of the *ApoE4* genotype in several studies [62–64]. For a more thorough compiling of evidence showing AD-associated oxidative damage across other domains of the nervous system, readers may refer to the following excellent articles [12,59,65].

The relationship between oxidative stress and AD has been extensively studied mainly through cellular and animal models [47,54]. However, these models may not fully capture key features of the disease. This limitation potentially leads to wrong conclusions about the pathogenic mechanisms and ultimately may dampen the development of effective therapies. Alternatively, patient-derived cells of neuronal lineage such as those from the olfactory epithelium may provide a convenient solution to this problem [5,9,42].

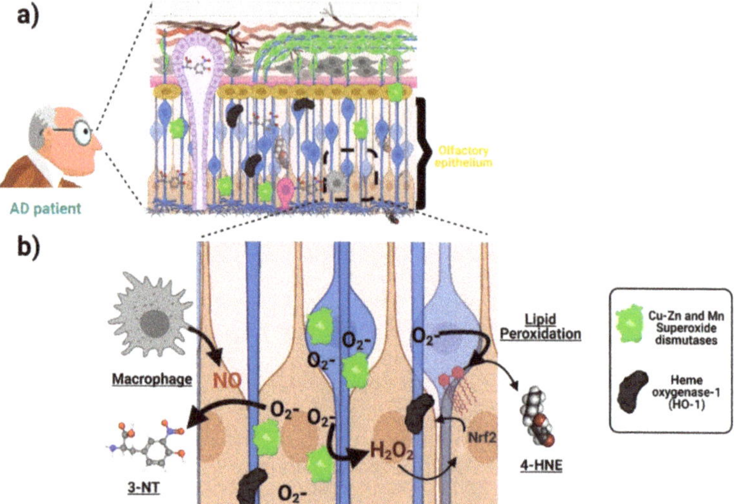

Figure 3. Oxidative stress associated with AD in the olfactory neuroepithelium. (**a**) ONPs and sustentacular cells in the olfactory epithelium (OE) show an increased antioxidant defense with elevated levels of manganese and copper-zinc superoxide dismutases as well as heme oxygenase-1 due to increased oxidative stress in AD patients compared with age-matched controls. Moreover, there is an increase in 3-nitrotyrosine (3-NT) and 4-hydroxynonenal (lipid peroxidation indicator) levels, suggesting AD-associated oxidative damage. (**b**) The increased generation of superoxide anion activates superoxide dismutases (SOD) as an antioxidant response. The generation of other reactive oxygen species (ROS), such as H_2O_2, induces the expression of other antioxidant enzymes (heme oxygenase-1). On the other hand, the accumulation of superoxide anion increases the levels of compounds such as 4-hydroxynonenal (4-HNE). Moreover, the increased levels of 3-NT are produced from the interaction of superoxide anion and nitric oxide (NO), whose probable source is located at activated macrophages in the OE of AD patients. Created with BioRender.com.

Interestingly, cultured patient-derived ONPs and other peripheral cells also manifest AD-associated oxidative stress. For example, an increase in the level of hydroxynonenal and Nε-(carboxymethyl)lysine) (indicating lipid peroxidation), as well as a higher content of heme oxygenase-1, has been found in ONPs isolated from AD patients compared with age-matched controls (Figure 3) [24]. Furthermore, ONPs from AD patients are also more susceptible to oxidative stress-induced cell death [25]. This is strikingly similar to what has been found by our group in blood-derived lymphocytes from AD patients [66,67]. Indeed, manifestations of oxidative stress associated with AD have been reported in different patient-derived peripheral cells ranging from blood cells to fibroblasts and iPSCs-derived neurons. These changes may include compensatory antioxidant responses and a rise in the concentration of oxidation by-products, as well as increased susceptibility to ROS-induced cell death, which has been demonstrated in different cellular types from AD patients. Many of those findings are summarized in the Table 1. In addition, Table 1 also summarizes similar evidence of other relevant pathogenic mechanisms proposed for AD pathogenesis, including Amyloid/Tau, mitochondria, and ER-stress. Thus, different cells throughout the body show signs of different proposed AD pathogenic mechanisms, including oxidative stress at early stages of the disease continuum. The robustness of this tendency highlights the potential of patient-derived cells, and in particular ONPs, for monitoring oxidative stress associated with AD.

Table 1. Signatures of oxidative stress and other AD mechanistic hypotheses are manifested in patient-derived peripheral cells, iPSCs and ONPs.

Pathogenic Mechanism	Main Finding	Cellular Type	Lineage	References
Amyloid/Tau	Platelets from AD patients reproduce the increased amyloidogenic processing of AβPP	Platelets	Non-neuronal	[68]
Amyloid/Tau	AD platelets harbor increased levels of a higher molecular weight tau isoform	Platelets	Non-neuronal	[69]
Amyloid/Tau	Alteration of AβPP, BACE, and ADAM 10 levels in early stages of the disease	Platelets	Non-neuronal	[70–72]
Amyloid/Tau	It is suggested a decreased non-amyloidogenic processing of AβPP by a lack of nicastrin mRNA expression in samples obtained from AD patients	Lymphocytes	Non-neuronal	[73]
Amyloid/Tau	Altered balance between Aβ-oligomers and PKCε levels in AD. Loss of PKCε-mediated inhibition of Aβ	Fibroblasts	Non-neuronal	[74]
Amyloid/Tau	Higher $A\beta_{42}/A\beta_{40}$ ratio compared to control cells	*PSEN1* iPSC-derived neural progenitors	Neuronal	[75]
Amyloid/Tau	Mutation alters the initial cleavage site of γ-secretase, resulting in an increased generation of $A\beta_{42}$, in addition to an increase in the levels of total and phosphorylated tau	Neuron-derived iPSCs from patients harboring the London FAD *AβPP* mutation V717I	Neuronal	[76]
Amyloid/Tau	Oligomeric forms of canonical Aβ impairs synaptic plasticity	Cortical neurons from three genetic forms of AD —*PSEN1* L113_I114insT, *AβPP* duplication (*AβPP*Dp), and Ts21— generated from iPSCs	Neuronal	[77]
Amyloid/Tau	Increase in the content and changes in the subcellular distribution of t-tau and p-tau in cells from AD patients compared to controls	Non-invasively isolated ONPs	Neuronal	[9]
Mitochondria	Compromise of mitochondrial COX from AD patients	Platelets	Non-neuronal	[78]
Mitochondria	Platelets isolated from AD patients show decreased ATP levels	Platelets	Non-neuronal	[79]
Mitochondria	AD lymphocytes exhibit impairment of total OXPHOS capacity	Lymphocytes	Non-neuronal	[80]
Mitochondria	AD skin fibroblasts show increased production of CO_2 and reduced oxygen uptake suggesting that mitochondrial electron transport chain might be compromised	Fibroblasts	Non-neuronal	[81]
Mitochondria	AD fibroblasts present reduction in mitochondrial length and a dysfunctional mitochondrial bioenergetics profile	Fibroblasts	Non-neuronal	[82]
Mitochondria	SAD fibroblasts exhibit aged mitochondria, and their recycling process is impaired	Fibroblasts	Non-neuronal	[83]
Mitochondria	Patient-derived cells show increased levels of oxidative phosphorylation chain complexes	Human induced pluripotent stem cell-derived neuronal cells (iN cells) from SAD patients	Neuronal	[84]
Mitochondria	Mitophagy failure as a consequence of lysosomal dysfunction	iPSC-derived neurons from FAD1 patients harboring *PSEN1* A246E mutation	Neuronal	[85]
Mitochondria	Neurons exhibit defective mitochondrial axonal transport	iPSC-derived neurons from an AD patient carrying *AβPP* -V715M mutation	Neuronal	[86]
Oxidative Stress	Increased activity of the antioxidant enzyme catalase in probable AD patients	Erythrocytes	Non-neuronal	[87]
Oxidative Stress	Increased production and content of thiobarbituric acid-reactive substances (TBARS), superoxide dismutase (SOD), and nitric oxide synthase (NOS)	Erythrocytes and Platelets	Non-neuronal	[88]
Oxidative Stress	Increase in the content of the unfolded version of p53 as well as reduced SOD activity	Peripheral blood mononuclear cells (PBMCs)	Non-neuronal	[89]
Oxidative Stress	Exacerbated response to NFKB pathway	PBMCs	Non-neuronal	[90]
Oxidative Stress	Increased ROS production in response to H_2O_2	PBMCs	Non-neuronal	[66]
Oxidative Stress	AD lymphocytes were more prone to cell death after a H_2O_2 challenge	Lymphocytes	Non-neuronal	[91]

Table 1. *Cont.*

Pathogenic Mechanism	Main Finding	Cellular Type	Lineage	References
Oxidative Stress	Reduced antioxidant capacity of FAD lymphocytes and fibroblasts together with increased lipid peroxidation on their plasma membrane	Lymphocytes and Fibroblasts	Non-neuronal	[92]
Oxidative Stress	Aβ peptides were better internalized and generated greater oxidative damage in FAD fibroblasts	Fibroblasts	Non-neuronal	[93]
Oxidative Stress	Aβ peptide caused a higher increase in the oxidation of HSP60	Fibroblasts	Non-neuronal	[94]
Oxidative Stress	Reduction in the levels of Vimentin in samples from AD patients	iPSCs-derived neurons from AD patient	Neuronal	[65]
Oxidative Stress	Increased levels of hydroxynonenal, Nε-(carboxymethyl)lysine), and heme oxygenase-1 in samples from AD patients	Biopsy-derived ONPs	Neuronal	[24]
Oxidative Stress	Increased susceptibility to oxidative-stress-induced cell death	Biopsy-derived ONPs	Neuronal	[25]
ER-Stress	Impaired ER Ca^{2+} and ER stress in PBMCs from MCIs and mild AD patients	PBMCs	Non-neuronal	[95]
ER-Stress	Accumulation of Aβ oligomers induced ER and oxidative stress	iPSC-derived neural cells from a patient carrying *APP*-E693Δ mutation and a sporadic AD patient	Neuronal	[96]
ER-Stress	Aβ-S8C dimer triggers an ER stress response more prominent in AD neuronal cultures where several genes from the UPR were upregulated	iPSC-derived neuronal cultures carrying the AD-associated *TREM2* R47H variant	Neuronal	[97]
ER-Stress	Accumulation of Aβ oligomers in iPSC-derived neurons from AD patients leads to increased ER stress	iPSC-derived neurons from patients with an *AβPP*-E693Δ mutation	Neuronal	[98]

4. The Role of NADH in Cell Metabolism and Antioxidant Defense

Metabolism is intimately associated with oxidative stress, since ATP production by mitochondria requires the reduction of oxygen to water, which is a major source of ROS. Enzymatic cofactors of energetic metabolism such as oxidized and reduced NAD (NAD+ and NADH, respectively), as well as their phosphorylated versions (NADP+ and NADPH), constitute key bridges between energy supply and the antioxidant defense of cells [30]. The availability of these cofactors is highly inter-related, and depending on the cellular context, their separate or combined measurement can be used to reveal redox homeostasis both in the cytosol and mitochondria [99]. We provide a brief overview of the main cellular sources and consumers of NAD+/NADH and their interplay with NADP+/NADPH levels with a special focus on neuronal cells.

The provision of NAD+ molecules in the body comes from de novo synthesis from tryptophan or via salvage pathways using nicotinamide (NAM) and nicotinamide riboside (NR) as precursors. The detailed pathways of NAD+ direct synthesis have been reviewed elsewhere [100]. In addition, the direct consumption of NAD+ is achieved mainly by the enzymatic activity of silent information regulator proteins or sirtuins (SIRTs) and poly (adenosine diphosphate-ribose) polymerases (PARPs). Sirtuins catalyze the deacetylation of target proteins by converting NAD+ into NAM and a *O*-Acyl ATP ribose. The activity of SIRTs has been profusely studied in the nucleus, where they control the function of different transcription factors and histone proteins to regulate cell senescence and neurodegeneration [101,102]. In addition, PARPs are enzymes that normally control DNA repair, whose overactivation under intense DNA oxidative damage may lead to cellular depletion of NAD+ and ATP. Both processes may promote cell death, potentially contributing to the pathogenesis of neurodegenerative disorders such as AD [103].

Different metabolic reactions determine the level and subcellular distribution of NADH. Accordingly, the synthesis of NADH from NAD+ in the cytosol is achieved by the glycolytic pathway, which generates two ATPs, two NADH, and two pyruvates as net yield per glucose. In addition, NADH is synthesized by two mitochondrial enzymes: pyruvate dehydrogenase (PDH), which produces acetyl-CoA entering to the tricarboxylic acid cycle (TCA), and malate dehydrogenase (MDH), which oxidates malate to generate oxaloacetate (part of TCA). The latter reaction may also occur in the cytosol in the opposite direction, leading to NADH consumption to sustain the malate shuttle towards mitochondria. Inside the mitochondria, NADH is oxidized to NAD+ by complex I (NADH: ubiquinone oxidoreductase) of the electron transport chain, donating its electrons to achieve oxidative phosphorylation and ATP synthesis. Importantly, in eukaryotic cells such as neurons, oxidation of NADH by complex I is the main source of ROS inside the cell [104]. In the cytosol, oxidation of NADH is produced by lactate dehydrogenase (LDH), which regenerates the NAD+ required for glycolysis to proceed. Indeed, the measurement of the NADH/NAD+ ratio may serve as an indicator of the balance between glycolysis and oxidative phosphorylation, which has been used for monitoring real time cellular metabolism [105]. Despite all these metabolic pathways that are present in astrocytes and neurons, both cell types differ in their metabolic profiles. For instance, astrocytes are richer in the expression of lactate dehydrogenase 5 (LDH5), which is better suited to produce lactate from pyruvate. On the contrary, neurons express more LDH1, which is more efficient at consuming lactate to produce pyruvate. These complementary molecular signatures are compatible with lines of evidence showing that neurons "outsource" glycolysis to astrocytes. As such, astrocytes behave as net sources of lactate, while neurons are net sinkers of this metabolite [106–109]. Importantly, cellular metabolism seems to be highly plastic and under some conditions, neurons can directly use glucose to perform glycolysis and all the subsequent metabolic steps [110,111].

The major cytosolic source of NADPH is the pentose phosphate pathway (PPP), which leads to the oxidative decarboxylation of glucose-6-phosphate (G6P) to produce NADPH and the ribose-5-phosphate sugar required for the synthesis of DNA and RNA [112]. The provision of NADPH obtained by neurons through PPP is relevant under oxidative stress. Indeed, it has been claimed that neurons may increase survival under oxidative stress conditions by diverting the metabolic flux of glucose from glycolysis to PPP in order to produce more NADPH and antioxidant power [113]. In addition, the subcellular levels of NADPH are replenished from the NADH pool by the action of the mitochondrial nicotinamide nucleotide transhydrogenase (NNT) [114]. Indeed, it has been estimated that half of the mitochondrial NADPH in the brain depends on the activity of NNT and interrupting its function may cause oxidative stress [99,115]. The abundance of NADPH is also partially determined by cytosolic as well as mitochondrial kinases (NAD kinases), which convert NAD+ into NADP+. In addition, two enzymes from the TCA cycle reduce NADP+ to NADPH inside the mitochondria, namely mitochondrial isocitrate dehydrogenase 2 (IDH2) and malic enzyme (ME1). Nevertheless, in the cytosol, there is another isocitrate dehydrogenase (IDH1) usually catalyzing the reaction in the opposite direction.

In general, while NADH levels are directly implicated in ATP and ROS synthesis, those of NADPH are directly involved in cellular antioxidant response and also in free radical generation by the enzyme NADPH oxidase [116]. However, given the metabolic conditions of brain cells, the role of NADPH would be predominantly antioxidant [99]. Accordingly, NADPH is used by glutathione reductase to reduce oxidized glutathione, and by thioredoxin reductase to reduce oxidized thioredoxin, which are major components of cellular ROS defense [117]. As both cytosolic and mitochondrial NADPH levels tightly depend on those of NADH, it follows that the concentration of both nucleotides determine ROS defense. Accordingly, it has been shown that the provision of NADH is required to support proper detoxification of peroxide from liver cells by mitochondria [117]. The main roles of NADH and NADPH in cell metabolism and antioxidant pathways are summarized in Figure 4.

Measuring NAD metabolism is of interest due to NAD's biological importance, and ties to human disease and normal aging. Different methods have been used to determine NAD metabolism. Some of them are highly sensitive, such as liquid chromatography mass spectrometry (LC/MS/MS). However, this approach only gives static information of a population of cells and is also invasive and destructive. Table 2 indicates some advantages and disadvantages of different methods for quantifying NAD metabolism, highlighting the relevant contribution of FLIM.

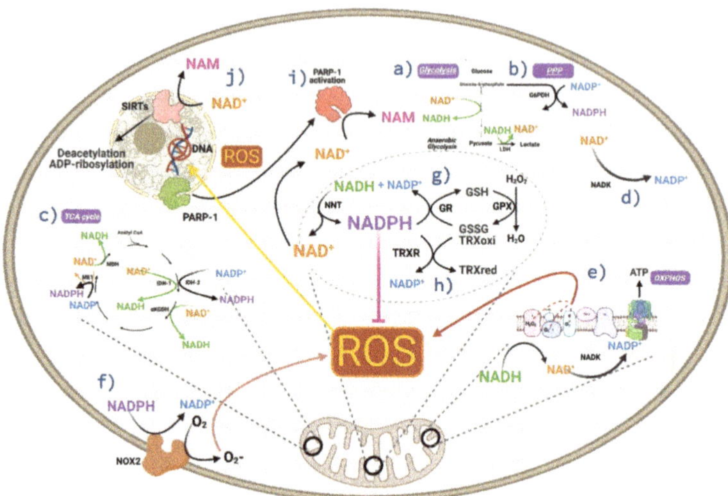

Figure 4. Roles of NADH and NADPH in metabolism and antioxidant pathways. (**a**–**c**) Synthesis of NADH from NAD+ in (**a**) glycolysis, and (**c**) TCA cycle; NADPH from NADP+ in (**b**) PPP and (**c**) TCA cycle. (**d**) Synthesis of NADP+ from NAD+ by NAD+ kinase both in cytosol and mitochondria. (**e**) Oxidation of NADH by complex I is the main source of ROS inside the cell in addition to (**f**) the activation NOX2 that generates ROS through a reduction of oxygen using NADPH as the source of donor electrons. In brain cells, the role of NADPH is predominantly antioxidant; for instance, (**g**) NADPH is used by glutathione reductase to reduce oxidized glutathione, and by (**h**) thioredoxin reductase to reduce oxidized thioredoxin. (**i**) Under oxidative stress and DNA damage, PARP-1 is activated, and this is manifested by an increase in the consumption of NAD+ by PARP. (**j**) On the other hand, the enzymatic activity of SIRTs consumes NAD+. SIRTs catalyze the deacetylation of target proteins by converting NAD+ into NAM. Created with BioRender.com.

Table 2. Methods for measuring NAD+ and derivatives.

Assay	Analyte	Advantages	Disadvantages	Ref
Luminometric analysis	NAD+, NADH, NADP+, and NADPH concentration	Method is reproducible and reported in tissues and cells.	Partial inactivation of luciferase system. Invasive and destructive.	[118]
Colorimetric Assay using thiazolyl blue	Intracellular NAD+ concentration	Identifies biological trends that are highly reproducible in the literature.	Indirect measurement affected by minor variations in temperature and pH. Cannot detect low picomolar levels. Invasive and destructive.	[119,120]
BRET-based biosensors	NAD+ concentration	Quantifies NAD+ levels in cell culture, tissue, and blood samples. The readout can be performed by a microplate reader or a simple digital camera. Minimum consumption of biological samples.	Invasive and destructive.	[121]
Reverse phase HPLC	Endogenous intracellular and extracellular levels of NAD+ and related metabolites	The method uses elements to increase sensitivity.	Limited to low micromolar detection levels. Since many NAD-related metabolites can be converted to one or more metabolites the identified concentrations may be fraught with inaccuracies. Invasive and destructive detection. Static information of a population of cells.	[122]

Table 2. *Cont.*

Assay	Analyte	Advantages	Disadvantages	Ref
LC-MS/MS	Endogenous intracellular and extracellular levels of NAD+ and related metabolites	High specificity and sensitivity.	The assay requires time, many preparations, and materials not readily available. Static information of a population of cells. Invasive and destructive detection.	[123,124]
LC-MS/MS (NAD metabolite isotopic labels)	Endogenous intracellular and extracellular levels of NAD+ and related metabolites	The method provides greater resolution and lower limit of detection.	Static information of a population cells. Invasive and destructive.	[125,126]
Fluorescent imaging with metabolite sensors	NADH, NAD+ concentrations, and their ratio	Metabolite sensors may be used to profile metabolic states of living cells in real-time and with single-cell or even subcellular resolution.	Invasive (metabolite sensors are introduced into any cell or organism). With some sensors, fluorescence is sensitive to pH. Other sensors have a limited dynamic range in fluorescence.	[127,128]
Novel MRI-based process	NAD+ and NADH concentrations	Non-invasive and non-destructive, measured in healthy aged human brains.	Only measures 2 analytes.	[129]
Fluorescence Lifetime Imaging (FLIM)	NAD+, NADH, NADP+, and NADPH	Non-invasive and non-destructive using autofluorescence intensity. May be used to profile metabolic states of living cells in real-time.	Requires an expensive equipment.	[99]

5. Analysis of NADH Autofluorescence by FLIM

It has been known for several decades that NADH emits autofluorescence and, in contrast, NAD+ does not [26]. It is important to notice that, as the spectral properties of NADH fully overlaps with those of NADPH, it is common to measure the fluorescent contribution of both components and denominate them as NAD(P)H. Conversely, reduced flavin adenine dinucleotide (FADH2) does not produce autofluorescence compared to its oxidized version (FAD) [130]. This inverse relationship has been used to measure a "redox ratio" defined as the total fluorescence intensity of FAD divided by the total fluorescence intensity of NADH [131]. As such, under relatively constant FAD, lower levels of NAD(P)H may indicate a larger redox ratio and may correlate with a more oxidative cellular environment.

Complementing the classical intensity-based fluorescence methods, the time-resolved decay of fluorescence by FLIM provides unique information about the environment of fluorophores, including changes in pH, viscosity, or binding state to enzymes [132–135]. Importantly, at least two configurations and fluorescence lifetimes of NADH can be distinguished with this approach, namely free NADH and protein-bound NADH [32]. This is possible because the fluorescence decay of NADH in solution markedly differs when binding to different proteins, i.e., enzymes. As such, when NADH is in solution (free NADH) it exists in a folded configuration, which causes quenching of the reduced nicotinamide by the adenine group and shortening of its fluorescent lifetime. On the contrary, protein-bound NADH has an extended configuration, favoring a prolonged decay of its fluorescence. As such, the reported lifetime of free NADH in solution is significantly lower (~0.4 ns) than the protein-bound conformation (the lifetime of NADH bound to LDH is 3.4 ns) [136]. Furthermore, taking advantage of their binding to different metabolic enzymes, it has been possible to measure the particular contribution of NADH and NADPH separately by FLIM [99,137]. This may constitute a great diagnostic tool to monitor oxidative stress as NADPH is an element directly involved in redox management.

Different methods can be used to calculate the fluorescence lifetime. For this purpose, data can be fitted into a single-exponential or multi-exponential decay function where the exponential factor tau (τ) corresponds to the fluorescence lifetime of the fluorophore. Nevertheless, it is often not possible to determine the best method to fit the data a priori. As a way to circumvent this limitation, data can be also analyzed by phasor approach. Phasor analysis is a fit-free technique in which the fluorescence decay from each pixel is transformed into a point in a two-dimensional (2-D) phasor space. As such, it works on the unbiased raw data without any approximation, and it does not require a priori knowledge of the sample being imaged, giving instantaneous results. Importantly, FLIM is compatible with confocal or multiphoton laser scanning microscopy as well as wide-field illumination. To obtain more details from each methodology, readers may refer to the following excellent publications [138–141].

6. Potential Monitoring of AD Progression through NADH FLIM

Cumulative evidence from patients, as well as cellular and animal models, have suggested that analyzing the content of NADH and NADPH may be useful to monitor AD progression and oxidative stress. Accordingly, mass spectrometry analysis of brains from triple-transgenic mice (3xTg-AD) showed that this AD model is associated with lower number of metabolites from NAD(P)+/NAD(P)H-dependent reactions [142]. In coherence with this result, it was reported that the brain cortex of 3xTgAD/Polβ+/− mice (in which DNA damage is further exacerbated) has reduced NAD+/NADH ratios [143]. The underlying cause of decreased NAD+/NADH ratio might be explained by an increase in oxidative stress due to PARP the activation. Accordingly, it is expected that the consumption of NAD+ by PARP rises under high oxidative stress and DNA damage [144]. The potential mechanistic relevance of PARP activation during AD pathogenesis has been partially supported by experiments in cultured hippocampal astrocytes treated with β-amyloid, which further activated PARP, while decreasing NAD(P)H autofluorescence as well as mitochondrial oxygen consumption [145]. Furthermore, exogenous treatment of AD patient-derived fibroblasts with NAD, which not only restores NAD+ levels but also inhibits PARP, decreased oxidative stress manifested as a rise in 8-Hydroxy-2′-deoxyguanosine (DNA oxidative damage) and mitochondrial ROS [143]. This is highly similar to what our group has previously reported by showing that the inhibition of PARP-1 reduces H_2O_2-induced cell death in MCI and AD lymphocytes [67]. Together, these results suggest that under high oxidative stress conditions manifested during AD, a PARP-mediated decrease in NAD+ content could be sensed by label-free microscopy as a drop in either free/protein-bound NADH or NADPH levels. In support of this possibility, it was determined by FLIM that cultured hippocampal neurons from both 3xTg-AD as well as aged mice have diminished cytoplasmic and mitochondrial concentrations of free NADH, which is the direct source of electrons for the mitochondrial complex I [146]

In a complementary approach that supports the potential relevance of a diagnostic tool based on FLIM, it has been shown that cultured neurons from 3xTg-AD mice manifest an early oxidized redox state and lower GSH defense before macromolecular ROS damage is evident [29]. Strikingly, this oxidative damage was reflected in lower resting levels of NAD(P)H/FAD fluorescence ratio and was fully reversible through treatment with NAM. Interestingly, it has been proposed that NAM, as well as other PARP-1 inhibitors, may be used as a treatment for AD at early stages [103]. In order to further test this therapeutic chance, it would be interesting to analyze the content of NADPH in AD patient-derived ONPs by FLIM during the treatment with PARP-1 inhibitors.

The content of NAD+ and NADH in the aging human brain have been non-invasively evaluated by means of magnetic resonance (MR)-based in vivo NAD assay [129]. In coherence with a progressive loss of mitochondrial activity and lower oxidative stress management during normal aging, an age-dependent decline in the content of NAD, NAD+, and NAD+/NADH ratio coupled to increased levels of NADH was revealed in healthy elderly subjects [129]. Interestingly, the decline in NAD+ levels during human aging has been linked to the development and progression of age-related diseases such as AD [147]. Thus, decreased NAD+ levels associated with aging and neurodegeneration are strikingly compatible with the results observed in AD transgenic mice (described above). The limited information from AD patients in this field, despite the promising results in animal models, stresses the need to improve our knowledge of the disease by using patient-derived cellular models.

We sustain that analyzing AD patient-derived ONPs through NADH FLIM is a valuable approach based on the following arguments. First, oxidative stress is an early feature of

AD which is manifested in the olfactory system as well as in cultured patient-derived ONPs. Accordingly, patient-derived ONPs are cells of neuronal lineage and can be easily cultured and non-invasively isolated, constituting a cost-effective way to obtain significant amounts of biological material. Second, the use of NADH and NADPH autofluorescence enables the non-invasive imaging of biological samples, minimizing the perturbation of normal physiological conditions and in a less time-consuming manner. With this approach, AD-related oxidative stress could be sensed as an increased FAD/NAD(P)H ratio or reduced levels of NADH or NADPH, which sustain the synthesis of cytosolic and mitochondrial antioxidant molecules. For all these measurements, FLIM not only provides the exclusive technology to discriminate between NADH and NADPH autofluorescence, but also enables obtaining a higher discrimination between the cytosolic and mitochondrial contribution [99]. Thus, we consider that analyzing AD patient-derived ONPs through NADH FLIM has a great translational potential.

7. Perspectives and Future Directions

Label-free monitoring of oxidative stress in patient-derived ONPs may accelerate the discovery of molecules for effectively targeting AD. In this sense, imaging the dynamics of NAD(P)H intrinsic fluorescence (e.g., by FLIM) may offer a readily available, less toxic, and comparatively richer lecture of drug effects compared to classic proteomic and cell-fixation methods. Interestingly, patient-derived ONPs have already been used for drug screening. In particular, these cells were used to test drugs that restored acetylated tubulin patient-derived stem cells with a variety of *SPAST* mutations in Hereditary Spastic Paraplegia (HSP) [48] and to perform a multidimensional phenotypic screening with different natural products in Parkinson's disease [148].

Different cellular AD models have been used for high-throughput screening (HTS) of therapeutic molecules [149–152]. For example, a search for inhibitors of calpain activity (to prevent Aβ-induced neurotoxicity) was performed on a library of approximately 120,000 compounds and tested on differentiated SH-SY5Y cells [153]. In another approach, the motility and proliferation of PC12 cells was assessed to test multiple drugs based on Chinese herbal compounds targeting Aβ42-induced apoptosis [154]. Similarly, it has been found that a combination of bromocriptine, cromolyn, and topiramate has a potent anti-Aβ effect on patient-derived iPSCs neurons [155]. Thus, it seems plausible to perform HTS of therapeutic molecules in patient-derived ONPs coupled to label free microscopy.

A potential therapeutic strategy, which could be monitored in patient-derived ONPs, is to delay the AD-associated depletion of free NADH. This is supported by the recent observation that imposed manipulation of cysteine/cystine (Cys/CySS) redox state was able to restore mitochondrial levels of free NADH to normal ranges in neurons from triple transgenic AD-like mice [28]. Given the relevance of free NADH inside cells—not only for redox management but also for metabolic supply, to sustain ATP levels—testing for antioxidant compounds capable of modulating free NADH deserves to be further studied. It is surprising to realize that the use of patient-derived ONPs to study the role of oxidative stress during AD has been, to some extent, neglected during the past decade. At least two reasons may have contributed to this delay; the first is that culturing patient-derived ONPs from biopsies is relatively more difficult and the second is the potential lack of technologies efficient enough to detect subtle changes of oxidative stress. Nevertheless, as highlighted in this article, both reasons can no longer be sustained.

Antioxidant therapies directed against AD have shown limited success; however, they still hold great promise and room for improvement. Some clinical trials in which AD patients were supplemented with antioxidants such as vitamins C and E, either alone or in combination with cholinesterase inhibitors, have failed to improve cognitive function [156,157]. However, other attempts have shown to be moderately effective; as is the case for polyphenols, a group of phytochemicals that showed a great antioxidant and anti-inflammatory potential together with neuroprotective properties [158,159]. As such, clinical trials have suggested that polyphenolic compounds such as curcumin, resveratrol, and green tea catechins may prevent and treat some forms of dementia [160–163]. Nevertheless, other reports show poor effects of antioxidants on cognitive function, which could be related to their low bioavailability [164–166]. Emerging evidence suggests that the combined intervention of different antioxidants may improve therapeutic efficacy. For example, some clinical trials have reported cognitive improvements in AD patients treated with a mix of antioxidant compounds harboring α-tocopherol, NAC, folate, acetyl-L-carnitine, vitamin B12, and S-adenosyl methionine [167]. In line with these findings, fibroblasts derived from AD patients have shown decreased mitochondrial oxidative stress after treatment with lipoic acid and N-acetyl-cysteine (NAC) [168]. It would be extremely interesting to monitor the intrinsic fluorescence of NADPH (reflecting the antioxidant capacity of the cell) in patient-derived ONPs in response to different mixes, proportions, and doses of these antioxidant compounds. Table 3 resumes some candidate natural and chemical compounds that could be successful in clinical trials evaluating them with AD-derived ONPs.

Table 3. Natural and chemical compounds that may target ONPs.

Compounds	Targeting	Mechanism	In Vitro/In Vivo Models	Comments	Refs
Incensole acetate (IA)	Oxidative stress induced by Aβ	Increased levels of the antioxidant enzyme HO-1	Human olfactory bulb neural stem cells (hOBNSCs)	The cellular model belongs to the olfactory system; therefore, we envision similar results in our proposed cellular model.	[169]
Curcumin loaded polymeric or lipid nanosuspensions	Oxidative stress	Elevation of total cellular glutathione levels and enhanced cell viability under oxidative stress	Normal and hypoxic olfactory ensheathing cells (OECs)	The use of OECs (non-myelinating glial cells that wrap olfactory neurons) in hypoxic conditions enables a roadmap to improve the delivery of antioxidants through the nose-to-brain route.	[170–172]
Saturated medium-chain fatty acid (MCFA) decanoic acid (10:0)	Oxidative stress	Upregulation of catalase activity and increase in mitochondrial citrate synthase	Neuroblastoma cells (SH-SY5Y cells)	MCFA decanoic acid has only been evaluated in human cell lines. These findings suggest it is worth testing them in AD patient-derived ONPs.	[173, 174]
Scutellarin (SC)	Oxidative stress and apoptosis	Enhances the levels of superoxide dismutase	L-Glu-treated HT22 cells/ AD mice induced by $AlCl_3$ and D-gal	SC has shown antioxidant and antiapoptotic properties only in induced models of AD; thus, it would be interesting to evaluate these properties in a cellular model derived from AD patients.	[175]
Curcumin and Vitamin D3	Oxidative stress	Increased SOD enzyme activity and catalase enzyme expression	Primary neuronal cortical culture from rats treated with Aβ	Antioxidant combinations show a synergistic effect that could be tested in an ONP model.	[176]
TM-10 (a ferulic acid derivative and a highly selective BuChE inhibitor)	Oxidative stress, Aβ aggregation, butyrylcholinesterase (BuChE) inhibition	Neuroprotective effect against $A\beta_{42}$- mediated SH-SY5Y neurotoxicity, and autophagy induction. In mice, improves scopolamine-inducedmemory impairment	SH-SY5Y cell, U87 cell, $AlCl_3$-induced zebrafish AD model, and mice treated with scopolamine	The search of Multi-Target-Directed-Ligands (MTDLs) has allowed fusing novel natural antioxidants derivatives and highly selective BuChE inhibitors. Thus, compounds with multiple biological activities are obtained, including ChE inhibitory activity, MAOs inhibitory potency, antioxidant activity, disaggregation effect on Aβ, and the ability to cross the blood−brain barrier. The use of AD patient-derived ONPs could be a valuable tool for validating these compounds in humans.	[177, 178]

Human embryonic stem cells (ESCs) and subsequently human induced pluripotent stem cells (iPSCs) have emerged as powerful tools due to their ability for modeling neurodegenerative diseases [179]. For instance, three-dimensional (3D) organoids using patient-derived induced pluripotent stem (iPS) cells can recapitulate microcephaly that has been difficult to model in mice [180]. On the other hand, 3D advanced culture models of the brain including blood–brain barrier (BBB) allow a precise study of candidate drugs by recapitulating the brain environment [181]. In this sense, the implementation of a human brain microvessel-on-a-chip that is amenable for quantitative live 3D fluorescence analysis with high-resolution will facilitate the monitoring of NADPH movement and permeability during oxidative stress [182]. Moreover, 3D models can be harnessed to perform cutting-edge super-resolution microscopy, including high resolution volumetric imaging using Focused Ion Beam Scanning Electron Microscopy (FIB-SEM) and also with the novel modality of expansion microscopy, which integrates with lattice light-sheet microscopy (Ex-LLSM) [183]. Thus, these emerging systems to model BBB will significantly improve drug discovery.

The olfactory system, the olfactory ensheathing cells (OECs), has been cultured in three dimensions to understand their behavior in a hampered environment, such as a spinal cord injury [184]. Although there is no evidence of ONPs cultured in 3D, the findings in other cells of the olfactory system suggest ONPs will have the same outcome. Thus, it would be extremely interesting to generate a 3D model for AD with ONPs from patients, incorporating a BBB microfluidic platform and analyzing cell metabolism by label-free microscopy in response to drug treatment like that which was reported in the organotypic microfluidic breast cancer model [185]. This approach will enable us to evaluate both the effect and the efficiency to traverse BBB of the drug candidate in an AD model.

In all, non-invasively isolated ONPs from AD patients coupled to real-time monitoring of relevant metabolic intermediaries may provide an unprecedent platform for early diagnosis and drug discovery. Furthermore, cellular models derived from patients might be sensitive enough to even develop personalized therapies, as has been suggested [186]. Proposed innovations are schematized in Figure 5. We envision that these strategies may generate large improvements required for the timely diagnosis and treatment of this devastating disease.

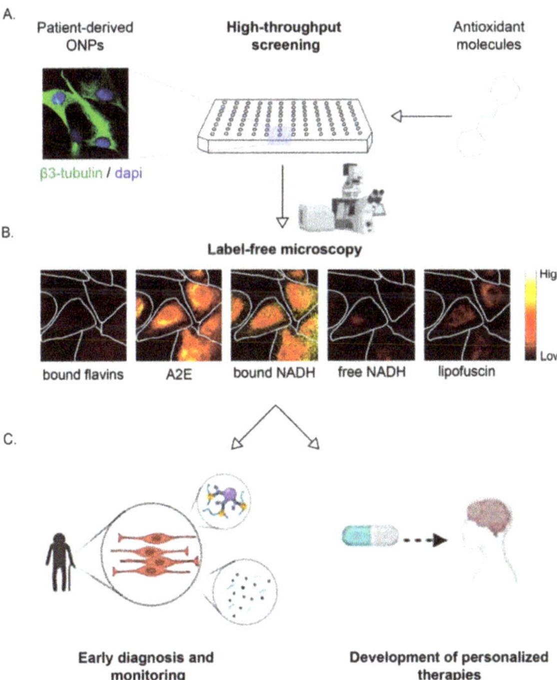

Figure 5. Isolation of patient-derived ONPs coupled to label-free microscopy offers relevant translational outcomes. (**A**) Schematic drawing of the high-throughput screening platform to study different

antioxidant molecules. (**B**) Cultured ONPs isolated from control patient-derived neurospheres were analyzed by label-free microscopy, using fluorescence hyperspectral analysis and the intensity of different intrinsic fluorophores was determined [187]. These fluorophores included: bound flavins, fluorescent retinoid derivative bis-retinoid N-retinylidene-N-retinyl ethanolamine (A2E), protein-bound NADH (bound NADH), free NADH, and lipofuscin. The original images from the publication of Gosnell et al. [187] (Figure 2) were adapted (cropped) with permission, following the guidelines of the creative commons license (CC BY 4.0, https://creativecommons.org/licenses/by/4.0/). (**C**) Analysis of intrinsic fluorophores such as protein-bound NADH or free NADH could provide relevant translational outcomes such detecting oxidative/metabolic signatures for early AD diagnosis and monitoring. In addition, those the subtle molecular profiling could settle the base for development of personalized therapies to treat AD. Created with BioRender.com.

Author Contributions: Writing—original draft preparation: L.G.-V., A.L.; writing—review and editing: L.G.-V., A.L., D.P.P., B.A.B., M.I.B.; supervision: M.I.B. All authors have read and agreed to the published version of the manuscript.

Funding: This work was supported by the grant: "Fondo Nacional de Desarrollo Científico y Tecnológico—Fondecyt" Grant Number: 1190958 (M.I.B.).

Acknowledgments: L.G.-V. was supported by Secretaría de Educación, Ciencia, Tecnología e Innovación de la Ciudad de México (SECTEI).

Conflicts of Interest: The authors declare no conflict of interest.

References

1. Haque, R.U.; Levey, A.I. Alzheimer's disease: A clinical perspective and future nonhuman primate research opportunities. *Proc. Natl. Acad. Sci. USA* **2019**. [CrossRef] [PubMed]
2. Serrano-Pozo, A.; Frosch, M.P.; Masliah, E.; Hyman, B.T. Neuropathological alterations in Alzheimer disease. *Cold Spring Harb. Perspect. Med.* **2011**, *1*, a006189. [CrossRef] [PubMed]
3. Perrin, R.J.; Fagan, A.M.; Holtzman, D.M. Multimodal techniques for diagnosis and prognosis of Alzheimer's disease. *Nature* **2009**, *461*, 916–922. [CrossRef]
4. Jack, C.R., Jr.; Knopman, D.S.; Jagust, W.J.; Shaw, L.M.; Aisen, P.S.; Weiner, M.W.; Petersen, R.C.; Trojanowski, J.Q. Hypothetical model of dynamic biomarkers of the Alzheimer's pathological cascade. *Lancet. Neurol.* **2010**, *9*, 119–128. [CrossRef]
5. Trushina, E. Alzheimer's disease mechanisms in peripheral cells: Promises and challenges. *Alzheimer's Dement.* **2019**, *5*, 652–660. [CrossRef] [PubMed]
6. Penney, J.; Ralvenius, W.T.; Tsai, L.H. Modeling Alzheimer's disease with iPSC-derived brain cells. *Mol. Psychiatry* **2020**, *25*, 148–167. [CrossRef] [PubMed]
7. Mackay-Sim, A. Concise review: Patient-derived olfactory stem cells: New models for brain diseases. *Stem Cells* **2012**, *30*, 2361–2365. [CrossRef]
8. Jimenez-Vaca, A.L.; Benitez-King, G.; Ruiz, V.; Ramirez-Rodriguez, G.B.; Hernandez-de la Cruz, B.; Salamanca-Gomez, F.A.; Gonzalez-Marquez, H.; Ramirez-Sanchez, I.; Ortiz-Lopez, L.; Velez-Del Valle, C.; et al. Exfoliated Human Olfactory Neuroepithelium: A Source of Neural Progenitor Cells. *Mol. Neurobiol.* **2018**, *55*, 2516–2523. [CrossRef]
9. Riquelme, A.; Valdes-Tovar, M.; Ugalde, O.; Maya-Ampudia, V.; Fernandez, M.; Mendoza-Duran, L.; Rodriguez-Cardenas, L.; Benitez-King, G. Potential Use of Exfoliated and Cultured Olfactory Neuronal Precursors for In Vivo Alzheimer's Disease Diagnosis: A Pilot Study. *Cell. Mol. Neurobiol.* **2020**, *40*, 87–98. [CrossRef]
10. Wolozin, B.; Zheng, B.; Loren, D.; Lesch, K.P.; Lebovics, R.S.; Lieberburg, I.; Sunderland, T. Beta/A4 domain of APP: Antigenic differences between cell lines. *J. Neurosci. Res.* **1992**, *33*, 189–195. [CrossRef]
11. Wolozin, B.; Bacic, M.; Merrill, M.J.; Lesch, K.P.; Chen, C.; Lebovics, R.S.; Sunderland, T. Differential expression of carboxyl terminal derivatives of amyloid precursor protein among cell lines. *J. Neurosci. Res.* **1992**, *33*, 163–169. [CrossRef]
12. Butterfield, D.A.; Halliwell, B. Oxidative stress, dysfunctional glucose metabolism and Alzheimer disease. *Nat. Rev. Neurosci.* **2019**, *20*, 148–160. [CrossRef]
13. Nunomura, A.; Perry, G.; Aliev, G.; Hirai, K.; Takeda, A.; Balraj, E.K.; Jones, P.K.; Ghanbari, H.; Wataya, T.; Shimohama, S.; et al. Oxidative damage is the earliest event in Alzheimer disease. *J. Neuropathol. Exp. Neurol.* **2001**, *60*, 759–767. [CrossRef] [PubMed]
14. Smith, M.A.; Nunomura, A.; Lee, H.G.; Zhu, X.; Moreira, P.I.; Avila, J.; Perry, G. Chronological primacy of oxidative stress in Alzheimer disease. *Neurobiol. Aging* **2005**, *26*, 579–580. [CrossRef]
15. Gonzalez-Reyes, R.E.; Nava-Mesa, M.O.; Vargas-Sanchez, K.; Ariza-Salamanca, D.; Mora-Munoz, L. Involvement of Astrocytes in Alzheimer's Disease from a Neuroinflammatory and Oxidative Stress Perspective. *Front. Mol. Neurosci.* **2017**, *10*, 427. [CrossRef]
16. Habas, A.; Hahn, J.; Wang, X.; Margeta, M. Neuronal activity regulates astrocytic Nrf2 signaling. *Proc. Natl. Acad. Sci. USA* **2013**, *110*, 18291–18296. [CrossRef]

17. Qiu, J.; Dando, O.; Febery, J.A.; Fowler, J.H.; Chandran, S.; Hardingham, G.E. Neuronal Activity and Its Role in Controlling Antioxidant Genes. *Int. J. Mol. Sci.* **2020**, *21*, 1933. [CrossRef]
18. Baxter, P.S.; Hardingham, G.E. Adaptive regulation of the brain's antioxidant defences by neurons and astrocytes. *Free Radic. Biol. Med.* **2016**, *100*, 147–152. [CrossRef]
19. Araque, A.; Parpura, V.; Sanzgiri, R.P.; Haydon, P.G. Tripartite synapses: Glia, the unacknowledged partner. *Trends Neurosci.* **1999**, *22*, 208–215. [CrossRef]
20. Perea, G.; Navarrete, M.; Araque, A. Tripartite synapses: Astrocytes process and control synaptic information. *Trends Neurosci.* **2009**, *32*, 421–431. [CrossRef] [PubMed]
21. Wang, Q.; Wang, Y.; Shimony, J.S.; Owen, C.J.; Liu, J.; Fagan, A.M.; Cairns, N.J.; Ances, B.; Morris, J.C.; Benzinger, T.L.S. IC-P-172: Simultaneous Quantification of White Matter Abnormalities and Vasogenic Edema in Early Alzheimer Disease. *Alzheimer's Dement.* **2016**, *12*, P125–P126. [CrossRef]
22. Kitchen, P.; Salman, M.M.; Halsey, A.M.; Clarke-Bland, C.; MacDonald, J.A.; Ishida, H.; Vogel, H.J.; Almutiri, S.; Logan, A.; Kreida, S.; et al. Targeting Aquaporin-4 Subcellular Localization to Treat Central Nervous System Edema. *Cell* **2020**, *181*, 784–799. [CrossRef]
23. Getchell, M.L.; Shah, D.S.; Buch, S.K.; Davis, D.G.; Getchell, T.V. 3-Nitrotyrosine immunoreactivity in olfactory receptor neurons of patients with Alzheimer's disease: Implications for impaired odor sensitivity. *Neurobiol. Aging* **2003**, *24*, 663–673. [CrossRef]
24. Ghanbari, H.A.; Ghanbari, K.; Harris, P.L.; Jones, P.K.; Kubat, Z.; Castellani, R.J.; Wolozin, B.L.; Smith, M.A.; Perry, G. Oxidative damage in cultured human olfactory neurons from Alzheimer's disease patients. *Aging Cell* **2004**, *3*, 41–44. [CrossRef] [PubMed]
25. Nelson, V.M.; Dancik, C.M.; Pan, W.; Jiang, Z.G.; Lebowitz, M.S.; Ghanbari, H.A. PAN-811 inhibits oxidative stress-induced cell death of human Alzheimer's disease-derived and age-matched olfactory neuroepithelial cells via suppression of intracellular reactive oxygen species. *J. Alzheimer's Dis.* **2009**, *17*, 611–619. [CrossRef]
26. Chance, B. Spectra and reaction kinetics of respiratory pigments of homogenized and intact cells. *Nature* **1952**, *169*, 215–221. [CrossRef]
27. Ghosh, D.; Levault, K.R.; Brewer, G.J. Relative importance of redox buffers GSH and NAD(P)H in age-related neurodegeneration and Alzheimer disease-like mouse neurons. *Aging Cell* **2014**, *13*, 631–640. [CrossRef] [PubMed]
28. Dong, Y.; Sameni, S.; Digman, M.A.; Brewer, G.J. Reversibility of Age-related Oxidized Free NADH Redox States in Alzheimer's Disease Neurons by Imposed External Cys/CySS Redox Shifts. *Sci. Rep.* **2019**, *9*, 11274. [CrossRef]
29. Ghosh, D.; LeVault, K.R.; Barnett, A.J.; Brewer, G.J. A reversible early oxidized redox state that precedes macromolecular ROS damage in aging nontransgenic and 3xTg-AD mouse neurons. *J. Neurosci. Off. J. Soc. Neurosci.* **2012**, *32*, 5821–5832. [CrossRef]
30. Bradshaw, P.C. Cytoplasmic and Mitochondrial NADPH-Coupled Redox Systems in the Regulation of Aging. *Nutrients* **2019**, *11*, 504. [CrossRef]
31. Schneckenburger, H.; Koenig, K. Fluorescence decay kinetics and imaging of NAD(P)H and flavins as metabolic indicators. *Opt. Eng.* **1992**, 31. [CrossRef]
32. Lakowicz, J.R.; Szmacinski, H.; Nowaczyk, K.; Johnson, M.L. Fluorescence lifetime imaging of free and protein-bound NADH. *Proc. Natl. Acad. Sci. USA* **1992**, *89*, 1271–1275. [CrossRef] [PubMed]
33. Jafek, B.W. Ultrastructure of human nasal mucosa. *Laryngoscope* **1983**, *93*, 1576–1599. [CrossRef]
34. Moran, D.T.; Rowley, J.C., 3rd; Jafek, B.W.; Lovell, M.A. The fine structure of the olfactory mucosa in man. *J. Neurocytol.* **1982**, *11*, 721–746. [CrossRef]
35. Morrison, E.E.; Costanzo, R.M. Morphology of the human olfactory epithelium. *J. Comp. Neurol.* **1990**, *297*, 1–13. [CrossRef]
36. Verhaagen, J.; Oestreicher, A.B.; Gispen, W.H.; Margolis, F.L. The expression of the growth associated protein B50/GAP43 in the olfactory system of neonatal and adult rats. *J. Neurosci. Off. J. Soc. Neurosci.* **1989**, *9*, 683–691. [CrossRef]
37. Holbrook, E.H.; Wu, E.; Curry, W.T.; Lin, D.T.; Schwob, J.E. Immunohistochemical characterization of human olfactory tissue. *Laryngoscope* **2011**, *121*, 1687–1701. [CrossRef] [PubMed]
38. Fletcher, R.B.; Das, D.; Gadye, L.; Street, K.N.; Baudhuin, A.; Wagner, A.; Cole, M.B.; Flores, Q.; Choi, Y.G.; Yosef, N.; et al. Deconstructing Olfactory Stem Cell Trajectories at Single-Cell Resolution. *Cell Stem Cell* **2017**, *20*, 817–830. [CrossRef]
39. Hahn, C.G.; Han, L.Y.; Rawson, N.E.; Mirza, N.; Borgmann-Winter, K.; Lenox, R.H.; Arnold, S.E. In vivo and in vitro neurogenesis in human olfactory epithelium. *J. Comp. Neurol.* **2005**, *483*, 154–163. [CrossRef] [PubMed]
40. Chen, X.; Fang, H.; Schwob, J.E. Multipotency of purified, transplanted globose basal cells in olfactory epithelium. *J. Comp. Neurol.* **2004**, *469*, 457–474. [CrossRef] [PubMed]
41. Dominici, M.; Le Blanc, K.; Mueller, I.; Slaper-Cortenbach, I.; Marini, F.; Krause, D.; Deans, R.; Keating, A.; Prockop, D.; Horwitz, E. Minimal criteria for defining multipotent mesenchymal stromal cells. The International Society for Cellular Therapy position statement. *Cytotherapy* **2006**, *8*, 315–317. [CrossRef] [PubMed]
42. Delorme, B.; Nivet, E.; Gaillard, J.; Haupl, T.; Ringe, J.; Deveze, A.; Magnan, J.; Sohier, J.; Khrestchatisky, M.; Roman, F.S.; et al. The human nose harbors a niche of olfactory ectomesenchymal stem cells displaying neurogenic and osteogenic properties. *Stem Cells Dev.* **2010**, *19*, 853–866. [CrossRef]
43. Murrell, W.; Feron, F.; Wetzig, A.; Cameron, N.; Splatt, K.; Bellette, B.; Bianco, J.; Perry, C.; Lee, G.; Mackay-Sim, A. Multipotent stem cells from adult olfactory mucosa. *Dev. Dyn. Off. Publ. Am. Assoc. Anat.* **2005**, *233*, 496–515. [CrossRef]
44. Tanos, T.; Saibene, A.M.; Pipolo, C.; Battaglia, P.; Felisati, G.; Rubio, A. Isolation of putative stem cells present in human adult olfactory mucosa. *PLoS ONE* **2017**, *12*, e0181151. [CrossRef]

45. Matigian, N.; Abrahamsen, G.; Sutharsan, R.; Cook, A.L.; Vitale, A.M.; Nouwens, A.; Bellette, B.; An, J.; Anderson, M.; Beckhouse, A.G.; et al. Disease-specific, neurosphere-derived cells as models for brain disorders. *Dis. Models Mech.* **2010**, *3*, 785–798. [CrossRef] [PubMed]
46. Feron, F.; Gepner, B.; Lacassagne, E.; Stephan, D.; Mesnage, B.; Blanchard, M.P.; Boulanger, N.; Tardif, C.; Deveze, A.; Rousseau, S.; et al. Olfactory stem cells reveal *MOCOS* as a new player in autism spectrum disorders. *Mol. Psychiatry* **2016**, *21*, 1215–1224. [CrossRef] [PubMed]
47. Stewart, R.; Wali, G.; Perry, C.; Lavin, M.F.; Feron, F.; Mackay-Sim, A.; Sutharsan, R. A Patient-Specific Stem Cell Model to Investigate the Neurological Phenotype Observed in Ataxia-Telangiectasia. *Methods Mol. Biol.* **2017**, *1599*, 391–400. [CrossRef]
48. Fan, Y.; Wali, G.; Sutharsan, R.; Bellette, B.; Crane, D.I.; Sue, C.M.; Mackay-Sim, A. Low dose tubulin-binding drugs rescue peroxisome trafficking deficit in patient-derived stem cells in Hereditary Spastic Paraplegia. *Biol. Open* **2014**, *3*, 494–502. [CrossRef]
49. Ayala-Grosso, C.A.; Pieruzzini, R.; Diaz-Solano, D.; Wittig, O.; Abrante, L.; Vargas, L.; Cardier, J. Amyloid-abeta Peptide in olfactory mucosa and mesenchymal stromal cells of mild cognitive impairment and Alzheimer's disease patients. *Brain Pathol.* **2015**, *25*, 136–145. [CrossRef]
50. Solis-Chagoyan, H.; Flores-Soto, E.; Valdes-Tovar, M.; Cercos, M.G.; Calixto, E.; Montano, L.M.; Barajas-Lopez, C.; Sommer, B.; Aquino-Galvez, A.; Trueta, C.; et al. Purinergic Signaling Pathway in Human Olfactory Neuronal Precursor Cells. *Stem Cells Int.* **2019**, *2019*, 2728786. [CrossRef]
51. Benitez-King, G.; Riquelme, A.; Ortiz-Lopez, L.; Berlanga, C.; Rodriguez-Verdugo, M.S.; Romo, F.; Calixto, E.; Solis-Chagoyan, H.; Jimenez, M.; Montano, L.M.; et al. A non-invasive method to isolate the neuronal linage from the nasal epithelium from schizophrenic and bipolar diseases. *J. Neurosci. Methods* **2011**, *201*, 35–45. [CrossRef]
52. Wolozin, B.; Sunderland, T.; Zheng, B.B.; Resau, J.; Dufy, B.; Barker, J.; Swerdlow, R.; Coon, H. Continuous culture of neuronal cells from adult human olfactory epithelium. *J. Mol. Neurosci. Mn* **1992**, *3*, 137–146. [CrossRef]
53. Gomez, G.; Rawson, N.E.; Hahn, C.G.; Michaels, R.; Restrepo, D. Characteristics of odorant elicited calcium changes in cultured human olfactory neurons. *J. Neurosci. Res.* **2000**, *62*, 737–749. [CrossRef]
54. Yazinski, S.; Gomez, G. Time course of structural and functional maturation of human olfactory epithelial cells in vitro. *J. Neurosci. Res.* **2014**, *92*, 64–73. [CrossRef]
55. Kagan, V.E.; Tyurina, Y.Y.; Sun, W.Y.; Vlasova, I.I.; Dar, H.; Tyurin, V.A.; Amoscato, A.A.; Mallampalli, R.; van der Wel, P.C.A.; He, R.R.; et al. Redox phospholipidomics of enzymatically generated oxygenated phospholipids as specific signals of programmed cell death. *Free Radic. Biol. Med.* **2020**, *147*, 231–241. [CrossRef] [PubMed]
56. Andersen, J.K. Oxidative stress in neurodegeneration: Cause or consequence? *Nat. Med.* **2004**, *10*, S18–S25. [CrossRef] [PubMed]
57. Sutherland, G.T.; Chami, B.; Youssef, P.; Witting, P.K. Oxidative stress in Alzheimer's disease: Primary villain or physiological by-product? *Redox Rep. Commun. Free Radic. Res.* **2013**, *18*, 134–141. [CrossRef]
58. Butterfield, D.A.; Boyd-Kimball, D. Oxidative Stress, Amyloid-beta Peptide, and Altered Key Molecular Pathways in the Pathogenesis and Progression of Alzheimer's Disease. *J. Alzheimer's Dis.* **2018**, *62*, 1345–1367. [CrossRef] [PubMed]
59. Tonnies, E.; Trushina, E. Oxidative Stress, Synaptic Dysfunction, and Alzheimer's Disease. *J. Alzheimer's Dis.* **2017**, *57*, 1105–1121. [CrossRef] [PubMed]
60. Kulkarni-Narla, A.; Getchell, T.V.; Schmitt, F.A.; Getchell, M.L. Manganese and copper-zinc superoxide dismutases in the human olfactory mucosa: Increased immunoreactivity in Alzheimer's disease. *Exp. Neurol.* **1996**, *140*, 115–125. [CrossRef]
61. Chuah, M.I.; Getchell, M.L. Metallothionein in olfactory mucosa of Alzheimer's disease patients and *apoE*-deficient mice. *Neuroreport* **1999**, *10*, 1919–1924. [CrossRef]
62. Calhoun-Haney, R.; Murphy, C. Apolipoprotein epsilon4 is associated with more rapid decline in odor identification than in odor threshold or Dementia Rating Scale scores. *Brain Cogn.* **2005**, *58*, 178–182. [CrossRef] [PubMed]
63. Gilbert, P.E.; Murphy, C. The effect of the *ApoE* epsilon4 allele on recognition memory for olfactory and visual stimuli in patients with pathologically confirmed Alzheimer's disease, probable Alzheimer's disease, and healthy elderly controls. *J. Clin. Exp. Neuropsychol.* **2004**, *26*, 779–794. [CrossRef] [PubMed]
64. Wang, Q.S.; Tian, L.; Huang, Y.L.; Qin, S.; He, L.Q.; Zhou, J.N. Olfactory identification and apolipoprotein E epsilon 4 allele in mild cognitive impairment. *Brain Res.* **2002**, *951*, 77–81. [CrossRef]
65. Chen, M.; Lee, H.K.; Moo, L.; Hanlon, E.; Stein, T.; Xia, W. Common proteomic profiles of induced pluripotent stem cell-derived three-dimensional neurons and brain tissue from Alzheimer patients. *J. Proteom.* **2018**, *182*, 21–33. [CrossRef]
66. Ponce, D.P.; Salech, F.; SanMartin, C.D.; Silva, M.; Xiong, C.; Roe, C.M.; Henriquez, M.; Quest, A.F.; Behrens, M.I. Increased susceptibility to oxidative death of lymphocytes from Alzheimer patients correlates with dementia severity. *Curr. Alzheimer Res.* **2014**, *11*, 892–898. [CrossRef]
67. Salech, F.; Ponce, D.P.; SanMartin, C.D.; Rogers, N.K.; Chacon, C.; Henriquez, M.; Behrens, M.I. PARP-1 and p53 Regulate the Increased Susceptibility to Oxidative Death of Lymphocytes from MCI and AD Patients. *Front. Aging Neurosci.* **2017**, *9*, 310. [CrossRef]
68. Tang, K.; Hynan, L.S.; Baskin, F.; Rosenberg, R.N. Platelet amyloid precursor protein processing: A bio-marker for Alzheimer's disease. *J. Neurol. Sci.* **2006**, *240*, 53–58. [CrossRef]
69. Neumann, K.; Farias, G.; Slachevsky, A.; Perez, P.; Maccioni, R.B. Human platelets tau: A potential peripheral marker for Alzheimer's disease. *J. Alzheimer's Dis. Jad* **2011**, *25*, 103–109. [CrossRef]

70. Colciaghi, F.; Marcello, E.; Borroni, B.; Zimmermann, M.; Caltagirone, C.; Cattabeni, F.; Padovani, A.; Di Luca, M. Platelet APP, ADAM 10 and BACE alterations in the early stages of Alzheimer disease. *Neurology* **2004**, *62*, 498–501. [CrossRef]
71. Colciaghi, F.; Borroni, B.; Pastorino, L.; Marcello, E.; Zimmermann, M.; Cattabeni, F.; Padovani, A.; Di Luca, M. [alpha]-Secretase ADAM10 as well as [alpha]APPs is reduced in platelets and CSF of Alzheimer disease patients. *Mol. Med.* **2002**, *8*, 67–74. [CrossRef]
72. Vignini, A.; Sartini, D.; Morganti, S.; Nanetti, L.; Luzzi, S.; Provinciali, L.; Mazzanti, L.; Emanuelli, M. Platelet amyloid precursor protein isoform expression in Alzheimer's disease: Evidence for peripheral marker. *Int. J. Immunopathol. Pharmacol.* **2011**, *24*, 529–534. [CrossRef]
73. Herrera-Rivero, M.; Soto-Cid, A.; Hernandez, M.E.; Aranda-Abreu, G.E. Tau, APP, NCT and BACE1 in lymphocytes through cognitively normal ageing and neuropathology. *An. Da Acad. Bras. De Cienc.* **2013**, *85*, 1489–1496. [CrossRef]
74. Khan, T.K.; Sen, A.; Hongpaisan, J.; Lim, C.S.; Nelson, T.J.; Alkon, D.L. PKCepsilon deficits in Alzheimer's disease brains and skin fibroblasts. *J. Alzheimer's Dis.* **2015**, *43*, 491–509. [CrossRef] [PubMed]
75. Sproul, A.A.; Jacob, S.; Pre, D.; Kim, S.H.; Nestor, M.W.; Navarro-Sobrino, M.; Santa-Maria, I.; Zimmer, M.; Aubry, S.; Steele, J.W.; et al. Characterization and molecular profiling of *PSEN1* familial Alzheimer's disease iPSC-derived neural progenitors. *PLoS ONE* **2014**, *9*, e84547. [CrossRef]
76. Muratore, C.R.; Rice, H.C.; Srikanth, P.; Callahan, D.G.; Shin, T.; Benjamin, L.N.; Walsh, D.M.; Selkoe, D.J.; Young-Pearse, T.L. The familial Alzheimer's disease *APP*V717I mutation alters APP processing and Tau expression in iPSC-derived neurons. *Hum. Mol. Genet.* **2014**, *23*, 3523–3536. [CrossRef]
77. Hu, N.W.; Corbett, G.T.; Moore, S.; Klyubin, I.; O'Malley, T.T.; Walsh, D.M.; Livesey, F.J.; Rowan, M.J. Extracellular Forms of Abeta and Tau from iPSC Models of Alzheimer's Disease Disrupt Synaptic Plasticity. *Cell Rep.* **2018**, *23*, 1932–1938. [CrossRef] [PubMed]
78. Bosetti, F.; Brizzi, F.; Barogi, S.; Mancuso, M.; Siciliano, G.; Tendi, E.A.; Murri, L.; Rapoport, S.I.; Solaini, G. Cytochrome c oxidase and mitochondrial F1F0-ATPase (ATP synthase) activities in platelets and brain from patients with Alzheimer's disease. *Neurobiol. Aging* **2002**, *23*, 371–376. [CrossRef]
79. Cardoso, S.M.; Proenca, M.T.; Santos, S.; Santana, I.; Oliveira, C.R. Cytochrome c oxidase is decreased in Alzheimer's disease platelets. *Neurobiol. Aging* **2004**, *25*, 105–110. [CrossRef]
80. Leuner, K.; Schulz, K.; Schutt, T.; Pantel, J.; Prvulovic, D.; Rhein, V.; Savaskan, E.; Czech, C.; Eckert, A.; Muller, W.E. Peripheral mitochondrial dysfunction in Alzheimer's disease: Focus on lymphocytes. *Mol. Neurobiol.* **2012**, *46*, 194–204. [CrossRef] [PubMed]
81. Sims, N.R.; Finegan, J.M.; Blass, J.P. Altered metabolic properties of cultured skin fibroblasts in Alzheimer's disease. *Ann. Neurol.* **1987**, *21*, 451–457. [CrossRef] [PubMed]
82. Perez, M.J.; Ponce, D.P.; Osorio-Fuentealba, C.; Behrens, M.I.; Quintanilla, R.A. Mitochondrial Bioenergetics Is Altered in Fibroblasts from Patients with Sporadic Alzheimer's Disease. *Front. Neurosci.* **2017**, *11*, 553. [CrossRef] [PubMed]
83. Martin-Maestro, P.; Gargini, R.; Garcia, E.; Perry, G.; Avila, J.; Garcia-Escudero, V. Slower Dynamics and Aged Mitochondria in Sporadic Alzheimer's Disease. *Oxidative Med. Cell. Longev.* **2017**, *2017*, 9302761. [CrossRef]
84. Birnbaum, J.H.; Wanner, D.; Gietl, A.F.; Saake, A.; Kundig, T.M.; Hock, C.; Nitsch, R.M.; Tackenberg, C. Oxidative stress and altered mitochondrial protein expression in the absence of amyloid-beta and tau pathology in iPSC-derived neurons from sporadic Alzheimer's disease patients. *Stem Cell Res.* **2018**, *27*, 121–130. [CrossRef] [PubMed]
85. Martin-Maestro, P.; Gargini, R.; Sproul, A.A.; Garcia, E.; Anton, L.C.; Noggle, S.; Arancio, O.; Avila, J.; Garcia-Escudero, V. Mitophagy Failure in Fibroblasts and iPSC-Derived Neurons of Alzheimer's Disease-Associated Presenilin 1 Mutation. *Front. Mol. Neurosci.* **2017**, *10*, 291. [CrossRef]
86. Li, L.; Roh, J.H.; Kim, H.J.; Park, H.J.; Kim, M.; Koh, W.; Heo, H.; Chang, J.W.; Nakanishi, M.; Yoon, T.; et al. The First Generation of iPSC Line from a Korean Alzheimer's Disease Patient Carrying *APP*-V715M Mutation Exhibits a Distinct Mitochondrial Dysfunction. *Exp. Neurobiol.* **2019**, *28*, 329–336. [CrossRef] [PubMed]
87. Repetto, M.G.; Reides, C.G.; Evelson, P.; Kohan, S.; de Lustig, E.S.; Llesuy, S.F. Peripheral markers of oxidative stress in probable Alzheimer patients. *Eur. J. Clin. Investig.* **1999**, *29*, 643–649. [CrossRef]
88. Kawamoto, E.M.; Munhoz, C.D.; Glezer, I.; Bahia, V.S.; Caramelli, P.; Nitrini, R.; Gorjao, R.; Curi, R.; Scavone, C.; Marcourakis, T. Oxidative state in platelets and erythrocytes in aging and Alzheimer's disease. *Neurobiol. Aging* **2005**, *26*, 857–864. [CrossRef] [PubMed]
89. Ihara, Y.; Hayabara, T.; Sasaki, K.; Kawada, R.; Nakashima, Y.; Kuroda, S. Relationship between oxidative stress and *apoE* phenotype in Alzheimer's disease. *Acta Neurol. Scand.* **2000**, *102*, 346–349. [CrossRef]
90. Ascolani, A.; Balestrieri, E.; Minutolo, A.; Mosti, S.; Spalletta, G.; Bramanti, P.; Mastino, A.; Caltagirone, C.; Macchi, B. Dysregulated NF-kappaB pathway in peripheral mononuclear cells of Alzheimer's disease patients. *Curr. Alzheimer Res.* **2012**, *9*, 128–137. [CrossRef]
91. Behrens, M.I.; Silva, M.; Salech, F.; Ponce, D.P.; Merino, D.; Sinning, M.; Xiong, C.; Roe, C.M.; Quest, A.F. Inverse susceptibility to oxidative death of lymphocytes obtained from Alzheimer's patients and skin cancer survivors: Increased apoptosis in Alzheimer's and reduced necrosis in cancer. *J. Gerontol. Ser. ABiol. Sci. Med Sci.* **2012**, *67*, 1036–1040. [CrossRef]
92. Cecchi, C.; Fiorillo, C.; Sorbi, S.; Latorraca, S.; Nacmias, B.; Bagnoli, S.; Nassi, P.; Liguri, G. Oxidative stress and reduced antioxidant defenses in peripheral cells from familial Alzheimer's patients. *Free Radic. Biol. Med.* **2002**, *33*, 1372–1379. [CrossRef]
93. Cecchi, C.; Fiorillo, C.; Baglioni, S.; Pensalfini, A.; Bagnoli, S.; Nacmias, B.; Sorbi, S.; Nosi, D.; Relini, A.; Liguri, G. Increased susceptibility to amyloid toxicity in familial Alzheimer's fibroblasts. *Neurobiol. Aging* **2007**, *28*, 863–876. [CrossRef] [PubMed]

94. Choi, J.; Malakowsky, C.A.; Talent, J.M.; Conrad, C.C.; Carroll, C.A.; Weintraub, S.T.; Gracy, R.W. Anti-apoptotic proteins are oxidized by Abeta25-35 in Alzheimer's fibroblasts. *Biochim. Et Biophys. Acta* **2003**, *1637*, 135–141. [CrossRef]
95. Mota, S.I.; Costa, R.O.; Ferreira, I.L.; Santana, I.; Caldeira, G.L.; Padovano, C.; Fonseca, A.C.; Baldeiras, I.; Cunha, C.; Letra, L.; et al. Oxidative stress involving changes in Nrf2 and ER stress in early stages of Alzheimer's disease. *Biochim. Et Biophys. Acta* **2015**, *1852*, 1428–1441. [CrossRef]
96. Piccini, A.; Fassio, A.; Pasqualetto, E.; Vitali, A.; Borghi, R.; Palmieri, D.; Nacmias, B.; Sorbi, S.; Sitia, R.; Tabaton, M. Fibroblasts from FAD-linked presenilin 1 mutations display a normal unfolded protein response but overproduce Abeta42 in response to tunicamycin. *Neurobiol. Dis.* **2004**, *15*, 380–386. [CrossRef] [PubMed]
97. Martins, S.; Muller-Schiffmann, A.; Erichsen, L.; Bohndorf, M.; Wruck, W.; Sleegers, K.; Van Broeckhoven, C.; Korth, C.; Adjaye, J. IPSC-Derived Neuronal Cultures Carrying the Alzheimer's Disease Associated *TREM2* R47H Variant Enables the Construction of an Abeta-Induced Gene Regulatory Network. *Int. J. Mol. Sci.* **2020**, *21*, 4516. [CrossRef]
98. Kondo, T.; Asai, M.; Tsukita, K.; Kutoku, Y.; Ohsawa, Y.; Sunada, Y.; Imamura, K.; Egawa, N.; Yahata, N.; Okita, K.; et al. Modeling Alzheimer's disease with iPSCs reveals stress phenotypes associated with intracellular Abeta and differential drug responsiveness. *Cell Stem Cell* **2013**, *12*, 487–496. [CrossRef]
99. Blacker, T.S.; Duchen, M.R. Investigating mitochondrial redox state using NADH and NADPH autofluorescence. *Free Radic. Biol. Med.* **2016**, *100*, 53–65. [CrossRef]
100. Xiao, W.; Wang, R.S.; Handy, D.E.; Loscalzo, J. NAD(H) and NADP(H) Redox Couples and Cellular Energy Metabolism. *Antioxid. Redox Signal.* **2018**, *28*, 251–272. [CrossRef]
101. Jesko, H.; Wencel, P.; Strosznajder, R.P.; Strosznajder, J.B. Sirtuins and Their Roles in Brain Aging and Neurodegenerative Disorders. *Neurochem. Res.* **2017**, *42*, 876–890. [CrossRef] [PubMed]
102. Manjula, R.; Anuja, K.; Alcain, F.J. SIRT1 and SIRT2 Activity Control in Neurodegenerative Diseases. *Front. Pharmacol.* **2020**, *11*, 585821. [CrossRef] [PubMed]
103. Salech, F.; Ponce, D.P.; Paula-Lima, A.C.; SanMartin, C.D.; Behrens, M.I. Nicotinamide, a Poly [ADP-Ribose] Polymerase 1 (PARP-1) Inhibitor, as an Adjunctive Therapy for the Treatment of Alzheimer's Disease. *Front. Aging Neurosci.* **2020**, *12*, 255. [CrossRef] [PubMed]
104. Holmstrom, K.M.; Finkel, T. Cellular mechanisms and physiological consequences of redox-dependent signalling. *Nat. Rev. Mol. Cell Biol.* **2014**, *15*, 411–421. [CrossRef] [PubMed]
105. Datta, R.; Heaster, T.M.; Sharick, J.T.; Gillette, A.A.; Skala, M.C. Fluorescence lifetime imaging microscopy: Fundamentals and advances in instrumentation, analysis, and applications. *J. Biomed. Opt.* **2020**, *25*, 1–43. [CrossRef]
106. Bordone, M.P.; Salman, M.M.; Titus, H.E.; Amini, E.; Andersen, J.V.; Chakraborti, B.; Diuba, A.V.; Dubouskaya, T.G.; Ehrke, E.; Espindola de Freitas, A.; et al. The energetic brain—A review from students to students. *J. Neurochem.* **2019**, *151*, 139–165. [CrossRef]
107. Pellerin, L.; Magistretti, P.J. Glutamate uptake into astrocytes stimulates aerobic glycolysis: A mechanism coupling neuronal activity to glucose utilization. *Proc. Natl. Acad. Sci. USA* **1994**, *91*, 10625–10629. [CrossRef] [PubMed]
108. Barros, L.F. Metabolic signaling by lactate in the brain. *Trends Neurosci.* **2013**, *36*, 396–404. [CrossRef] [PubMed]
109. Vergara, R.C.; Jaramillo-Riveri, S.; Luarte, A.; Moenne-Loccoz, C.; Fuentes, R.; Couve, A.; Maldonado, P.E. The Energy Homeostasis Principle: Neuronal Energy Regulation Drives Local Network Dynamics Generating Behavior. *Front. Comput. Neurosci.* **2019**, *13*, 49. [CrossRef] [PubMed]
110. Diaz-Garcia, C.M.; Mongeon, R.; Lahmann, C.; Koveal, D.; Zucker, H.; Yellen, G. Neuronal Stimulation Triggers Neuronal Glycolysis and Not Lactate Uptake. *Cell Metab.* **2017**, *26*, 361–374. [CrossRef]
111. Bak, L.K.; Walls, A.B.; Schousboe, A.; Ring, A.; Sonnewald, U.; Waagepetersen, H.S. Neuronal glucose but not lactate utilization is positively correlated with NMDA-induced neurotransmission and fluctuations in cytosolic Ca2+ levels. *J. Neurochem.* **2009**, *109* Suppl 1, 87–93. [CrossRef]
112. Chen, L.; Zhang, Z.; Hoshino, A.; Zheng, H.D.; Morley, M.; Arany, Z.; Rabinowitz, J.D. NADPH production by the oxidative pentose-phosphate pathway supports folate metabolism. *Nat. Metab.* **2019**, *1*, 404–415. [CrossRef]
113. Bolanos, J.P.; Almeida, A. The pentose-phosphate pathway in neuronal survival against nitrosative stress. *IUBMB Life* **2010**, *62*, 14–18. [CrossRef] [PubMed]
114. Rydstrom, J. Mitochondrial NADPH, transhydrogenase and disease. *Biochim. Biophys. Acta* **2006**, *1757*, 721–726. [CrossRef]
115. Lopert, P.; Patel, M. Nicotinamide nucleotide transhydrogenase (Nnt) links the substrate requirement in brain mitochondria for hydrogen peroxide removal to the thioredoxin/peroxiredoxin (Trx/Prx) system. *J. Biol. Chem.* **2014**, *289*, 15611–15620. [CrossRef]
116. Ying, W. NAD+/NADH and NADP+/NADPH in cellular functions and cell death: Regulation and biological consequences. *Antioxid. Redox Signal.* **2008**, *10*, 179–206. [CrossRef] [PubMed]
117. Sies, H.; Berndt, C.; Jones, D.P. Oxidative Stress. *Annu. Rev. Biochem.* **2017**, *86*, 715–748. [CrossRef] [PubMed]
118. Brolin, S.E.; Naeser, P. Bioluminescence analysis of NAD(P)H and NAD(P)+ preventing mutual interference by selective nucleotide destruction and enhanced specific light emission. *J. Biochem. Biophys. Methods* **1992**, *25*, 149–162. [CrossRef]
119. Bernofsky, C.; Swan, M. An improved cycling assay for nicotinamide adenine dinucleotide. *Anal. Biochem.* **1973**, *53*, 452–458. [CrossRef]
120. Grant, R.S.; Kapoor, V. Murine glial cells regenerate NAD, after peroxide-induced depletion, using either nicotinic acid, nicotinamide, or quinolinic acid as substrates. *J. Neurochem.* **1998**, *70*, 1759–1763. [CrossRef]

121. Yu, Q.; Pourmandi, N.; Xue, L.; Gondrand, C.; Fabritz, S.; Bardy, D.; Patiny, L.; Katsyuba, E.; Auwerx, J.; Johnsson, K. A biosensor for measuring NAD(+) levels at the point of care. *Nat. Metab.* **2019**, *1*, 1219–1225. [CrossRef] [PubMed]
122. Casabona, G.; Sturiale, L.; L'Episcopo, M.R.; Raciti, G.; Fazzio, A.; Sarpietro, M.G.; Genazzani, A.A.; Cambria, A.; Nicoletti, F. HPLC analysis of cyclic adenosine diphosphate ribose and adenosine diphosphate ribose: Determination of NAD+ metabolites in hippocampal membranes. *Ital. J. Biochem.* **1995**, *44*, 258–268.
123. Trammell, S.A.; Brenner, C. Targeted, LCMS-based Metabolomics for Quantitative Measurement of NAD(+) Metabolites. *Comput. Struct. Biotechnol. J.* **2013**, *4*, e201301012. [CrossRef]
124. Evans, J.; Wang, T.C.; Heyes, M.P.; Markey, S.P. LC/MS analysis of NAD biosynthesis using stable isotope pyridine precursors. *Anal. Biochem.* **2002**, *306*, 197–203. [CrossRef] [PubMed]
125. Yamada, K.; Hara, N.; Shibata, T.; Osago, H.; Tsuchiya, M. The simultaneous measurement of nicotinamide adenine dinucleotide and related compounds by liquid chromatography/electrospray ionization tandem mass spectrometry. *Anal. Biochem.* **2006**, *352*, 282–285. [CrossRef]
126. Bustamante, S.; Jayasena, T.; Richani, D.; Gilchrist, R.B.; Wu, L.E.; Sinclair, D.A.; Sachdev, P.S.; Braidy, N. Quantifying the cellular NAD+ metabolome using a tandem liquid chromatography mass spectrometry approach. *Metab. Off. J. Metab. Soc.* **2017**, *14*, 15. [CrossRef]
127. Zhao, Y.; Yang, Y. Profiling metabolic states with genetically encoded fluorescent biosensors for NADH. *Curr. Opin. Biotechnol.* **2015**, *31*, 86–92. [CrossRef] [PubMed]
128. Cohen, M.S.; Stewart, M.L.; Goodman, R.H.; Cambronne, X.A. Methods for Using a Genetically Encoded Fluorescent Biosensor to Monitor Nuclear NAD. *Methods Mol. Biol.* **2018**, *1813*, 391–414. [CrossRef]
129. Zhu, X.H.; Lu, M.; Lee, B.Y.; Ugurbil, K.; Chen, W. In vivo NAD assay reveals the intracellular NAD contents and redox state in healthy human brain and their age dependences. *Proc. Natl. Acad. Sci. USA* **2015**, *112*, 2876–2881. [CrossRef]
130. Reinert, K.C.; Dunbar, R.L.; Gao, W.; Chen, G.; Ebner, T.J. Flavoprotein autofluorescence imaging of neuronal activation in the cerebellar cortex in vivo. *J. Neurophysiol.* **2004**, *92*, 199–211. [CrossRef] [PubMed]
131. Chance, B.; Schoener, B.; Oshino, R.; Itshak, F.; Nakase, Y. Oxidation-reduction ratio studies of mitochondria in freeze-trapped samples. NADH and flavoprotein fluorescence signals. *J. Biol. Chem.* **1979**, *254*, 4764–4771. [CrossRef]
132. Yaseen, M.A.; Sakadzic, S.; Wu, W.; Becker, W.; Kasischke, K.A.; Boas, D.A. In vivo imaging of cerebral energy metabolism with two-photon fluorescence lifetime microscopy of NADH. *Biomed. Opt. Express* **2013**, *4*, 307–321. [CrossRef] [PubMed]
133. Becker, W.; Bergmann, A.; Biscotti, G.; Rueck, A. *Advanced Time-Correlated Single Photon Counting Techniques for Spectroscopy and Imaging in Biomedical Systems*; SPIE: Bellingham, WA, USA, 2004; Volume 5340.
134. Sharick, J.T.; Favreau, P.F.; Gillette, A.A.; Sdao, S.M.; Merrins, M.J.; Skala, M.C. Protein-bound NAD(P)H Lifetime is Sensitive to Multiple Fates of Glucose Carbon. *Sci. Rep.* **2018**, *8*, 5456. [CrossRef] [PubMed]
135. Gomez, C.A.; Fu, B.; Sakadzic, S.; Yaseen, M.A. Cerebral metabolism in a mouse model of Alzheimer's disease characterized by two-photon fluorescence lifetime microscopy of intrinsic NADH. *Neurophotonics* **2018**, *5*, 045008. [CrossRef] [PubMed]
136. Ranjit, S.; Malacrida, L.; Stakic, M.; Gratton, E. Determination of the metabolic index using the fluorescence lifetime of free and bound nicotinamide adenine dinucleotide using the phasor approach. *J. Biophotonics* **2019**, *12*, e201900156. [CrossRef]
137. Blacker, T.S.; Mann, Z.F.; Gale, J.E.; Ziegler, M.; Bain, A.J.; Szabadkai, G.; Duchen, M.R. Separating NADH and NADPH fluorescence in live cells and tissues using FLIM. *Nat. Commun.* **2014**, *5*, 3936. [CrossRef] [PubMed]
138. Ranjit, S.; Malacrida, L.; Jameson, D.M.; Gratton, E. Fit-free analysis of fluorescence lifetime imaging data using the phasor approach. *Nat. Protoc.* **2018**, *13*, 1979–2004. [CrossRef]
139. Evers, M.; Salma, N.; Osseiran, S.; Casper, M.; Birngruber, R.; Evans, C.L.; Manstein, D. Enhanced quantification of metabolic activity for individual adipocytes by label-free FLIM. *Sci. Rep.* **2018**, *8*, 8757. [CrossRef]
140. Blacker, T.S.; Sewell, M.D.E.; Szabadkai, G.; Duchen, M.R. Metabolic Profiling of Live Cancer Tissues Using NAD(P)H Fluorescence Lifetime Imaging. *Methods Mol. Biol.* **2019**, *1928*, 365–387. [CrossRef]
141. Stringari, C.; Nourse, J.L.; Flanagan, L.A.; Gratton, E. Phasor fluorescence lifetime microscopy of free and protein-bound NADH reveals neural stem cell differentiation potential. *PLoS ONE* **2012**, *7*, e48014. [CrossRef]
142. Dong, Y.; Brewer, G.J. Global Metabolic Shifts in Age and Alzheimer's Disease Mouse Brains Pivot at NAD+/NADH Redox Sites. *J. Alzheimer's Dis.* **2019**, *71*, 119–140. [CrossRef] [PubMed]
143. Hou, Y.; Lautrup, S.; Cordonnier, S.; Wang, Y.; Croteau, D.L.; Zavala, E.; Zhang, Y.; Moritoh, K.; O'Connell, J.F.; Baptiste, B.A.; et al. NAD(+) supplementation normalizes key Alzheimer's features and DNA damage responses in a new AD mouse model with introduced DNA repair deficiency. *Proc. Natl. Acad. Sci. USA* **2018**, *115*, E1876–E1885. [CrossRef]
144. Alano, C.C.; Garnier, P.; Ying, W.; Higashi, Y.; Kauppinen, T.M.; Swanson, R.A. NAD+ depletion is necessary and sufficient for poly(ADP-ribose) polymerase-1-mediated neuronal death. *J. Neurosci. Off. J. Soc. Neurosci.* **2010**, *30*, 2967–2978. [CrossRef]
145. Abeti, R.; Abramov, A.Y.; Duchen, M.R. Beta-amyloid activates PARP causing astrocytic metabolic failure and neuronal death. *Brain: A J. Neurol.* **2011**, *134*, 1658–1672. [CrossRef] [PubMed]
146. Dong, Y.; Digman, M.A.; Brewer, G.J. Age- and AD-related redox state of NADH in subcellular compartments by fluorescence lifetime imaging microscopy. *GeroScience* **2019**, *41*, 51–67. [CrossRef]
147. Covarrubias, A.J.; Perrone, R.; Grozio, A.; Verdin, E. NAD(+) metabolism and its roles in cellular processes during ageing. *Nat. Rev. Mol. Cell Biol.* **2021**, *22*, 119–141. [CrossRef]

148. Vial, M.L.; Zencak, D.; Grkovic, T.; Gorse, A.D.; Mackay-Sim, A.; Mellick, G.D.; Wood, S.A.; Quinn, R.J. A Grand Challenge. 2. Phenotypic Profiling of a Natural Product Library on Parkinson's Patient-Derived Cells. *J. Nat. Prod.* **2016**, *79*, 1982–1989. [CrossRef] [PubMed]
149. Aldewachi, H.; Al-Zidan, R.N.; Conner, M.T.; Salman, M.M. High-Throughput Screening Platforms in the Discovery of Novel Drugs for Neurodegenerative Diseases. *Bioengineering* **2021**, *8*, 30. [CrossRef]
150. Perez Del Palacio, J.; Diaz, C.; de la Cruz, M.; Annang, F.; Martin, J.; Perez-Victoria, I.; Gonzalez-Menendez, V.; de Pedro, N.; Tormo, J.R.; Algieri, F.; et al. High-Throughput Screening Platform for the Discovery of New Immunomodulator Molecules from Natural Product Extract Libraries. *J. Biomol. Screen.* **2016**, *21*, 567–578. [CrossRef]
151. Choi, S.H.; Kim, Y.H.; Quinti, L.; Tanzi, R.E.; Kim, D.Y. 3D culture models of Alzheimer's disease: A road map to a "cure-in-a-dish". *Mol. Neurodegener.* **2016**, *11*, 75. [CrossRef] [PubMed]
152. LaBarbera, K.M.; Limegrover, C.; Rehak, C.; Yurko, R.; Izzo, N.J.; Knezovich, N.; Watto, E.; Waybright, L.; Catalano, S.M. Modeling the mature CNS: A predictive screening platform for neurodegenerative disease drug discovery. *J. Neurosci. Methods* **2021**. [CrossRef]
153. Seyb, K.I.; Schuman, E.R.; Ni, J.; Huang, M.M.; Michaelis, M.L.; Glicksman, M.A. Identification of small molecule inhibitors of beta-amyloid cytotoxicity through a cell-based high-throughput screening platform. *J. Biomol. Screen.* **2008**, *13*, 870–878. [CrossRef] [PubMed]
154. Hou, X.Q.; Yan, R.; Yang, C.; Zhang, L.; Su, R.Y.; Liu, S.J.; Zhang, S.J.; He, W.Q.; Fang, S.H.; Cheng, S.Y.; et al. A novel assay for high-throughput screening of anti-Alzheimer's disease drugs to determine their efficacy by real-time monitoring of changes in PC12 cell proliferation. *Int. J. Mol. Med.* **2014**, *33*, 543–549. [CrossRef] [PubMed]
155. Kondo, T.; Imamura, K.; Funayama, M.; Tsukita, K.; Miyake, M.; Ohta, A.; Woltjen, K.; Nakagawa, M.; Asada, T.; Arai, T.; et al. iPSC-Based Compound Screening and In Vitro Trials Identify a Synergistic Anti-amyloid beta Combination for Alzheimer's Disease. *Cell Rep.* **2017**, *21*, 2304–2312. [CrossRef]
156. Kryscio, R.J.; Abner, E.L.; Caban-Holt, A.; Lovell, M.; Goodman, P.; Darke, A.K.; Yee, M.; Crowley, J.; Schmitt, F.A. Association of Antioxidant Supplement Use and Dementia in the Prevention of Alzheimer's Disease by Vitamin E and Selenium Trial (PREADViSE). *Jama Neurol.* **2017**, *74*, 567–573. [CrossRef]
157. Masaki, K.H.; Losonczy, K.G.; Izmirlian, G.; Foley, D.J.; Ross, G.W.; Petrovitch, H.; Havlik, R.; White, L.R. Association of vitamin E and C supplement use with cognitive function and dementia in elderly men. *Neurology* **2000**, *54*, 1265–1272. [CrossRef]
158. Choi, D.Y.; Lee, Y.J.; Hong, J.T.; Lee, H.J. Antioxidant properties of natural polyphenols and their therapeutic potentials for Alzheimer's disease. *Brain Res. Bull.* **2012**, *87*, 144–153. [CrossRef] [PubMed]
159. Karunaweera, N.; Raju, R.; Gyengesi, E.; Munch, G. Plant polyphenols as inhibitors of NF-kappaB induced cytokine production-a potential anti-inflammatory treatment for Alzheimer's disease? *Front. Mol. Neurosci.* **2015**, *8*, 24. [CrossRef]
160. Baum, L.; Lam, C.W.; Cheung, S.K.; Kwok, T.; Lui, V.; Tsoh, J.; Lam, L.; Leung, V.; Hui, E.; Ng, C.; et al. Six-month randomized, placebo-controlled, double-blind, pilot clinical trial of curcumin in patients with Alzheimer disease. *J. Clin. Psychopharmacol.* **2008**, *28*, 110–113. [CrossRef]
161. Turner, R.S.; Thomas, R.G.; Craft, S.; van Dyck, C.H.; Mintzer, J.; Reynolds, B.A.; Brewer, J.B.; Rissman, R.A.; Raman, R.; Aisen, P.S.; et al. A randomized, double-blind, placebo-controlled trial of resveratrol for Alzheimer disease. *Neurology* **2015**, *85*, 1383–1391. [CrossRef] [PubMed]
162. Kuriyama, S.; Hozawa, A.; Ohmori, K.; Shimazu, T.; Matsui, T.; Ebihara, S.; Awata, S.; Nagatomi, R.; Arai, H.; Tsuji, I. Green tea consumption and cognitive function: A cross-sectional study from the Tsurugaya Project 1. *Am. J. Clin. Nutr.* **2006**, *83*, 355–361. [CrossRef] [PubMed]
163. Molino, S.; Dossena, M.; Buonocore, D.; Ferrari, F.; Venturini, L.; Ricevuti, G.; Verri, M. Polyphenols in dementia: From molecular basis to clinical trials. *Life Sci.* **2016**, *161*, 69–77. [CrossRef] [PubMed]
164. Ringman, J.M.; Frautschy, S.A.; Teng, E.; Begum, A.N.; Bardens, J.; Beigi, M.; Gylys, K.H.; Badmaev, V.; Heath, D.D.; Apostolova, L.G.; et al. Oral curcumin for Alzheimer's disease: Tolerability and efficacy in a 24-week randomized, double blind, placebo-controlled study. *Alzheimer's Res. Ther.* **2012**, *4*, 43. [CrossRef] [PubMed]
165. de Vries, K.; Strydom, M.; Steenkamp, V. Bioavailability of resveratrol: Possibilities for enhancement. *J. Herb. Med.* **2018**, *11*, 71–77. [CrossRef]
166. Broman-Fulks, J.J.; Canu, W.H.; Trout, K.L.; Nieman, D.C. The effects of quercetin supplementation on cognitive functioning in a community sample: A randomized, placebo-controlled trial. *Ther. Adv. Psychopharmacol.* **2012**, *2*, 131–138. [CrossRef]
167. Remington, R.; Bechtel, C.; Larsen, D.; Samar, A.; Doshanjh, L.; Fishman, P.; Luo, Y.; Smyers, K.; Page, R.; Morrell, C.; et al. A Phase II Randomized Clinical Trial of a Nutritional Formulation for Cognition and Mood in Alzheimer's Disease. *J. Alzheimer's Dis.* **2015**, *45*, 395–405. [CrossRef]
168. Moreira, P.I.; Harris, P.L.; Zhu, X.; Santos, M.S.; Oliveira, C.R.; Smith, M.A.; Perry, G. Lipoic acid and N-acetyl cysteine decrease mitochondrial-related oxidative stress in Alzheimer disease patient fibroblasts. *J. Alzheimer's Dis.* **2007**, *12*, 195–206. [CrossRef] [PubMed]
169. El-Magd, M.A.; Khalifa, S.F.; FA, A.A.; Badawy, A.A.; El-Shetry, E.S.; Dawood, L.M.; Alruwaili, M.M.; Alrawaili, H.A.; Risha, E.F.; El-Taweel, F.M.; et al. Incensole acetate prevents beta-amyloid-induced neurotoxicity in human olfactory bulb neural stem cells. *Biomed. Pharmacother. Biomed. Pharmacother.* **2018**, *105*, 813–823. [CrossRef]

170. Bonaccorso, A.; Pellitteri, R.; Ruozi, B.; Puglia, C.; Santonocito, D.; Pignatello, R.; Musumeci, T. Curcumin Loaded Polymeric vs. Lipid Nanoparticles: Antioxidant Effect on Normal and Hypoxic Olfactory Ensheathing Cells. *Nanomaterials* **2021**, *11*, 159. [CrossRef]
171. Bonferoni, M.C.; Rassu, G.; Gavini, E.; Sorrenti, M.; Catenacci, L.; Giunchedi, P. Nose-to-Brain Delivery of Antioxidants as a Potential Tool for the Therapy of Neurological Diseases. *Pharmaceutics* **2020**, *12*, 1246. [CrossRef]
172. Mythri, R.B.; Jagatha, B.; Pradhan, N.; Andersen, J.; Bharath, M.M. Mitochondrial complex I inhibition in Parkinson's disease: How can curcumin protect mitochondria? *Antioxid. Redox Signal.* **2007**, *9*, 399–408. [CrossRef]
173. Mett, J.; Muller, U. The medium-chain fatty acid decanoic acid reduces oxidative stress levels in neuroblastoma cells. *Sci. Rep.* **2021**, *11*, 6135. [CrossRef] [PubMed]
174. Hughes, S.D.; Kanabus, M.; Anderson, G.; Hargreaves, I.P.; Rutherford, T.; O'Donnell, M.; Cross, J.H.; Rahman, S.; Eaton, S.; Heales, S.J. The ketogenic diet component decanoic acid increases mitochondrial citrate synthase and complex I activity in neuronal cells. *J. Neurochem.* **2014**, *129*, 426–433. [CrossRef]
175. Hu, X.; Teng, S.; He, J.; Sun, X.; Du, M.; Kou, L.; Wang, X. Pharmacological basis for application of scutellarin in Alzheimer's disease: Antioxidation and antiapoptosis. *Mol. Med. Rep.* **2018**, *18*, 4289–4296. [CrossRef] [PubMed]
176. Alamro, A.A.; Alsulami, E.A.; Almutlaq, M.; Alghamedi, A.; Alokail, M.; Haq, S.H. Therapeutic Potential of Vitamin D and Curcumin in an In Vitro Model of Alzheimer Disease. *J. Cent. Nerv. Syst. Dis.* **2020**, *12*, 1179573520924311. [CrossRef] [PubMed]
177. Bacci, A.; Runfola, M.; Sestito, S.; Rapposelli, S. Beyond Antioxidant Effects: Nature-Based Templates Unveil New Strategies for Neurodegenerative Diseases. *Antioxidants* **2021**, *10*, 367. [CrossRef]
178. Sang, Z.; Wang, K.; Han, X.; Cao, M.; Tan, Z.; Liu, W. Design, Synthesis, and Evaluation of Novel Ferulic Acid Derivatives as Multi-Target-Directed Ligands for the Treatment of Alzheimer's Disease. *Acs Chem. Neurosci.* **2019**, *10*, 1008–1024. [CrossRef]
179. Keller, J.M.; Frega, M. Past, Present, and Future of Neuronal Models In Vitro. *Adv. Neurobiol.* **2019**, *22*, 3–17. [CrossRef]
180. Lancaster, M.A.; Renner, M.; Martin, C.A.; Wenzel, D.; Bicknell, L.S.; Hurles, M.E.; Homfray, T.; Penninger, J.M.; Jackson, A.P.; Knoblich, J.A. Cerebral organoids model human brain development and microcephaly. *Nature* **2013**, *501*, 373–379. [CrossRef]
181. Ahn, S.I.; Kim, Y. Human Blood-Brain Barrier on a Chip: Featuring Unique Multicellular Cooperation in Pathophysiology. *Trends Biotechnol.* **2021**. [CrossRef]
182. Salman, M.M.; Marsh, G.; Kusters, I.; Delince, M.; Di Caprio, G.; Upadhyayula, S.; de Nola, G.; Hunt, R.; Ohashi, K.G.; Gray, T.; et al. Design and Validation of a Human Brain Endothelial Microvessel-on-a-Chip Open Microfluidic Model Enabling Advanced Optical Imaging. *Front. Bioeng. Biotechnol.* **2020**, *8*, 573775. [CrossRef] [PubMed]
183. Oddo, A.; Peng, B.; Tong, Z.; Wei, Y.; Tong, W.Y.; Thissen, H.; Voelcker, N.H. Advances in Microfluidic Blood-Brain Barrier (BBB) Models. *Trends Biotechnol.* **2019**, *37*, 1295–1314. [CrossRef] [PubMed]
184. Vadivelu, R.K.; Ooi, C.H.; Yao, R.Q.; Tello Velasquez, J.; Pastrana, E.; Diaz-Nido, J.; Lim, F.; Ekberg, J.A.; Nguyen, N.T.; St John, J.A. Generation of three-dimensional multiple spheroid model of olfactory ensheathing cells using floating liquid marbles. *Sci. Rep.* **2015**, *5*, 15083. [CrossRef] [PubMed]
185. Ayuso, J.M.; Gillette, A.; Lugo-Cintron, K.; Acevedo-Acevedo, S.; Gomez, I.; Morgan, M.; Heaster, T.; Wisinski, K.B.; Palecek, S.P.; Skala, M.C.; et al. Organotypic microfluidic breast cancer model reveals starvation-induced spatial-temporal metabolic adaptations. *EBioMedicine* **2018**, *37*, 144–157. [CrossRef]
186. Liang, N.; Trujillo, C.A.; Negraes, P.D.; Muotri, A.R.; Lameu, C.; Ulrich, H. Stem cell contributions to neurological disease modeling and personalized medicine. *Prog. Neuro-Psychopharmacol. Biol. Psychiatry* **2018**, *80*, 54–62. [CrossRef] [PubMed]
187. Gosnell, M.E.; Anwer, A.G.; Cassano, J.C.; Sue, C.M.; Goldys, E.M. Functional hyperspectral imaging captures subtle details of cell metabolism in olfactory neurosphere cells, disease-specific models of neurodegenerative disorders. *Biochim. Et Biophys. Acta* **2016**, *1863*, 56–63. [CrossRef] [PubMed]

International Journal of
Molecular Sciences

Article

Discovery of a Necroptosis Inhibitor Improving Dopaminergic Neuronal Loss after MPTP Exposure in Mice

Sara R. Oliveira, Pedro A. Dionísio, Maria M. Gaspar, Maria B. T. Ferreira, Catarina A. B. Rodrigues, Rita G. Pereira, Mónica S. Estevão, Maria J. Perry, Rui Moreira, Carlos A. M. Afonso, Joana D. Amaral and Cecília M. P. Rodrigues *

Research Institute for Medicines (iMed.ULisboa), Faculty of Pharmacy, Universidade de Lisboa, 1649-003 Lisbon, Portugal; sararoliveira@ff.ulisboa.pt (S.R.O.); pedelandionisio@gmail.com (P.A.D.); mgaspar@ff.ulisboa.pt (M.M.G.); mariabtferreira@gmail.com (M.B.T.F.); cataxana@gmail.com (C.A.B.R.); ragpereira@gmail.com (R.G.P.); monica.estevao@gmail.com (M.S.E.); mjprocha@ff.ulisboa.pt (M.J.P.); rmoreira@ff.ulisboa.pt (R.M.); carlosafonso@ff.ulisboa.pt (C.A.M.A.); jamaral@ff.ulisboa.pt (J.D.A.)

* Correspondence: cmprodrigues@ff.ulisboa.pt; Tel.: +351-217946490

Abstract: Parkinson's disease (PD) is the second most common neurodegenerative disorder, mainly characterized by motor deficits correlated with progressive dopaminergic neuronal loss in the substantia nigra pars compacta (SN). Necroptosis is a caspase-independent form of regulated cell death mediated by the concerted action of receptor-interacting protein 3 (RIP3) and the pseudokinase mixed lineage domain-like protein (MLKL). It is also usually dependent on RIP1 kinase activity, influenced by further cellular clues. Importantly, necroptosis appears to be strongly linked to several neurodegenerative diseases, including PD. Here, we aimed at identifying novel chemical inhibitors of necroptosis in a PD-mimicking model, by conducting a two-step screening. Firstly, we phenotypically screened a library of 31 small molecules using a cellular model of necroptosis and, thereafter, the hit compound effect was validated in vivo in a sub-acute 1-methyl-1-4-phenyl-1,2,3,6-tetrahydropyridine hydrochloride (MPTP) PD-related mouse model. From the initial compounds, we identified one hit—Oxa12—that strongly inhibited necroptosis induced by the pan-caspase inhibitor zVAD-fmk in the BV2 murine microglia cell line. More importantly, mice exposed to MPTP and further treated with Oxa12 showed protection against MPTP-induced dopaminergic neuronal loss in the SN and striatum. In conclusion, we identified Oxa12 as a hit compound that represents a new chemotype to tackle necroptosis. Oxa12 displays in vivo effects, making this compound a drug candidate for further optimization to attenuate PD pathogenesis.

Keywords: MPTP; necroptosis; neurodegeneration; Parkinson's disease; small molecules

Citation: Oliveira, S.R.; Dionísio, P.A.; Gaspar, M.M.; Ferreira, M.B.T.; Rodrigues, C.A.B.; Pereira, R.G.; Estevão, M.S.; Perry, M.J.; Moreira, R.; Afonso, C.A.M.; et al. Discovery of a Necroptosis Inhibitor Improving Dopaminergic Neuronal Loss after MPTP Exposure in Mice. *Int. J. Mol. Sci.* **2021**, *22*, 5289. https://doi.org/10.3390/ijms22105289

Academic Editor: Anne Vejux

Received: 16 April 2021
Accepted: 14 May 2021
Published: 18 May 2021

1. Introduction

Parkinson's disease (PD) is the second most common neurodegenerative disorder worldwide. PD is pathologically defined by the progressive dysfunction of the nigrostriatal pathway, which culminates in the loss of dopaminergic neurons in the substantia nigra pars compacta (SN) and depletion of dopaminergic enervation in the striatum [1,2]. Of note, degeneration of dopaminergic neurons in the SN precedes the first motor deficits afflicting PD patients [2].

The pathological mechanisms causing PD are thought to stimulate a cascade of events that activate regulated cell death (RCD) pathways, which are responsible for neuronal death [3,4]. Recent evidence has shown that necroptosis, a type of regulated necrosis, plays crucial pathogenic roles in several human diseases, including neurodegenerative diseases, such as PD, while holding high potential for clinical targeting [5]. Necroptosis is a caspase-independent type of RCD commonly executed after RIP1 and RIP3 kinase activation. Typically, this type of cell death is initiated following cell death transmembrane receptor stimulation, with tumor necrosis factor (TNF) receptor 1 (TNFR1) being the most well-studied example [6,7]. In conditions where caspase-8 activation is genetically or pharmacologically prevented, RIP1 and RIP3 are not cleaved and accumulate in the so-called necrosome complex [8]. Then, activated RIP3 recruits and phosphorylates pseudokinase mixed lineage domain-like protein (MLKL), which further translocates to the plasma membrane, inducing membrane disruption and necroptosis execution [5]. Importantly, necroptosis has already been associated with neuronal death induced by the neurotoxin 1-methyl-4-phenyl-1,2,3,6-tetrahydropyridine (MPTP), a PD-mimicking neurotoxin, in both in vivo and in vitro rodent models, as well as in PD human samples [9,10].

Importantly, the first necroptosis inhibitors were identified back in 2005 through a phenotypic screening for chemical inhibitors of necroptosis induced by TNF and zVAD-fmk in human monocytic U937 cells. This led to the identification of necrostatin-1 (Nec-1) and its optimized derivative necrostatin-1 stable (Nec-1s), later confirmed to be RIP1 kinase inhibitors [11–13]. Additional studies regarding Nec-1 properties pointed to off-target activity and limited metabolic stability in mice, while Nec-1s presented higher activity as a necroptosis inhibitor, along with no nonspecific cytotoxicity and reasonable pharmacokinetic properties. However, Nec-1s still presents a short in vivo half-life [11,14–17]. Of note, Nec-1s identification proved that RIP1 inhibition is beneficial in several diseases [9,10,13,18]. Moreover, molecules targeting other components of necroptotic signaling pathway, such as RIP3 or MLKL, have also been proposed [19,20]. However, none of the compounds discovered so far reached the expectation of clinical application, which has led researchers to unceasingly screen new drugs with better selectivity and potency.

In this study, we phenotypically screened a library of 31 new molecules to discover novel inhibitors of necroptotic cell death with structural novelty. Although target-based drug discovery has been the dominant approach to drug discovery in the past decades, there has been a recent reawakening interest in phenotypic drug discovery approaches, based on their potential to address the complexity of only partially understood diseases and their promise of delivering first-in-class drugs, in parallel with major advances in cell-based phenotypic screening tools. In addition, prioritized hits from the primary screen were further evaluated for in vivo proof-of-concept using an MPTP PD-mimicking mouse model. We identified a potential hit, Oxa12, that significantly inhibited zVAD-fmk-induced necroptotic cell death in the BV2 microglial cell line, with a half-maximal effective concentration (EC_{50}) of ~1 μM. Importantly, Oxa12 showed to protect dopaminergic neuronal cells from death induced by MPTP in vivo, in the SN and striatum, which highlights the potential benefits of discovering and optimizing new molecules with mechanisms of action that affect disease-relevant pathways, such as necroptosis.

2. Results

2.1. Phenotypic Screening for Hit Selection

To identify novel necroptosis inhibitors, we performed a cell-based phenotypic screening assay to select hit compounds that strongly inhibit necroptotic cell death (Figure 1A).

Previous studies have demonstrated that the pan-caspase inhibitor zVAD-fmk induces necroptosis in different cellular models, including in the L929 fibrosarcoma and in the BV2 murine microglial cell lines, by a mechanism that most likely depends on TNF autocrine secretion [12,21,22]. Here, we used BV2 cells incubated with zVAD-fmk as an in vitro model of necroptosis and screened a total of 31 small compounds potentially inhibitors of this type of cell death. These compounds are part of a larger in-house library containing mainly the heterocyclic core of 4-methylene-2-aryloxazol-5(4H)-one and are represented in Figure 2. The remaining compounds have been tested in previous papers [12,21,22]. Cells were incubated with 25 μM zVAD-fmk alone or in combination with 30 μM test compounds for 24 h. BV2 cells treated with DMSO were used for data normalization. BV2 cells co-incubated with zVAD-fmk and Nec-1, a well-known RIP1 kinase inhibitor, were used as positive control of necroptosis inhibition (Figure 1B). As expected, treatment with Nec-1 almost completely rescued zVAD-fmk-induced decrease in cell viability ($p < 0.001$) (Figure 1B), as assessed by MTS metabolism. More importantly, we identified one hit compound—Oxa12—that significantly protected cells from zVAD-fmk-mediated necroptosis by ~70% ($p < 0.01$) (Figure 1B). Importantly, RIP1 kinase activity inhibitory potential of Oxa12 was determined using radiometric binding-based assays. The results show that Oxa12 at 5 μM inhibits RIP1 activity by ~15%, which confirms RIP1 as a possible target. The anti-necroptotic effects, however, may not be solely dependent on this target. These results are in line with our previous work, where we showed the sequestration of key necroptosis mediators, RIP1, RIP3, and p-MLKL, in the insoluble fraction in zVAD-fmk-treated BV2 cells, while Oxa12 abolished necrosome assembly and MLKL phosphorylation [22]. Moreover, Oxa12 strongly rescued zVAD-fmk-induced cell death, as observed by microscopy analysis of cell morphology [22]. Furthermore, our previous in silico molecular docking calculations for Oxa12 inside the RIP1 kinase domain demonstrated that Oxa12 is occupying a region similar to the co-crystallized inhibitor, suggesting a similar interaction pattern and indicating that Oxa12 most likely targets RIP1 to some extent [22].

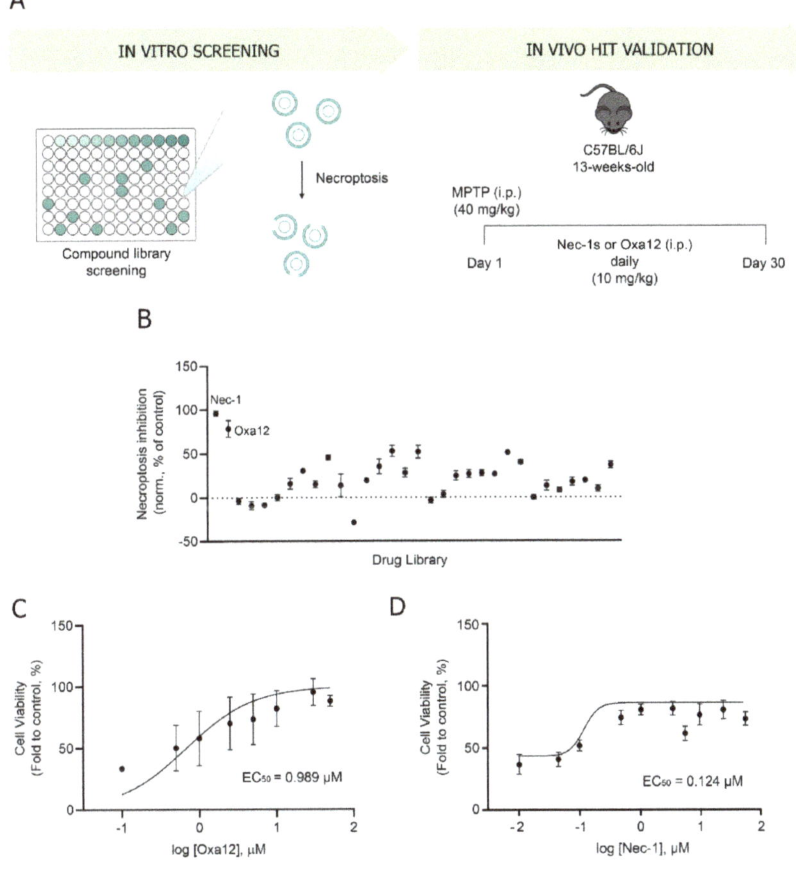

Figure 1. A cell-based phenotypic screening identifies Oxa12 as a necroptosis inhibitor. (**A**) Schematic overview of the two-step screening workflow. (**B**) Determination of the ability of the compound to inhibit necroptosis. Cell metabolic activity is depicted as a percentage of the control (DMSO = 0; Nec-1 at 30 μM) for compounds at 30 μM tested in BV2 murine microglia cells exposed to 25 μM zVAD-fmk for 24 h. (**C**) Half-maximal effective concentration (EC_{50}) determination in BV2 murine microglia cells in a dose-response concentration (0.1 to 50 μM Oxa12) plus 25 μM zVAD-fmk for 24 h. (**D**) Half-maximal effective concentration (EC_{50}) determination in BV2 murine microglia cells in a dose-response concentration (0.01 to 50 μM Nec-1) plus 25 μM zVAD-fmk for 24 h.

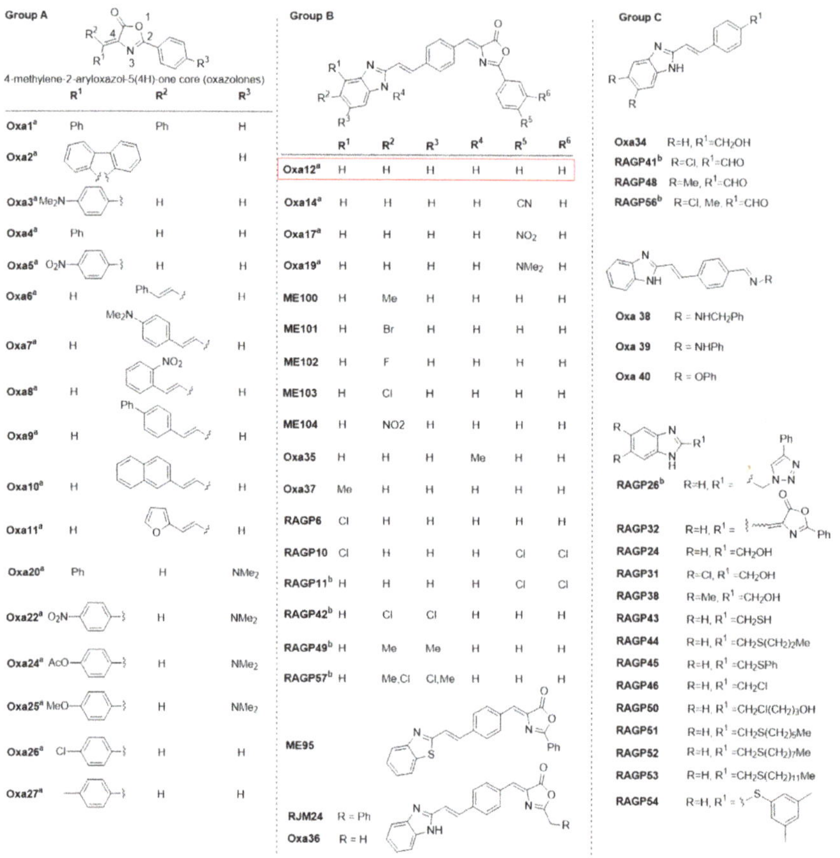

Group A

4-methylene-2-aryloxazol-5(4H)-one core (oxazolones)

	R^1	R^2	R^3
Oxa1[a]	Ph	Ph	H
Oxa2[a]			H
Oxa3[a]	Me_2N–	H	H
Oxa4[a]	Ph	H	H
Oxa5[a]	O_2N–	H	H
Oxa6[a]	H	Ph	H
Oxa7[a]	H	Me_2N	H
Oxa8[a]	H	NO_2	H
Oxa9[a]	H	Ph	H
Oxa10[a]	H		H
Oxa11[a]	H		H
Oxa20[a]	Ph	H	NMe_2
Oxa22[a]	O_2N–	H	NMe_2
Oxa24[a]	AcO–	H	NMe_2
Oxa25[a]	MeO–	H	NMe_2
Oxa26[a]	Cl–	H	H
Oxa27[a]		H	H

Group B

	R^1	R^2	R^3	R^4	R^5	R^6
Oxa12[a]	H	H	H	H	H	H
Oxa14[a]	H	H	H	H	CN	H
Oxa17[a]	H	H	H	H	NO_2	H
Oxa19[a]	H	H	H	H	NMe_2	H
ME100	H	Me	H	H	H	H
ME101	H	Br	H	H	H	H
ME102	H	F	H	H	H	H
ME103	H	Cl	H	H	H	H
ME104	H	NO2	H	H	H	H
Oxa35	H	H	H	Me	H	H
Oxa37	Me	H	H	H	H	H
RAGP6	Cl	H	H	H	H	H
RAGP10	Cl	H	H	H	Cl	Cl
RAGP11[b]	H	H	H	H	Cl	Cl
RAGP42[b]	H	Cl	Cl	H	H	H
RAGP49[b]	H	Me	Me	H	H	H
RAGP57[b]	H	Me,Cl	Cl,Me	H	H	H

ME95

RJM24 R = Ph

Oxa36 R = H

Group C

Oxa34 R=H, R^1=CH_2OH

RAGP41[b] R=Cl, R^1=CHO

RAGP48 R=Me, R^1=CHO

RAGP56[b] R=Cl, Me, R^1=CHO

Oxa 38 R = $NHCH_2Ph$

Oxa 39 R = NHPh

Oxa 40 R = OPh

RAGP26[b] R=H, R^1 =

RAGP32 R=H, R^1 =

RAGP24 R=H, R^1 =CH_2OH

RAGP31 R=Cl, R^1 =CH_2OH

RAGP38 R=Me, R^1 =CH_2OH

RAGP43 R=H, R^1 =CH_2SH

RAGP44 R=H, R^1 =$CH_2S(CH_2)_2Me$

RAGP45 R=H, R^1 =CH_2SPh

RAGP46 R=H, R^1 =CH_2Cl

RAGP50 R=H, R^1 =$CH_2Cl(CH_2)_3OH$

RAGP51 R=H, R^1 =$CH_2S(CH_2)_5Me$

RAGP52 R=H, R^1 =$CH_2S(CH_2)_7Me$

RAGP53 R=H, R^1 =$CH_2S(CH_2)_{11}Me$

RAGP54 R=H, R^1 =

Figure 2. Structures of the synthesized small-molecule library compounds based on the 4-methylene-2-aryloxazol-5(4H)-one core (oxazolones). Group A includes oxazolones containing substituted aryl and aromatic heterocyclic attached to the 4-methylene position. Group B consists of extended oxazolones containing terminal fused heterocycles. Group C includes additional compounds without the oxazolone core or the 4-methylene side chain. [a] Synthesized compounds evaluated in a previous paper [22]. [b] Compounds not evaluated due to insoluble properties at the tested concentration.

To further characterize Oxa12 activity, we determined the EC_{50}, a pharmacologic parameter commonly used as a measure of compound potency, which here indicates 50% of compound maximal potency to inhibit necroptosis, by performing a dose-response study. BV2 cells were incubated with zVAD-fmk along with Oxa12 at a range of concentrations (0.1–50 μM) for 24 h and the EC_{50} was calculated. Our results showed that Oxa12 presented an EC_{50} value of 0.989 μM, comparable with 0.124 μM of Nec-1, thus indicating its effectiveness as a necroptosis inhibitor (Figure 1C,D).

We next found that Oxa12 met stringent criteria on activity, chemical tractability and structural novelty, including a low chemical similarity with Nec-1 and its analogues, as measured by Tanimoto's index. Moreover, we analyzed whether Oxa12 was included in the central nervous system (CNS) drug property space, using a CNS multiparameter optimization (MPO) approach, a prospective design tool and a widely utilized predictor for ADME and safety properties suitable for CNS targeting [23]. CNS MPO utilizes a set of six physicochemical descriptors to calculate a score, between 0 and 6. A higher score represents the optimal chemical space possessing alignment of key drug properties for CNS therapeutic agents. Generally, a CNS MPO score ≥ 4 is desired, although there are drugs active on the CNS with lower score, with a range of 3–4 being still acceptable [24]. Here, we observed that Oxa12 has a value of 3.6. While being lower than the CNS MPO score of 5 calculated for Nec-1, the value for Oxa12 is similar to that determined for several well-known CNS drugs, suggesting good brain exposure (Table 1).

Table 1. Central nervous system MPO calculations for Oxa12 and Nec-1.

Oxa12				Nec-1			
Physiochemical Descriptor		Individual Score	Score CNS MPO	Physiochemical Descriptor		Individual Score	Score CNS MPO
MW	391.43	0.8	3.6	MW	259.33	1.0	5.5
clogP	5.29	0		clogP	1.66	1.0	
$clogD_{7.4}$	5.69	0		$clogD_{7.4}$	1.66	1.0	
TPSA	67.34	1.0		TPSA	48.13	1.0	
HBD	1	0.8		HBD	2	0.5	
pK_a	5.34	1.0		pK_a	0	1.0	

MW, molecular weight; clogP, calculated logP (*n*-octanol/buffer distribution coefficient) using Molinspiration desktop property calculator (https://www.molinspiration.com/cgi-bin/properties, accessed on 15 April 2021); $clogD_{7.4}$, distribution coefficient at pH 7.4; TPSA, topological polar surface area; HDB, number of hydrogen bond donor atoms; pK_a, acidity constant of ionizable function.

Finally, we also determined the extent of Oxa12 metabolism in mouse liver microsomes. Using our medium-throughput drug metabolism platform, we observed that 20% of Oxa12 was detected by HPLC after incubation for 30 min in mouse liver microsomes, which corresponds to an estimated half-life of 13 min, a liability that negatively impacts the systemic and brain exposure to Oxa-12. These results allowed us to calculate a clearance Clint in vitro of 0.99 mL/min/mg protein and a predicted hepatic extraction ratio of 0.76. Of note, although the original Nec-1 has a half-life of < 5 min, chemical optimization of this compound has led to commonly used derivatives with a half-life of about 1 h in mouse microsomal assays [25]. Thus, a combined analysis of the CNS MPO score and metabolic data confirms that Oxa12 is a novel addition to the chemical toolbox of CNS-targeting anti-necroptotic compounds.

2.2. *In Vivo Efficacy of Oxa12 in the Sub-Acute MPTP Mouse Model*

To determine if our identified hit compound was effective in vivo, we used a sub-acute MPTP mouse model. MPTP is the most widely used neurotoxin to mimic PD-related nigrostriatal degeneration, since it can be easily administered systemically, and human exposure to MPTP has been associated with severe parkinsonism, which reinforces the relevance of MPTP animal models [26,27]. Mechanistically, MPTP can rapidly cross the blood–brain barrier and, once in the brain, it is mostly oxidized to an intermediate, which then diffuses to the extracellular space and converts to the active toxic metabolite, 1-methyl-4-phenylpyridinium (MPP^+) [28,29]. Then, dopaminergic neurons selectively uptake MPP^+. In mitochondria, MPP^+ inhibits the complex I of the respiratory chain, leading to decreased ATP generation along with the production of reactive oxygen species (ROS), and ultimately cell death [29,30]. In this regard, several studies have already demonstrated that necroptosis is involved in neuronal death induced by MPTP and that Nec-1/Nec-1s administration is neuroprotective in multiple MPTP exposure regimens [9,10].

Here, we used a sub-acute MPTP regimen consisting of a single dose of MPTP (40 mg/kg), or vehicle, intraperitoneally injected in 13-week-old mice. Then, 10 mg/kg of Oxa12 or Nec-1s was administered 1 h after MPTP and then once every day for 30 days. As expected, we observed a significant reduction of approximately ~50% in TH-positive staining, a marker of dopaminergic neurons, in the SN and in the striatum of MPTP-exposed animals (Figure 3A,B), while no differences were observed in mice injected with Oxa12 or Nec-1s alone (Figure 3A,B).

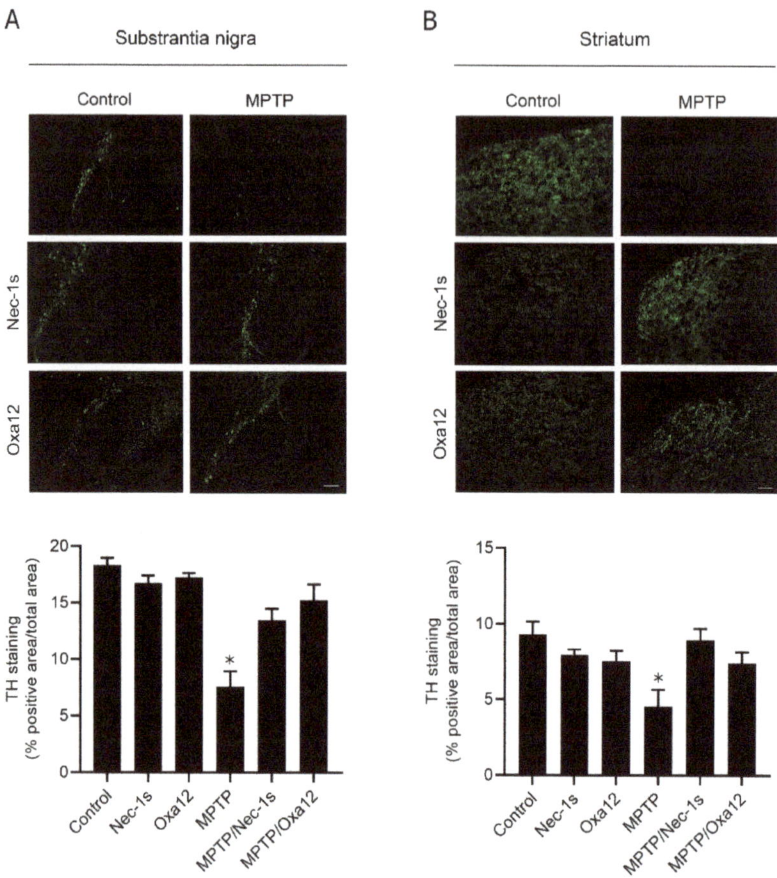

Figure 3. Oxa12 protects from MPTP-driven dopaminergic neuronal loss. Representative images of TH-positive immunostaining from control- and MPTP-injected mice treated with vehicle, Nec-1s or Oxa12 in the SN (**A**) and the striatum (**B**), and respective quantification. Scale bar, 100 µm. * $p < 0.05$ vs. control mice.

Importantly, treatment with Oxa12 or Nec-1s partially protected cells from MPTP-induced dopaminergic cell death, in the SN (Oxa12, $p = 0.05$; Nec-1s, $p = 0.12$) and striatum (Oxa12, $p = 0.06$; Nec-1s, $p = 0.08$), as observed by the rescue of TH-positive staining in comparison with MPTP-treated animals (Figure 3A,B).

In accordance, exposure to MPTP significantly reduced TH protein levels in the SN, while Oxa12 and Nec-1s treatment alone did not alter TH levels (Figure 4). Notably, Oxa12 significantly rescued TH protein levels by ~30% ($p < 0.05$), while Nec-1s showed a ~10% difference (Figure 4). The challenging dissection of SN-enriched midbrain sections, which include other dopaminergic structures, may account for the less significant rescue of TH in the SN by western blot as compared with immunofluorescence. Our results show that Oxa12 reveals a propensity to protect neuronal cells from MPTP-induced cell loss, with the potential to be explored as a new chemotype to tackle necroptosis. Nevertheless, careful examination of target engagement, including RIP1 and MLKL sequestration in insoluble fractions, a strong indicator of necrosome assembly and, therefore, of necroptosis commitment [28], deserves further investigation using appropriate antibodies.

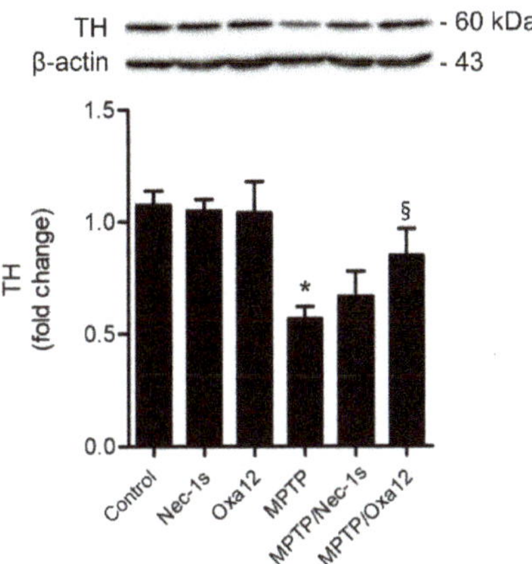

Figure 4. Oxa12 protects dopaminergic neurons from MPTP-induced cell death. Representative western blot of TH protein levels from control and MPTP-injected mice untreated or treated with Nec-1s or Oxa12 in the SN, and the respective densitometric analysis. β-actin was used as loading control. Values are expressed as mean ± SEM of three independent experiments. * $p < 0.05$ vs. control mice; § $p < 0.05$ vs. MPTP mice. TH, tyrosine hydroxylase.

Finally, MPTP mice models normally develop an inflammatory response that initiates in the striatum, with astrogliosis developing later than microgliosis and being sustained for a longer period of time [29,30]. Concordantly, we observed an increase in GFAP immunostaining and GFAP protein levels in MPTP-treated mice, thus suggesting a prolonged astrogliosis in these animals (data not shown). However, no significant differences were observed in mice treated with Oxa12 or Nec-1s.

3. Discussion

In the present study, a two-step screening workflow was performed to identify novel necroptosis inhibitors from a small-molecule library of 31 compounds. In a first step, we conducted an in vitro phenotypic screening to identify hit compounds that strongly inhibited zVAD-fmk-induced necroptotic cell death in the murine BV2 microglial cell line. In a second step, the ability of our hit molecule to protect against dopaminergic neuronal cell loss was determined in vivo using the sub-acute MPTP mouse model of PD.

As a strategy for the first step, we initially screened a range of diverse heterocyclic functionalized compounds collected from different developed synthetic methodologies. From this preliminary screening, several existing compounds containing the 4-methylene-2-aryloxazol-5(4H)-one (oxazolones) core were identified as bioactive [31]. Furthermore, we also synthesized a range of oxazolones containing diverse substituents at the 4-methylene and 2-aryl positions (Groups A and B) and by replacing the oxazolone core (Group C) as well as by inclusion of longer phenylmethylene spacer and additional structural tuning on the attached fused heterocycles (Group B). Importantly, we identified one molecule—Oxa12—that strongly inhibited necroptosis in the zVAD-fmk-treated BV2 microglia cells, and this effect was further demonstrated in vivo. Oxa12 tended to protect dopaminergic neuronal cells from MPTP-induced cell death, thus suggesting the potential of this compound in ameliorating PD pathogenesis.

Since the discovery of Nec-1 as the first necroptosis inhibitor, several other studies were performed and new inhibitors for RIP1, RIP3 and MLKL were identified. However, the therapeutic potential of all these inhibitors is restricted by low potency or reduced selectivity. In fact, so far, there is no compound that reached the prospects of clinical application, which has led to the continuous screening of new molecules with higher selectivity and potency. Despite its strongness at inhibiting necroptosis, Nec-1 is a far-from-ideal drug due to its short in vivo half-life of approximately 1 h, as well as its off-target effects, including inhibitory binding to indoleamine 2,3-dioxygenase (IDO), a protein involved in inflammation [32]. Nec-1s presents higher specificity and improved pharmacokinetics properties; however, it still has poor in vivo half-life [33]. Thus, Nec-1s represents a better choice for in vivo experiments. Other RIP1 inhibitors, including GSK'481, GSK'963 and GSK'772, demonstrate similar limitations [34–36]. RIP3 inhibitors, such as GSK'872, have also been studied; however, while they inhibit necroptosis, they may also promote apoptosis [20]. Thus, considering the limitations of all known RIP1 and RIP3 inhibitors, the development of novel and potent small compounds able to specifically attenuate necroptosis continues to be of strong need.

Here, we used the BV2 murine microglia cell line treated with the pan-caspase inhibitor, zVAD-fmk, as a model of necroptosis execution. In fact, we and others already demonstrated that zVAD-fmk strongly induces necroptotic cell death in both L929 fibrosarcoma and BV2 murine microglia cells [12,21,22]. Importantly, the pharmacological relevance of the in vitro phenotypic screening was guaranteed by using a well-established necroptosis inhibitor. Here, Nec-1, which inhibits RIP1 kinase activity, was used as positive control throughout the in vitro screening and demonstrated to be a strong necroptosis inhibitor. To identify novel necroptosis inhibitors, an in-house library of 31 small compounds were screened at a concentration of 30 μM, which allowed us to filter new necroptosis inhibitor scaffolds with potential to be further modified chemically. We identified one hit—Oxa12—that strongly inhibited necroptosis in the BV2 microglia cells by ~70%. To further characterize Oxa12 potency in vitro, dose-response curves were performed, where Oxa12 showed an EC_{50} value of 0.989 μM, thus reflecting its potential to be tested in vivo.

Throughout the years, MPTP mouse models have been widely used due to their specific and reproducible neurotoxic effect on the nigrostriatal system, being considered a convenient model of dopaminergic neurodegeneration to study therapeutic interventions. However, these models do not fully reproduce the human condition, and behavioral abnormalities are still a challenging question [36]. Moreover, previous studies have already demonstrated the involvement of necroptosis in the acute and sub-chronic MPTP mouse model of PD, where Nec-1/Nec-1s administration attenuated dopaminergic neurodegeneration [9,10]. Here, we used a sub-acute regimen of MPTP exposure, consisting of a single injection of MPTP (40 mg/kg) and still observed a significant increase in dopaminergic neurodegeneration in the SN and the striatum, while Nec-1s administration showed a tendency to protect cells from MPTP-induced dopaminergic cell death [27,37,38]. Importantly, Oxa12 alone did not induce neuronal toxicity, but it was able to protect dopaminergic neurons from MPTP-induced cell death in both SN and striatum. This experimental observation together with the MPO value suggest that Oxa-12 can cross the BBB, at least to some extent. Medicinal chemistry optimization of Oxa-12 should tackle issues such as solubility and be followed by extensive in vivo characterization of a more advanced lead compound, including by evaluating BBB penetration, half-life and target engagement. The role for inhibiting necroptosis deserves further exploitation [39].

Overall, these results indicate that Oxa12 has reasonable potency, associated with an apparent lack of toxicity, which makes this compound fit for additional chemistry optimization. Therapies using small molecules targeting key components of dopaminergic neuron cell death could evolve as a potential therapeutic approach for ameliorating PD progression.

4. Materials and Methods

4.1. Cell Culture and Reagents

The BV2 murine microglia cell line (gently provided by Dr. Elsa Rodrigues, University of Lisbon) was cultured in RPMI 1640 medium (GIBCO® Life Technologies, Inc., Grand Island, NY, USA), supplemented with 10% heat inactivated fetal bovine serum (FBS), 1% antibiotic/antimycotic (A/A) solution and 1% GlutaMAX™ (GIBCO). During the experiments, culture media was substituted by RPMI supplemented with 1% A/A, 1% insulin-transferrin-selenium (RPMI/ITS) and 1 mg/mL bovine serum albumin (BSA;

GIBCO). Cells were maintained at 37 °C in a humidified atmosphere of 5% CO_2. Chemicals used were Nec-1 (Sigma-Aldrich, St. Louis, MO, USA), dimethyl sulfoxide (DMSO; Sigma-Aldrich) and Z-Val-Ala-Asp-fluoromethylketone (zVAD-fmk) pan-caspase inhibitor (Enzo Life Sciences, Farmingdale, NY, USA).

4.2. Chemical Synthesis and Analysis

The oxazol-5-(4H)-ones (Oxas) and the precursors aldehydes were synthesized following a general synthetic reported route: a mixture of the corresponding aldehyde (30–600 µmol scale, 1 equiv.), hippuric acid or derivatives (1 equiv.), sodium acetate (1 equiv.) and acetic anhydride (3 equiv.) in a round-bottom flask was heated under magnetic stirring at 110 °C for 2 h [31]. Then, the reaction mixture was cooled to room temperature and the obtained solid was washed with distilled water (2×) and MeOH (4×) and dried under vacuum. If further purification was needed, some of the compounds were recrystallized in a suitable solvent. The purity and structural identification were performed by ^{1}H, ^{13}C NMR and mass spectrometry analyses.

4.3. Screening of Necroptosis Inhibitors

BV2 cells were plated in 96-well plates at 7×10^3 cells/well and after 24 h, necroptosis was induced by adding 25 µM zVAD-fmk in RPMI/ITS. Test compounds or Nec-1 (positive control of necroptosis inhibition) were incubated along with zVAD-fmk at a final concentration of 30 µM for 24 h. DMSO was used as vehicle control. Cell viability was determined based on measurement of MTS metabolism using the CellTiter 96® Aqueous Non-Radioactive Cell Proliferation (MTS) Assay (Promega, Madison, WI, USA). Differences in absorbance were measured at 490 nm using GloMax® Multi Detection System (Sunnyvale, CA, USA).

4.4. EC_{50} Determination

To quantitatively determine the effective potency in BV2 cells, the compound half maximal effective concentration or EC_{50} was determined by a dose-response curve. In brief, BV2 cells were plated in 96-well plates as described above and treated with increasing concentrations of compound Oxa12 (from 0.1 to 50 µM) and Nec-1 (from 0.01 to 50 µM) using DMSO as vehicle control. Following 24 h of incubation, cell viability was determined based on the measurement of MTS, as previously described.

4.5. RIP1 and RIP3 Kinase Activity Assays

Oxa12 were tested at 5 µM for RIP1 kinase activity by a radiometric-binding assay using myelin basic protein as substrate (Eurofins, France) as before [39].

4.6. Microsomal Stability Assay

In the microsomal stability assay, an analytical high-performance liquid chromatography (HPLC) system was used with the following condition: column, Merck Lichrospher 100 RP18 125 mm 4.6 mm (5 µm); mobile phase A = 0.1% trifluoroacetic acid in water, B = 0.1% trifluoroacetic acid in acetonitrile, isocratic; flow rate, 1 mL/min; detection, UV at 400 nm injection, 20 µL; column temperature, ambient. The metabolic stability assays were carried out using the cosolvent method, appropriate for assessing the metabolic stability of compounds poorly soluble in aqueous medium [40]. For the cosolvent method, a 0.5 mM DMSO stock solution of Oxa12 was prepared. Then, a diluted solution of the compound was prepared by adding 50 µL of the previous 0.05 mM solution with 200 µL of acetonitrile, to make a 0.1 mM solution of Oxa12 in 20% DMSO/80% acetonitrile. Cosolvent assay conditions were: substrate concentration, 1 µM; microsomal protein, 0.5 mg/mL; organic solvents, 0.2% DMSO, 0.8% acetonitrile; incubation time, 30 min; number of assays, duplicates for T0 and T30 min. Time 0 and time 30 batches, after quenching with acetonitrile, were centrifuged at 11,000× *g* for 5 min and the supernatants were analyzed by HPLC, in order to quantify compound Oxa12.

4.7. MPTP Mouse Model

Animal studies were performed according to the animal welfare of the Faculty of Pharmacy, University of Lisbon, and approved by the competent national authority Direção-Geral de Alimentação e Veterinária (DAGV) and in accordance with the EU Directive (2010/63/UE), Portuguese laws (DR 113/2013, 2880/2015 and 260/2016) and all relevant legislation. To evaluate the neuroprotective effect of our hit compound—Oxa12—in the sub-acute MPTP mouse model, male 13-week-old C57BL/6N wild-type (wt) mice (Charles River Laboratories, Wilmington, MA, USA) were injected intraperitoneally (i.p.) with a unique dose of MPTP-HCl (40 mg/kg; Sigma Aldrich, St Louis, MO, USA), dissolved in sterile 0.9% saline, or vehicle only (control group). One hour after MPTP administration, mice were intraperitoneally injected with 10 mg/kg Oxa12 solubilized in 1% DMSO (Sigma-Aldrich) and 30% 2-hydroxypropyl-beta-cyclodextrin (Sigma-Aldrich) or 10 mg/kg Nec-1s (Focus Biomolecules, Plymouth Meeting, PA, USA) solubilized in 1% DMSO, 4% 2-hydroxypropyl-beta-cyclodextrin (Sigma-Aldrich), in PBS [38,41,42]. Oxa12 and Nec-1s injections were administered once every day for 30 days. Oxa12 dosage and regimen of administration were selected based on published protocols for Nec-1s [9,33]. Seven animal per group were used. After 30 days, mice were sacrificed in a CO_2 chamber followed by transcardiac perfusion with ice-cold PBS. Brains were then excised, and one hemisphere was used to isolate the midbrain region, containing the SN, and the striatum, as previously described [38,41], which was rapidly frozen in liquid nitrogen and stored at −80 °C until processing for protein extraction. The other hemisphere was fixed in 4% paraformaldehyde for 48 h and then stored in 20% sucrose/PBS and 0.025% sodium azide, at 4 °C, for further immunohistochemistry analyses.

4.8. Immunohistochemistry

Hemispheres previously fixed in paraformaldehyde were cryoprotected in 20% sucrose/PBS and embedded in gelatin. Then, sequential coronal brain sections (8 µm thick) near the midstriatum (Bregma 1.00) and SN (Bregma −3.20) were obtained by cryostat sectioning and mounted on SuperFrost-Plus glass slides (Thermo Fisher Scientific). Afterward, sections were incubated in warm PBS at 37 °C for 15 min, followed by two washes in PBS, to remove gelatin. Then, sections were blocked in Tris-buffered saline (TBS) containing 10% (*v*/*v*) normal donkey serum (Jackson ImmunoResearch Laboratories Inc., West Grove, PA, USA) and 0.1% (*v*/*v*) Triton X-100 (Sigma-Aldrich) for 1 h. Subsequentially, to stain dopaminergic neurons, sections were incubated with primary rabbit polyclonal anti-tyrosine hydroxylase (TH) antibody (#ab112; Abcam, Cambridge, UK, 1:700), overnight at 4 °C. After several washes with PBS, anti-TH primary antibody was detected with diluted (1:200) Alexa Fluor 488 (anti-rabbit) conjugated secondary antibody (Invitrogen—Thermo Fisher Scientific) for 2 h at room temperature. After extensive rinsing, sections were counterstained with Hoechst 33258 (Sigma-Aldrich) and mounted on Mowiol 4-88 (Sigma-Aldrich).

4.9. Image Analysis

Images were obtained by an Axioskop fluorescence microscope (Carl Zeiss GmbH, Hamburg, Germany). Images were captured from six region-matched sections for nigral and striatal regions for each animal and converted into a gray scale with an 8-bit format using the ImageJ software version 1.48 (National Institute of Health, Bethesda, MD, USA). A threshold optical density was determined for each staining. Areas occupied by positive staining were quantified in thresholded images, normalized to the total area of interest region and calculated as percentage of total area.

4.10. Protein Isolation

For total protein isolation, tissues obtained from dissected midbrains and striata were homogenized in radio-immunoprecipitation assay (RIPA) buffer (50 nM Tris/HCl, pH 8; 150 nM NaCl; 1% NP-40; 0.5% sodium deoxycholate; 0.1% SDS) and 1x Halt Protease and Phosphatase Inhibitor Cocktail (Pierce, Thermo Fisher Scientific) with a moto-driven Bio-vortexer (No 1083; Biospec Products, Bartlesfield, UK). Lysates were maintained on ice for 30 min and were then sonicated and centrifuged at 10,000× *g* for 10 min. Supernatants were collected and used as total protein extracts. Protein concentrations were determined by using the Bio-Rad protein assay kit, according to the manufacturer's instructions.

4.11. Western Blot

Equal amounts of total protein extracts were electrophoretically resolved on 8% SDS-PAGE. Resolved proteins were then transferred onto nitrocellulose membranes and blocked with a 5% milk solution in Tris-buffered saline (TBS). Then, membranes were incubated with primary antibody rabbit polyclonal TH (#ab112, Abcam), overnight at 4 °C. After washing with TBS/0.2% Tween 20 (TBS-T), membranes were incubated with secondary goat anti-rabbit IgG antibody conjugated with horseradish peroxidase (Bio-Rad Laboratories) for 2 h at room temperature. Membranes were processed for protein detection using Immobilon^{TM} Western (Millipore). β-actin (AC-15) (#A5441, Sigma-Aldrich) was used as loading control. Densitometric analysis was performed using the Image Lab Software version 5.1 Beta (Bio-Rad).

4.12. Statistical Analysis

All data are presented as mean ± standard error of the mean (SEM). Data analysis were performed with a one-way analysis of variance (ANOVA) followed by a post hoc Bonferroni test. Analysis and graphical presentation were conducted with the GraphPad Prim Software version 8 (GraphPad Software Inc., San Diego, CA, USA). Statistical significance was achieved when $p < 0.05$.

Author Contributions: Conceptualization, P.A.D., J.D.A., C.A.M.A. and C.M.P.R.; methodology, S.R.O., P.A.D., M.M.G., M.B.T.F., C.A.B.R., R.G.P., M.S.E. and M.J.P.; investigation, S.R.O., P.A.D., M.M.G., M.B.T.F., C.A.B.R., R.G.P., M.S.E. and M.J.P.; data curation, S.R.O., P.A.D., J.D.A., R.M. and C.A.M.A.; writing—original draft preparation, S.R.O.; writing—review and editing, S.R.O., P.A.D., M.M.G., M.B.T.F., C.A.B.R., R.G.P., M.S.E., M.J.P., R.M., C.A.M.A., J.D.A. and C.M.P.R.; visualization, S.R.O., J.D.A. and C.A.M.A.; supervision, J.D.A., C.A.M.A. and C.M.P.R.; project administration, J.D.A. and C.M.P.R.; funding acquisition, C.M.P.R. All authors have read and agreed to the published version of the manuscript.

Funding: This research received funding from European Structural & Investment Funds through the COMPETE Program—Programa Operacional Regional de Lisboa—Program Grant LISBOA-01-0145-FEDER-016405, and from National Funds through the FCT—Fundação para a Ciência e a Tecnologia—Program Grant SAICTPAC/0019/2015. S.R.O. received a PhD fellowship (PD/BD/128332//2017) from the FCT.

Institutional Review Board Statement: The study was conducted according to the guidelines of the Declaration of Helsinki and approved by the institutional Review Board and by the national authority Direção-Geral de Alimentação e Veterinária (DGAV), initiated on 1 August 2018 (Necroptosis in dopaminergic neurodegeneration in Parkinson's disease), in accordance with the EU Directive (2010/63/UE), Portuguese laws (DL 113/2013, 2880/2015, 260/2016 and 1/2019) and all relevant legislation.

Informed Consent Statement: Not applicable.

Acknowledgments: We thank all members of the laboratory for advice and technical assistance.

Conflicts of Interest: The authors declare no conflict of interest. The funders had no role in the design of the study; in the collection, analyses or interpretation of data; in the writing of the manuscript, or in the decision to publish the results.

References

1. Poewe, W.; Seppi, K.; Tanner, C.M.; Halliday, G.M.; Brundin, P.; Volkmann, J.; Schrag, A.E.; Lang, A.E. Parkinson disease. *Nat. Rev. Dis. Primers.* **2017**, *3*, 17013. [CrossRef]
2. Kordower, J.H.; Olanow, C.W.; Dodiya, H.B.; Chu, Y.; Beach, T.G.; Adler, C.H.; Halliday, G.M.; Bartus, R.T. Disease duration and the integrity of the nigrostriatal system in Parkinson's disease. *Brain* **2013**, *136*, 2419–2431. [CrossRef] [PubMed]
3. Levy, O.A.; Malagelada, C.; Greene, L.A. Cell death pathways in Parkinson's disease: Proximal triggers, distal effectors, and final steps. *Apoptosis* **2009**, *14*, 478–500. [CrossRef] [PubMed]
4. Venderova, K.; Park, D.S. Programmed cell death in Parkinson's disease. *Cold Spring Harb. Perspect. Med.* **2012**, *2*, a009365. [CrossRef]
5. Conrad, M.; Angeli, J.P.; Vandenabeele, P.; Stockwell, B.R. Regulated necrosis: Disease relevance and therapeutic opportunities. *Nat. Rev. Drug. Discov.* **2016**, *15*, 348–366. [CrossRef] [PubMed]
6. He, S.; Wang, L.; Miao, L.; Wang, T.; Du, F.; Zhao, L.; Wang, X. Receptor interacting protein kinase-3 determines cellular necrotic response to TNF-alpha. *Cell* **2009**, *137*, 1100–1111. [CrossRef] [PubMed]
7. Bonnet, M.C.; Preukschat, D.; Welz, P.S.; van Loo, G.; Ermolaeva, M.A.; Bloch, W.; Haase, I.; Pasparakis, M. The adaptor protein FADD protects epidermal keratinocytes from necroptosis in vivo and prevents skin inflammation. *Immunity* **2011**, *35*, 572–582. [CrossRef] [PubMed]
8. Grootjans, S.T.; Vanden Berghe, T.; Vandenabeele, P. Initiation and execution mechanisms of necroptosis: An overview. *Cell Death Differ.* **2017**, *24*, 1184–1195. [CrossRef]
9. Iannielli, A.; Bido, S.; Folladori, L.; Segnali, A.; Cancellieri, C.; Maresca, A.; Massimino, L.; Rubio, A.; Morabito, G.; Caporali, L.; et al. Pharmacological Inhibition of Necroptosis Protects from Dopaminergic Neuronal Cell Death in Parkinson's Disease Models. *Cell Rep.* **2018**, *22*, 2066–2079. [CrossRef]
10. Lin, Q.S.; Chen, P.; Wang, W.X.; Lin, C.C.; Zhou, Y.; Yu, L.H.; Lin, Y.X.; Xu, Y.F.; Kang, D.Z. RIP1/RIP3/MLKL mediates dopaminergic neuron necroptosis in a mouse model of Parkinson disease. *Lab. Investig.* **2020**, *100*, 503–511. [CrossRef]
11. Wang, L.; Du, F.; Wang, X. TNF-alpha induces two distinct caspase-8 activation pathways. *Cell* **2008**, *133*, 693–703. [CrossRef] [PubMed]
12. Christofferson, D.E.; Li, Y.; Hitomi, J.; Zhou, W.; Upperman, C.; Zhu, H.; Gerber, S.A.; Gygi, S.; Yuan, J. A novel role for RIP1 kinase in mediating TNFα production. *Cell Death Dis.* **2012**, *3*, e320. [CrossRef] [PubMed]
13. Degterev, A.; Huang, Z.; Boyce, M.; Li, Y.; Japtag, P.; Mizushima, N.; Cuny, G.D.; Mitchison, T.J.; Moskowitz, M.A.; Yuan, J. Chemical inhibitor of nonapoptotic cell death with therapeutic potential for ischemic brain injury. *Nat. Chem. Biol.* **2005**, *1*, 112–119. [CrossRef]
14. Declercq, W.; Vanden Berghe, T.; Vandenabeele, P. RIP kinases at the crossroads of cell death and survival. *Cell* **2009**, *138*, 229–232. [CrossRef] [PubMed]
15. Xie, T.; Peng, W.; Liu, Y.; Yan, C.; Maki, J.; Degterev, A.; Yuan, J.; Shi, Y. Structural basis of RIP1 inhibition by necrostatins. *Structure* **2013**, *21*, 493–499. [CrossRef]
16. Brenner, D.; Blaser, H.; Mak, T.W. Regulation of tumour necrosis factor signalling: Live or let die. *Nat. Rev. Immunol.* **2015**, *15*, 362–374. [CrossRef]
17. Teng, X.; Degterev, A.; Japtag, P.; Xing, X.; Choi, S.; Denu, R.; Yuan, J.; Cuny, G.D. Structure-activity relationship study of novel necroptosis inhibitors. *Bioorganic Med. Chem. Lett.* **2005**, *15*, 5039–5044. [CrossRef]
18. Caccamo, A.; Branca, C.; Piras, I.S.; Ferreira, E.; Huentelman, M.J.; Liang, W.S.; Readhead, B.; Dudley, J.T.; Spangenberg, E.E.; Green, K.N.; et al. Necroptosis activation in Alzheimer's disease. *Nat. Neurosci.* **2017**, *20*, 1236–1246. [CrossRef]
19. Sun, L.; Wang, H.; Wang, Z.; He, S.; Chen, S.; Liao, D.; Wang, L.; Yan, J.; Liu, W.; Lei, X.; et al. Mixed lineage kinase domain-like protein mediates necrosis signaling downstream of RIP3 kinase. *Cell* **2012**, *148*, 213–227. [CrossRef] [PubMed]
20. Mandal, P.; Berger, S.B.; Pillay, S.; Moriwaki, K.; Huang, C.; Guo, H.; Lich, J.D.; Finger, J.; Kasparcova, V.; Votta, B.; et al. RIP3 induces apoptosis independent of pronecrotic kinase activity. *Mol. Cell* **2014**, *56*, 481–495. [CrossRef] [PubMed]
21. Wu, Y.T.; Peng, W.; Liu, Y.; Yan, C.; Maki, J.; Degterev, A.; Yuan, J.; Shi, Y. zVAD-induced necroptosis in L929 cells depends on autocrine production of TNFα mediated by the PKC-MAPKs-AP-1 pathway. *Cell Death Differ.* **2011**, *18*, 26–37. [CrossRef]
22. Oliveira, S.R.; Dionísio, P.A.; Brito, H.; Franco, L.; Rodrigues, C.A.B.; Guedes, R.C.; Afonso, C.A.M.; Amaral, J.D.; Rodrigues, C.M.P. Phenotypic screening identifies a new oxazolone inhibitor of necroptosis and neuroinflammation. *Cell Death Discov.* **2018**, *4*, 10. [CrossRef] [PubMed]
23. Wager, T.T.; Hou, X.; Verhoest, P.R.; Villalobos, A. Moving beyondrules: The development of a central nervous system multiparameter optimization (CNS MPO) approach to enable alignment of druglike properties. *ACS Chem. Neurosci.* **2010**, *1*, 435–449. [CrossRef] [PubMed]
24. Wager, T.T.; Hou, X.; Verhoest, P.R.; Villalobos, A. Central nervous system multiparameter optimization desirable application in drug discovery. *ACS Chem. Neurosci.* **2016**, *7*, 767–775. [CrossRef]
25. Degterev, A.; Maki, J.L.; Yuan, J. Activity and specificity of necrostatin-1, small-molecule inhibitor of RIP1 kinase. *Cell Death Differ.* **2013**, *20*, 366. [CrossRef] [PubMed]
26. Meredith, G.E.; Rademacher, D.J. MPTP mouse models of Parkinson's disease: An update. *J. Parkinson's Dis.* **2011**, *1*, 19–33. [CrossRef] [PubMed]
27. Langston, J.W. The MPTP Story. *J. Parkinson's Dis.* **2017**, *7*, S11–S19. [CrossRef] [PubMed]

28. Ransom, B.R.; Kunis, D.M.; Irwin, I.; Langston, J.W. Astrocytes convert the parkinsonism inducing neurotoxin, MPTP, to its active metabolite, MPP+. *Neurosci. Lett.* **1987**, *75*, 323–328. [CrossRef]
29. Schildknecht, S.; Pape, R.; Meiser, J.; Karreman, C.; Strittmatter, T.; Odermatt, M.; Cirri, E.; Friemel, A.; Ringwald, M.; Pasquarelli, N.; et al. Preferential Extracellular Generation of the Active Parkinsonian Toxin MPP+ by Transporter-Independent Export of the Intermediate MPDP+. *Antioxid. Redox Signal.* **2015**, *23*, 1001–1016. [CrossRef]
30. Dauer, W.; Przedborski, S. Parkinson's disease: Mechanisms and models. *Neuron* **2003**, *39*, 889–909. [CrossRef]
31. Rodrigues, C.A.B.; Mariz, I.F.A.; Maçôas, E.M.S.; Afonso, C.A.M.; Martinho, J.M.G. Unsaturated oxazolones as nonlinear fluorophores. *Dyes Pigment.* **2013**, *99*, 642–652. [CrossRef]
32. Vandenabeele, P.; Grootjans, S.; Callewaert, N.; Takahashi, N. Necrostatin-1 blocks both RIPK1 and IDO: Consequences for the study of cell death in experimental disease models. *Cell Death Differ.* **2013**, *20*, 185–187. [CrossRef] [PubMed]
33. Ofengeim, D.; Ito, Y.; Najafov, A.; Zhang, Y.; Shan, B.; DeWitt, J.P.; Ye, J.; Zhang, X.; Chang, A.; Vakifahmetoglu-Norberg, H.; et al. Activation of necroptosis in multiple sclerosis. *Cell Rep.* **2015**, *10*, 1836–1849. [CrossRef] [PubMed]
34. Berger, S.B.; Harris, P.; Nagilla, R.; Kasparcova, V.; Hoffman, S.; Swift, B.; Dare, L.; Schaeffer, M.; Capriotti, C.; Ouellette, M.; et al. Characterization of GSK'963: A structurally distinct, potent and selective inhibitor of RIP1 kinase. *Cell Death Discov.* **2015**, *1*, 15009. [CrossRef] [PubMed]
35. Harris, P.A.; Bandyopadhyay, D.; Berger, S.B.; Campobasso, N.; Capriotti, C.A.; Cox, J.A.; Dare, L.; Finger, J.N.; Hoffman, S.J.; Kahler, K.M.; et al. Discovery of Small Molecule RIP1 Kinase Inhibitors for the Treatment of Pathologies Associated with Necroptosis. *ACS Med. Chem. Lett.* **2013**, *4*, 1238–1243. [CrossRef]
36. Harris, P.A.; Berger, S.B.; Jeong, J.U.; Nagilla, R.; Bandyopadhyay, D.; Campobasso, N.; Capriotti, C.A.; Cox, J.A.; Dare, L.; Dong, X.; et al. Discovery of a First-in-Class Receptor Interacting Protein 1 (RIP1) Kinase Specific Clinical Candidate (GSK2982772) for the Treatment of Inflammatory Diseases. *J. Med. Chem.* **2017**, *60*, 1247–1261. [CrossRef] [PubMed]
37. Pong, K.; Doctrow, S.R.; Huffman, K.; Adinolfi, C.A.; Baudry, M. Attenuation of staurosporine-induced apoptosis, oxidative stress, and mitochondrial dysfunction by synthetic superoxide dismutase and catalase mimetics, in cultured cortical neurons. *Exp. Neurol.* **2001**, *171*, 84–97. [CrossRef]
38. Dionísio, P.A.; Oliveira, S.R.; Gaspar, M.M.; Gama, M.J.; Castro-Caldas, M.; Amaral, J.D.; Rodrigues, C.M.P. Ablation of RIP3 protects from dopaminergic neurodegeneration in experimental Parkinson's disease. *Cell Death Dis.* **2019**, *10*, 840. [CrossRef]
39. Brito, H.; Marques, V.; Afonso, M.B.; Brown, D.G.; Börjesson, U.; Selmi, N.; Smith, D.M.; Roberts, I.O.; Fitzek, M.; Aniceto, N.; et al. Phenotypic high-throughput screening platform identifies novel chemotypes for necroptosis inhibition. *Cell Death Discov.* **2020**, *6*, 6. [CrossRef]
40. Di, L.; Kerns, E.H.; Li, S.Q.; Petusky, S.L. High throughtput microsomal stability assay for insoluble compounds. *Int. J. Pharm.* **2006**, *317*, 54–60. [CrossRef]
41. Castro-Caldas, M.; Neves Carvalho, A.; Peixeiro, I.; Rodrigues, E.; Lechner, M.C.; Gama, M.J. GSTpi expression in MPTP-induced dopaminergic neurodegeneration of C57BL/6 mouse midbrain and striatum. *J. Mol. Neurosci.* **2009**, *38*, 114–127. [CrossRef] [PubMed]
42. Saporito, M.S.; Thomas, B.A.; Scott, R.W. MPTP activates c-Jun NH(2)-terminal kinase (JNK) and its upstream regulatory kinase MKK4 in nigrostriatal neurons in vivo. *J. Neurochem.* **2000**, *75*, 1200–1208. [CrossRef] [PubMed]

International Journal of
Molecular Sciences

Article

Epigallocatechin-3-Gallate-Loaded Liposomes Favor Anti-Inflammation of Microglia Cells and Promote Neuroprotection

Chun-Yuan Cheng [1,2], Lassina Barro [2], Shang-Ting Tsai [2,3], Tai-Wei Feng [2,3], Xiao-Yu Wu [2], Che-Wei Chao [4], Ruei-Siang Yu [2], Ting-Yu Chin [4,*] and Ming Fa Hsieh [2,3,*]

1 Division of Neurosurgery, Department of Surgery, Changhua Christian Hospital, 135 Nanxiao St., Changhua City, Changhua County 500, Taiwan; 83998@cch.org.tw
2 Department of Biomedical Engineering, Chung Yuan Christian University, No. 200, Zhongbei Rd., Zhongli Dist., Taoyuan City 320314, Taiwan; lassinabarro_dieron@yahoo.fr (L.B.); shant7707@gmail.com (S.-T.T.); taverenspirit85@gmail.com (T.-W.F.); ggg2004aaa@yahoo.com.tw (X.-Y.W.); shung804@yahoo.com.tw (R.-S.Y.)
3 Center for Minimally-Invasive Medical Devices and Technologies, Chung Yuan Christian University, No. 200, Zhongbei Rd., Zhongli Dist., Taoyuan City 320314, Taiwan
4 Department of Bioscience Technology, Chung Yuan Christian University, No. 200, Zhongbei Rd., Zhongli Dist., Taoyuan City 320314, Taiwan; oe4213@gmail.com
* Correspondence: tychin@cycu.edu.tw (T.-Y.C.); mfhsieh@cycu.edu.tw (M.F.H.)

Abstract: Microglia-mediated neuroinflammation is recognized to mainly contribute to the progression of neurodegenerative diseases. Epigallocatechin-3-gallate (EGCG), known as a natural antioxidant in green tea, can inhibit microglia-mediated inflammation and protect neurons but has disadvantages such as high instability and low bioavailability. We developed an EGCG liposomal formulation to improve its bioavailability and evaluated the neuroprotective activity in in vitro and in vivo neuroinflammation models. EGCG-loaded liposomes have been prepared from phosphatidylcholine (PC) or phosphatidylserine (PS) coated with or without vitamin E (VE) by hydration and membrane extrusion method. The anti-inflammatory effect has been evaluated against lipopolysaccharide (LPS)-induced BV-2 microglial cells activation and the inflammation in the substantia nigra of Sprague Dawley rats. In the cellular inflammation model, murine BV-2 microglial cells changed their morphology from normal spheroid to activated spindle shape after 24 h of induction of LPS. In the in vitro free radical 2,2-diphenyl-1-picrylhydrazyl (DPPH) assay, EGCG scavenged 80% of DPPH within 3 min. EGCG-loaded liposomes could be phagocytized by BV-2 cells after 1 h of cell culture from cell uptake experiments. EGCG-loaded liposomes improved the production of BV-2 microglia-derived nitric oxide and TNF-α following LPS. In the in vivo Parkinsonian syndrome rat model, simultaneous intra-nigral injection of EGCG-loaded liposomes attenuated LPS-induced pro-inflammatory cytokines and restored motor impairment. We demonstrated that EGCG-loaded liposomes exert a neuroprotective effect by modulating microglia activation. EGCG extracted from green tea and loaded liposomes could be a valuable candidate for disease-modifying therapy for Parkinson's disease (PD).

Keywords: neuroprotection; neuroinflammation; Parkinson's disease; catechin; L-α-phosphatidylcholine; phosphatidylserine

Citation: Cheng, C.-Y.; Barro, L.; Tsai, S.-T.; Feng, T.-W.; Wu, X.-Y.; Chao, C.-W.; Yu, R.-S.; Chin, T.-Y.; Hsieh, M.F. Epigallocatechin-3-Gallate-Loaded Liposomes Favor Anti-Inflammation of Microglia Cells and Promote Neuroprotection. *Int. J. Mol. Sci.* **2021**, *22*, 3037. https://doi.org/10.3390/ijms22063037

Academic Editor: Anne Vejux

Received: 8 February 2021
Accepted: 12 March 2021
Published: 16 March 2021

1. Introduction

When the nervous system is damaged or infected, microglia cells are activated and transformed to be branched, resulting in excessive expression of a large amount of pro-inflammatory cytokines such as tumor necrosis factor-α (TNF-α), interleukin-1β (IL-1β), interleukin 6 (IL-6), and inflammatory mediators such as nitric oxide (NO) and reactive oxygen species (ROS). Finally, nerve cells are damaged, degenerated, or die from these inflammatory mediators. It is recently found that in the brain of patients with neurodegenerative diseases such as Parkinson's disease (PD), Alzheimer's disease, Huntington's disease, and Creutzfeldt-Jakob disease, large amounts of microglia cells are activated and over-expressed [1–3]. Epidemiologically, PD's cause is mostly linked to the neuroinflammatory reaction. The inflammatory mediators resulting, such as TNF-α, IL-1β, IL-6, NO, and ROS, are found in the striatum of the brain [1,4–7]. The degradation of dopaminergic neurons can be regulated by microglia cells [8].

The neuroinflammation process that causes PD is preceded by primary damage of neurons caused by environmental toxins, including rotenone [9], lipopolysaccharide (LPS) [5,7], and the effects of abnormal protein accumulation [10]. The damage will cause lesions and even apoptosis of dopaminergic neurons. Then microglia cells are activated to release cytokines, resulting in inflammation and death of the neurons and finally leading to PD.

When the microglia cells are stimulated by LPS, LPS binds to the surface receptor CD14 binding site of microglia cells. The LPS-CD14 complex is linked with MD2 linker through toll-like receptor-4 (TLR4) transmembrane proteins and then involved in multiple message transmission pathways generated by mitogen-activated protein kinases (MAPK) and activating transcription factors (nuclear factor-kappa B, NF-κB). After gene transcription [5,11,12], microglial cells release cytokines such as TNF-α and IL-1β, or express genes of inducible nitric oxide synthase (iNOS) and cyclo-oxygenase-2 (COX-2), resulting in the release of prostaglandins or NO. In addition, destructive ONOO free radicals are generated by combining superoxide anions produced from nicotinamide adenine dinucleotide phosphate (NADPH) oxidase with NO produced from iNOS, leading to the death of dopaminergic neurons [13]. Therefore, in this study, LPS was used to induce neuroinflammation of microglia cells as an in vitro model of PD.

Catechins are natural antioxidants that can prevent cell damage and provide many pharmacological benefits such as anti-tumor, anti-cancer, anti-aging, anti-pharmacological radiation, and free radical scavenging [14]. Green tea contains about 10% polyphenols by weight, including large amounts of a catechin called epigallocatechin gallate (EGCG). EGCG has the highest antioxidant activity and free radical scavenging capacity in all green tea catechins and can capture ROS to protect cells from the damage of oxidative stress [15]. EGCG also has high anti-inflammatory efficacy, which can effectively inhibit secretion of cytokines (TNF-α, IL-2, and IL-8) by macrophages [16], phosphorylation of Akt signaling proteins and IκB proteins in inflammatory pathways to reduce NF-κB expression, or AP-1 transcription by inhibiting phosphorylation of upstream MAPK proteins to balance COX-2 expression and reduce the production of pro-inflammatory cytokines [17].

Recently, EGCG was reported to be potentially therapeutic or prophylactic for PD due to suppressing active oligomers of α-synuclein (αS) [18]. EGCG also prevents αS aggregation in vitro [19–21], and cytoplasmic αS aggregation in dopaminergic neurons is one possible pathogenesis of PD leading to damage of dopaminergic neurons in substantia nigra [22]. Furthermore, EGCG can recover 1-methyl-4-phenyl-1,2,3,6-tetrahydropyridine (MPTP)-induced neurochemical or functional damage and regulate ferroportin in substantia nigra and reduce oxidative stress [23]. EGCG also has neuroprotective and immune-protective effects in MPTP-treated mice and can modulate neuroinflammation and protect dopaminergic neuron loss in MPTP-induced PD [24].

The anti-inflammation effects of EGCG were investigated. EGCG suppressed LPS-induced NO production and expression of iNOS in BV-2 microglial cells. EGCG can effectively inhibit the expressions of pro-inflammatory cytokines such as TNF-α and IL-1β in BV-2 cells [25]. EGCG pretreatment of human macrophages significantly inhibited LPS-induced expression of pro-inflammatory cytokines such as TNF-α, IL-1β, and IL-6 [26]. In addition, post-treatment of EGCG on LPS-injured mice decreased production of a pro-inflammatory cytokine through modulating the TLR4-NF-κB pathway [27]. Moreover, poly(lactide-co-glycolide) (PLGA) microspheres loaded with EGCG and optimized by the addition of β-cyclodextrin (β-CD) could effectively suppress NO production from BV-2 cells in the in vitro model of murine BV-2 microglial cells stimulated by LPS, indicating the microspheres can suppress inflammation of activated microglial cells [28].

Although green tea is a daily drink, the efficacy of catechins is ineffective due to low oral bioavailability; thus, effective pharmaceutical dosage forms are required. Nano drug carrier has the benefits of avoiding premature metabolism, extending drug action time, and targeting drug delivery. Therefore, this study intends to develop liposomes containing phosphatidylcholine (PC) and phosphatidylserine (PS), similar components to the cell membrane, as anti-inflammatory dosage forms. EGCG extracted from green tea leaves was loaded in liposomes to slow down inflammatory reaction in microglia cells induced by LPS. The therapeutic effect of EGCG-loaded liposomes on the in vivo model of PD for neuroprotection was also evaluated.

2. Results

2.1. Extraction of EGCG

2.1.1. EGCG Extract

The epigallocatechin-3-gallate characterization data can be found in the Supplementary File (Figures S1 and S2).

2.1.2. Various Formulations of EGCG

In Table 1, the average particle diameters of placebo PS-, PS-EGCE-, and PS-EGCG-VE-liposomes were smaller than that of placebo PC-, PC-EGCE-, and PC-EGCG-VE-liposomes, respectively. Because PC is neutral and PS is negatively charged [29], this indicates that the additive surface potential affected the particle size. Polydispersity index (PDI) of all liposomes was less than 0.22, indicating that the structure of liposomes in solution is stable. The negative charges of PS led to a repulsive force on the surface potential between PS-liposomes, avoiding aggregation and reducing the size of PS-liposomes.

Table 1. Characteristics of the formulation of EGCG-loaded liposomes.

Samples	Molar Ratio				Diameter (nm)	Encapsulation Efficiency(%)	PDI
	PC	PS	CH	VE			
PC-liposome	0.73	-	0.27	-	184.4 ± 1.59	-	0.218
PS-liposome	0.24	0.49	0.27	-	117.47 ± 1.12	-	0.092
PC-EGCG-liposome	0.73	-	0.27	-	155.2 ± 1.23	55.4%	0.121
PS-EGCG-liposome	0.24	0.49	0.27	-	132.86 ± 2.05	70.4%	0.094
PC-EGCG-VE-liposome	0.73	-	0.27	0.07	161.5 ± 0.56	60.2%	0.058
PS-EGCG-VE-liposome	0.24	0.49	0.27	0.07	142.9 ± 0.36	76.8%	0.101

PC: L-α-phosphatidylcholine; PS: phosphatidylserine; CH: cholesterol; VE: vitamin E (α-tocopherol); EGCG: (−)-epigallocatechin-3-gallate; PDI: polydispersity index.

The encapsulation efficiency/size of PS-containing liposomes described in Table 1 was larger/smaller than that of corresponding PC-containing liposomes [30]. The encapsulation efficiency of PC-EGCG-VE-liposomes and PS-EGCG-VE-liposomes was more extensive than that of PC-EGCG-liposomes and PS-EGCG-liposomes, respectively. This is because vitamin E is fat-soluble, embedded in a phospholipid bilayer membrane, and provides an antioxidant protectant for EGCG.

2.2. In Vitro Cell Analysis

2.2.1. Cell Viability

The cell viability of cells treated with EGCG from 50 to 400 μM was significantly decreased compared to the control group (Figure 1A). In contrast, the viability of cells receiving 5 to 25 μM is similar to the control group cells' viability. The concentration of EGCG used in this study was determined to be 25 μM. Similarly, the concentration of

LPS used for induction of cell inflammation was determined to be 50 ng/mL according to Figure 1B. In Figure 1C, the cell viability of cells co-cultured with placebo liposomes of all concentrations was not statistically significant compared to the control group. Therefore, placebo liposomes are not cytotoxic to microglia cells.

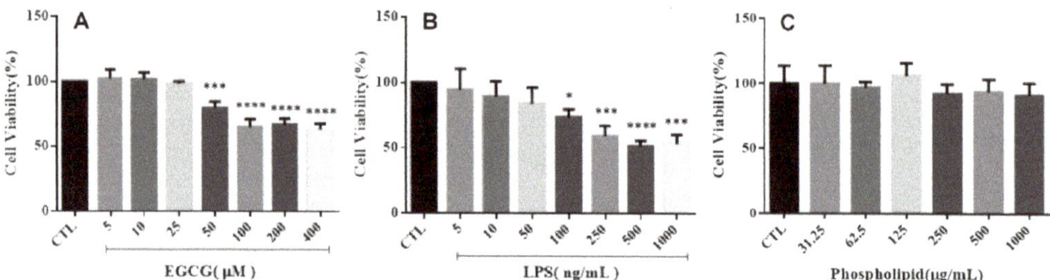

Figure 1. Cell viability of BV-2 cells treated by (**A**) EGCG, (**B**) lipopolysaccharide (LPS), and (**C**) placebo liposome (compared with control group (CTL), * $p < 0.05$; *** $p < 0.001$, **** $p < 0.0001$, $n = 3$).

2.2.2. Cell Morphology

Cell morphology of the control group (Figure 2A) and the morphology of cells treated with 25 μM EGCG (Figure 2B) were spherical, while the morphology of cells treated with 50 ng/mL LPS (Figure 2C) was spindle-shaped. However, the morphology of cells treated with 25 μM EGCG and activated by LPS (Figure 2D) was spherical. This denoted that EGCG can inhibit the activation induced by LPS. Therefore, the pretreatment of EGCG has an inhibitory effect on neuroinflammation, protecting the microglial cells from activation.

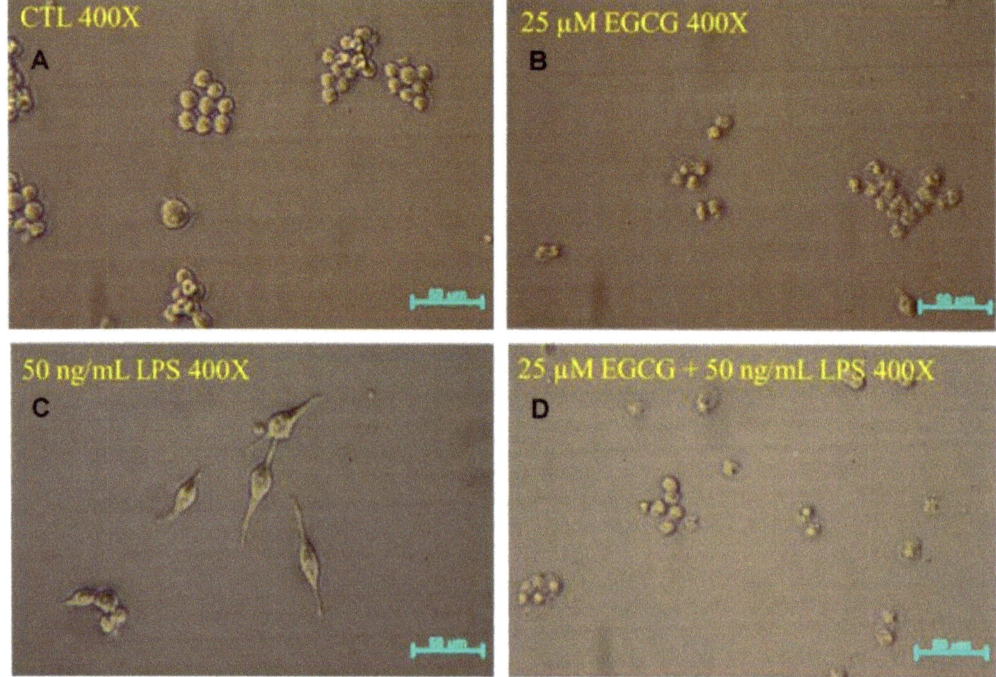

Figure 2. Morphology of BV-2 cells (**A**) of the control group, (**B**) treated with EGCG, (**C**) treated with LPS, and (**D**) treated with EGCG and then LPS.

2.2.3. NO Release

In Figure 3A, the NO release from BV-2 cells induced with 5–1000 ng/mL LPS during 24 h incubation was statistically significant compared to the NO release from the control group. The concentration of LPS used for activation of cell inflammation to work was determined to be 50 ng/mL.

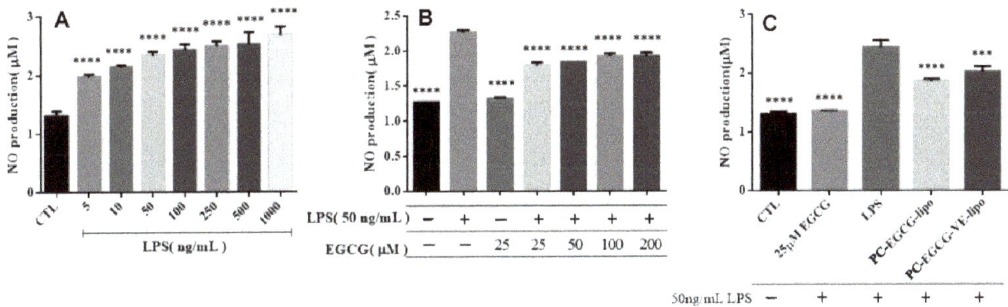

Figure 3. Nitric oxide (NO) production from BV-2 cells (*** $p < 0.001$; **** $p < 0.0001$, $n = 3$). (**A**) Inflammation induced by various concentration of LPS (compared with control group), (**B**) cells treated with EGCG followed by LPS induction (compared with LPS (+) and EGCG (−)), (**C**) cells treated with EGCG, PC-EGCG-liposomes, and EGCG-VE-liposomes then LPS (compared with LPS (+)).

NO release from BV-2 cells treated with 25 μM EGCG was not statistically significant compared to the control group, as shown in Figure 3B. However, the cell inflammation induced with LPS for 24 h showed a significant increase compared to the control group. Those cells treated with 25–200 μM EGCG for 1 h and then activated with LPS showed a statistically significant decrease compared to the group of cells activated with LPS only. NO release did not decrease when EGCG increased from 50–200 μM since the cell viability decreased when EGCG had risen from 50–200 μM, according to Figure 1A.

The NO production of the cells treated with 25 μM EGCG followed by the inflammation induced with 50 ng/mL LPS was not statistically significant compared with the control group (Figure 3C). However, the NO released in the group of cells treated with PC-EGCG-liposomes or PC-EGCG-VE-liposomes followed by LPS activation with 50 ng/mL showed a significant decrease compared to the group of cells treated only with LPS. The NO release from cells pretreated with PC-EGCG-liposomes or PC-EGCG-VE-liposomes was higher than the NO release from the group of cells pretreated by EGCG, which should be explained by the slow release of EGCG from the liposomes.

2.2.4. Cytokine Analysis

In Figure 4A, the concentration of TNF-α of cells treated with LPS after 24 h showed a statistically significant increase compared to that of the control group or the cell culture medium (DMEM). Placebo PC-liposomes were close to that of cells treated with LPS. However, the TNF-α concentration in the group of cells pretreated with PS-EGCG-liposomes or PS-EGCG-VE-liposomes followed by the LPS activation showed a statistically significant decrease compared to that of cells treated with LPS. This decreasing concentration of TNF-α indicates that EGCG-loaded liposomes can reduce the activation of microglial cells induced by LPS. The inhibitory effect of PS-EGCG-VE-liposomes was better than that of PS-EGCG-liposomes.

The phospholipids on the cell membrane can be hydrolyzed by cytosolic phospholipase A2 ($cPLA_2$) to produce arachidonic acid. The cyclo-oxygenase (COX) is a key enzyme that converts arachidonic acid to prostaglandin. COX-2 and $cPLA_2$ are often generated from inflammation or malignant disease [31–34]. In Figure 4B, the inflammation was induced by LPS of 5–50 ng/mL for 24 h, and the expression of $cPLA_2$ increased when the

LPS concentration increased. Figure 4C showed the increased activities of COX-2, induced by LPS (5–50 ng/mL) for 24 h. The expression of COX-2 increased from 5–25 ng/mL LPS, while it decreased at 50 ng/mL. In Figure 4D, the $cPLA_2$ expression was reduced when BV-2 cells were pretreated with EGCG and induced by LPS. The expression of COX-2 when LPS activated BV-2 cells was increased compared to the control group (Figure 4E). There was a significant decrease of COX-2 in the group of cells pretreated with EGCG, placebo PS-liposomes, PS-EGCG-liposomes, and PS-EGCG-VE-liposomes, followed by LPS inflammatory induction. Especially, expression of COX-2 of cells pretreated with placebo PS-liposomes had a statistically significant difference from that of cells induced by LPS.

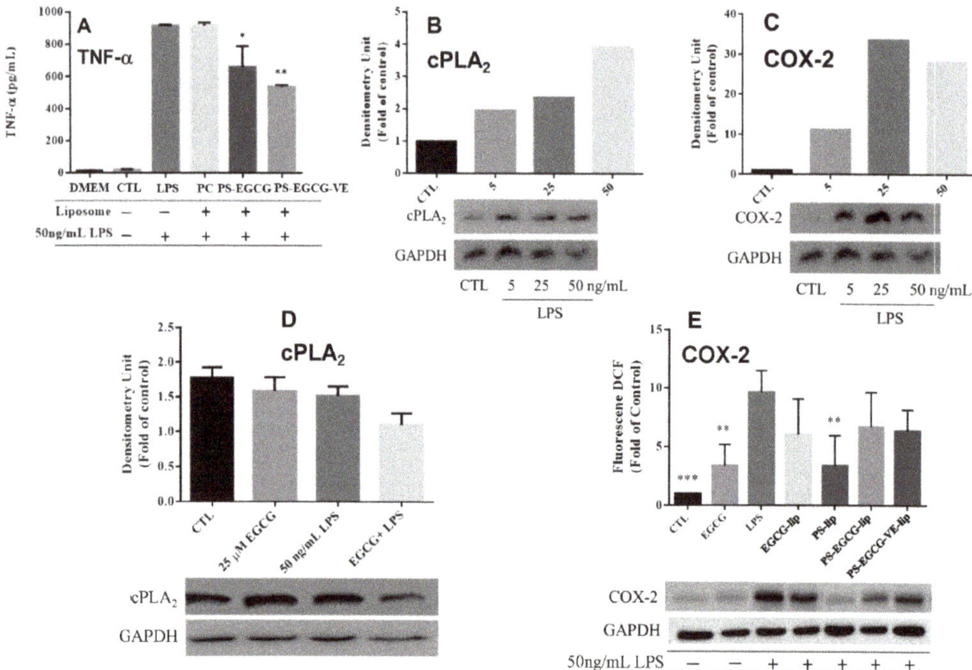

Figure 4. In LPS-induced BV-2 cells, expression of (**A**) TNF-α (compared with the liposome (−) and LPS (+) group, * $p < 0.05$; ** $p < 0.01$, $n = 3$), (**B**) cytosolic phospholipase A2 ($cPLA_2$), and (**C**) cyclo-oxygenase-2 (COX-2). In BV-2 cells pretreated by EGCG or liposomes and then induced by LPS, expression of (**D**) $cPLA_2$ and (**E**) COX-2 (compared with LPS (+) group ** $p < 0.01$, *** $p < 0.001$, $n = 3$).

2.3. *In Vivo Animal Test*

2.3.1. Animal Behavioral Test

The number of circles completed by the Parkinsonian rats after amphetamine administration (shown in Figure 5A) was significantly increased compared to the number of circles completed by the control group rats. Parkinsonian rats' behavior treated with the various formulation of EGCG was similar to the control group rats. These data indicated that EGCG attenuated LPS-induced unilateral lesions of the nigrostriatal system.

2.3.2. Inflammatory Markers Analysis

The ratio of TNF-α/GAPDH in the treated group was significantly lower compared to the ratio found in syndromic rats. The IL-1β trend was similar to the TNF-α trend.

The in vivo results are consistent with the results of in vitro studies. The expression of brain-derived neurotrophic factor (BDNF) in the treated group was similar to the LPS-induced group (Figure 5D). However, this outcome could indicate that the improvement of limb coordination in Parkinsonian syndrome rats by PC-EGCG-loaded liposomes is caused by reducing neuroinflammation response, but not by increasing expression of BDNF.

It also fits previous reports that reduction of TNF-α and NO production by pretreatment of rats with EGCG (10 mg/kg) for 24 h and induction with LPS after 7 days decreased comparing to that of LPS-treated rats, and concluded that EGCG has a potential therapeutic effect for LPS-induced neurotoxicity due to reduction of TNF-α and NO release.

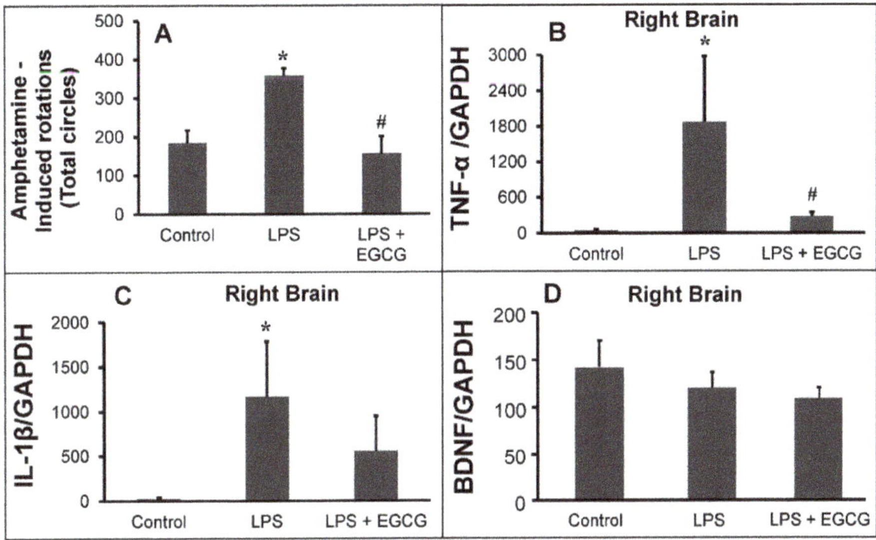

Figure 5. Animal study analysis; (**A**) effects of epigallocatechin-3-gallate (EGCG) liposomes on Parkinsonian syndrome rat rotation behavioral test ($n = 5$ for the control group (injected with 5 μL PBS); $n = 4$ for the LPS-induced group (with 15 μg LPS); $n = 5$ for the LPS-induced and then the PC-EGCG-VE-liposome improvement group (with 15 μg LPS and PC-EGCG-VE-liposomes with 25 μM EGCG), for mean ± SEM, * $p < 0.05$ represents the comparison between the control group and LPS. No difference between the control group and the LPS+EGCG group, # $p < 0.05$ represents the comparison between the LPS group and the LPS+EGCG group). Expression of (**B**) TNF-α, (**C**) IL-1β, and (**D**) brain-derived neurotrophic factor (BDNF) in substantia nigra area of rats sacrificed after rotation behavioral test, $n = 3$, for mean ± SEM, (* $p < 0.05$ represents the comparison between the control group and LPS. No difference between the control group and the LPS+EGCG group, # $p < 0.05$ represents the comparison between the LPS group and the LPS+EGCG group).

3. Discussion

In this study, the purity of EGCG extracted from green tea was detected to be 90.5%, and the free radical scavenging activity of EGCG was more extensive than 80% within 3 min. It was increasing with a higher concentration of EGCG or longer reaction time. The particle size of PC-EGCG-VE- and PS-EGCG-VE-liposomes were 161.5 and 142.9 nm, which were smaller than that of EGCG loaded in PLGA microspheres [28] with additional β-cyclodextrin (ranging from 1–14 μm). The encapsulation efficiency of PC-EGCG-VE- and PS-EGCG-VE-liposomes were 60.2% and 76.8%, respectively. These results showed that PS-containing liposomes were smaller, more stable, and had higher encapsulation efficiency due to the charge on PS and resulted in a repulsive force between liposomes to avoid aggregation.

Expression of TNF-α in cells pretreated with PS-EGCG- and PS-EGCG-VE-liposomes and then induced by LPS had statistically significant differences compared with that of cells induced with LPS, and similar results were observed in the pretreatment of EGCG on LPS-induced TNF-α expression in BV-2 cells [25] and human macrophages [26]. In summary, the cell morphology and expression of TNF-α of the group treated with EGCG showed an inhibition effect of the inflammation induced by LPS.

In the present study, EGCG pretreatment can reduce the release of NO. Suppression of NO production from LPS-induced BV-2 cells by pretreatment of EGCG-loaded PLGA microspheres was also investigated in our previous study [28].

In the present study, Parkinsonian syndrome in rats is created by the damage to the unilateral substantia nigra region induced by LPS. Another important finding is that, through statistical quantitative analysis, EGCG-loaded liposomes can alleviate the syndrome due to damage to the unilateral midbrain nigrosome region of the rat brain induced by LPS in the rotational test, and production of neuroinflammatory factors TNF-α in the substantia nigra area of the rat brain can also be reduced by EGCG-loaded liposomes. This study indicates that improvement of limb coordination and reduction of neuroinflammation in LPS-induced Parkinsonian syndrome is caused by locally administering EGCG-loaded liposomes but not by increasing expression of BDNF. However, the anti-neuroinflammation results are necessary for a neuroprotective effect [35]. BV-2 activation releases pro-inflammatory factors, which are neurotoxic and lead to cell damage. By preventing BV-2 activation, EGCG provides a neuroprotective effect. Some post-treatment of EGCG for improving proliferation, survival rate, and neuronal differentiation of adult neural stem cells in dentate gyrus induced by LPS indicates that EGCG may be a potential therapeutic agent for neuroinflammatory diseases [27].

Previous pharmacokinetic studies showed that exogenous PS can cross the blood-brain barrier (BBB), in which it appears to have an affinity for the hypothalamus [6], and the oral administration results in peak levels in 1–4 hours. In addition, it was found that PS-containing liposomes mimic apoptotic cells to promote the secretion of anti-inflammatory mediators, such as transforming growth factor-β1 (TGF-β1) (to downregulate NO produced from macrophages) [36] and prostaglandin E_2 (PGE_2) by macrophages and microglia cells in vitro [6,37], and also promote the alleviation of inflammation in vivo [38]. Accordingly, PS-containing EGCG-loaded liposomes demonstrated in this study have the advantage of smaller particle size, higher encapsulation efficiency, and inhibiting activation of Parkinsonian syndrome both in microglia cells and in the in vivo rat model, showing an improving anti-inflammatory function and neuroprotection.

The limitation is that our study was carried out with some missing analyses, such as in the case of the neuroprotective effect, and the NeuN staining was not investigated. We also analyzed only BDNF as a neurotrophic factor. FGF2 and IGF2 are also neurotrophic factors that should be considered [39].

4. Material and Methods

4.1. The Source of Epigallocatechin-3-Gallate

4.1.1. Epigallocatechin-3-Gallate Extraction

The extraction process and purity measurement of extracted EGCG had been published elsewhere [40]. Briefly, we dispersed five grams of green tea powder (Ten Ren Tea Co., Ltd., Taiwan) in 65% ethanol and refluxed at 100 °C for 15 min. The rotary evaporator (N-1000, Eyela, Tokyo, Japan) was used to remove ethanol. We then purified the dried solid in two consecutive extraction processes. The solid dispersed in distilled water was extracted against an equal volume of chloroform. The resulting water phase was then extracted against ethyl acetate. Next, the extracts were obtained from an ethyl acetate phase followed by rotary evaporation and then stored in the containers at low humidity before use.

4.1.2. Free Radical Scavenging Activity Analysis

Most in vitro antioxidant activity evaluation uses 2,2-diphenyl-1-picrylhydrazyl (DPPH) to determine the ability of antioxidants for scavenging free radicals [41]. A total of 200 μM DPPH solution dissolved in methanol (80%, HPLC, Sigma-Aldrich, USA) was prepared, then mixed with EGCG solutions at various concentrations from 0.0625 to 1 mg/mL dissolved in H_2O (Purelab, ELGA LabWater), respectively. The absorbance was measured at 517 nm using a UV/VIS spectrophotometer (Shimadzu cooperation, Japan). The DPPH scavenging activity of EGCG was calculated using Equations (1).

$$\text{DPPH scavenging activity } (\%) = \left(1 - \frac{\text{absorbance of DPPH related with catechin exact}}{\text{absorbance of DPPH}}\right) \times 100\% \quad (1)$$

4.2. Liposome Preparation

Solutions for the fabrication of liposome dosage form including 10 mg/mL of L-α-phosphatidylcholine (PC), phosphatidylserine (PS), and cholesterol (CH) solutions dissolved in chloroform, 43.6 mM EGCG solution dissolved in 1 mL H_2O, and 10 mg/mL α-tocopherol (vitamin E, VE) solution dissolved in ethanol were prepared and then stored in a 4 °C refrigerator.

PC, PS, CH, and VE solutions were mixed in the molar ratio listed in Table 1 to a total volume of 1 mL and 3 μL PKH-26 cell linker kit (Sigma-Aldrich) for cellular membrane labeling fluorescent dye was added and mixed homogeneously. The organic solvent was then removed by a rotary evaporator and a vacuum dryer to obtain a layer of translucent lipid film. In preparation for placebo PC- and PS-liposomes, 1 mL H_2O was added, and for PC-EGCG-, PS-EGCG-, PC-EGCG-VE-, and PS-EGCG-VE-liposomes, 0.5 mL H_2O and 0.5 mL EGCG solution was added to the lipid film. The solution was shaken by a vortex mixer at high speed for 10 min to obtain large unilamellar vesicles. A mini-extruder (Avanti Polar Lipid Inc., Alabama, USA) with two polycarbonate filter membranes with a pore size of 400 and 200 nm was used to squeeze the sample 11 times repeatedly. Thus, the liposome was forced to pass through the filter membrane and self-assembled to obtain liposomes with uniform particle size.

Analysis of Fabricated Liposome

Particle size and particle size distribution (Polydispersity index, PDI) of liposomes were analyzed by a dynamic light scattering (DLS) apparatus (Zetasizer 3000 HSA, Malvern, U.K.) at 25 °C, and the light scattering angle was 90°. To determine the amount of EGCG encapsulated in liposomes, EGCG-loaded liposomes were placed in a sealed dialysis membrane (MWCO = 1000). The membrane was placed in an isotonic aqueous solution at 4 °C for removing the EGCG molecule that was not encapsulated. Then, dialyzed liposomes and ethanol were reacted at 4 °C for 20 min for rupturing liposomes, and the solution was centrifuged at 10,000 rpm for 20 min at 4 °C in a high-speed centrifuge. The supernatant of the centrifuged solution was collected, and the amount of EGCG loaded in liposomes was estimated by comparing its absorption value at 274 nm to that of the calibration curve. The encapsulation efficiency of EGCG in liposomes was calculated by Equations (2). Surface morphology and liposome structure were observed with a negative staining transmission electron microscope (TEM) using phosphotungstic acid (PTA) dye because liposomes are transparent and colorless.

$$\text{Encapsulat ion efficiency } (\%) = \left(\frac{\text{Total amount of encapsulat ed EGCG}}{\text{Total amont of EGCG in supernatan t}}\right) \times 100\% \quad (2)$$

In this study, the concentration of EGCG in EGCG-loaded liposomes was calculated from 43.6 mM (the EGCG stock solution) × 0.5 mL/1 mL × corresponding encapsulation efficiency × the dilution factor in use. All the concentrations describing the EGCG-loaded liposomes in this study denoted the concentration of EGCG in EGCG-loaded liposomes.

4.3. In Vitro Cell Analysis

4.3.1. Cell Culture

BV-2 cells were used for in vitro cell experiments. We cultured the cells with a culture medium of Dulbecco's Modified Eagle Medium (DMEM) (Gibco, Grand Island, NY, USA), supplemented with 2.2 g sodium bicarbonate, 4.8 g HEPES (4-(2-hydroxyethyl)-1-

piperazineethanesulfonic acid) buffer adjusted at pH 7.2–7.4, antibiotic-antimycotic 100× (Biowest, France), L-glutamine, and 10% (v/v) of fetal bovine serum (FBS) (Gibco, Grand Island, USA). The BV-2 cells were cultured in 6-well plates at the density of 10^5 cells/cm^2 and incubated under 5% CO_2 at 37 °C. The activation was induced by lipopolysaccharide (LPS).

The shape of BV-2 cells was round and irregular spindle when they were not stimulated and stimulated by LPS.

4.3.2. Cell Viability

EGCG or EGCG-loaded liposomes with different concentrations and LPS solution were added to the BV-2 cells and incubated for 24 h. The medium was then removed, rinsed once with PBS, and replaced by the medium containing the 3-(4,5-dimethylthiazolyl-2)-2, 5-diphenyltetrazolium bromide (MTT) solutions (3-(4,5-dimethylthiazol-2-yl)-2,5-diphenyl tetrazolium bromide, ab146345, Lot. No. GR3203144-11, Abcam, Cambridge, MA, USA) at the volume ratio of 9: 1. After incubating for three hours, the medium of each well was removed, and dimethyl sulfoxide (DMSO) was added to dissolve the purple crystals for 5 min. The enzyme-linked immunosorbent assay (ELISA) reader (Multiskan FC, Thermo Scientific, Shanghai, China) was used to detect the absorbance at 570 nm of each well. The cell viability was calculated using Equations (3).

$$\text{Cell viability } (\%) = \left(\frac{\text{Absorbance of cells cultured with EGCG}}{\text{Absorbance of control group}} \right) \times 100\% \quad (3)$$

4.3.3. Cell Morphology

BV-2 cells were cultured in a 6-well plate (10^5 cells/ cm^2) for 24 h to stabilize cell attachments. We established various culture conditions to monitor cell morphology: (i) cells treated with 25 μM EGCG; (ii) cells treated with 50 ng/mL LPS; and (iii) cells pretreated with EGCG for 1 h, followed by 50 ng/mL LPS treatment. The inverted microscope (Nikon Eclipse 50i, Tokyo, Japan) was used.

4.3.4. Nitric Oxide Release

We performed the nitric oxide (NO) release assessment by seeding the cells with various concentrations of EGCG-loaded liposomes in one hand, and on the other hand, cells were seeded without EGCG. After 24 h incubation, cells were treated with 50 ng/mL LPS for both conditions and incubated for 24 h. Next, we collected 50 μL of culture media from each culture condition and transferred them into 96-well plates. 50 μL of Griess reagent was used to evaluate NO concentration in cells and was then added into different wells. The disposal was protected from light and incubated at 25 °C for 20 min. ELISA reader assessed the optic density (OD) values at 540 nm wavelength for determining the concentration of NO released by cells.

4.3.5. Cytokine Analysis

TNF-α Assessment

We cultured BV-2 cells in 6-well plates. Cells were then treated with 125 μg/mL EGCG-loaded liposomes or placebo and incubated for one hour. We induced the inflammation by treating the cells with 50 ng/mL LPS and incubating them for 24 h. The culture media were collected for TNF-α secreted assessment using TNF-α mouse ELISA assay kit (R&D Systems' biotechnology brand).

COX-2 and cPLA2 Analysis by Western Blotting

To evaluate the COX-2 and cPLA2 secretion, BV-2 cells were in 6-well plates as aforementioned. When the cells reached 90% confluently, the medium was discarded. We then collected the cells in cold conditions to avoid protein degradation. We lysed the cell by adding 100 μL lysis buffer and protease inhibitor into them at 4 °C for 45 min. The tubes were centrifuged at 15,000 rpm for 15 min at 4 °C, and the supernatant was collected and stored for further analysis. The protein content was measured with a Pierce bicinchoninic acid protein assay kit (Pierce Biotechnology). Proteins (20 μg) from each sample were suspended in Laemmli buffer, heated for 5 min at 100 °C, and separated by sodium dodecyl sulfate-polyacrylamide gel electrophoresis (SDS-PAGE) on a 4~12% gradient. Proteins were electroblotted onto a hydrophobic polyvinylidene difluoride membrane (Pall Corporation, Port Washington, NY, USA). Following transfer, membranes were blocked in 5% bovine serum albumin (BSA) in Tris-buffered saline solution (TBST) with 0.1% Tween-20 for 45 min and then incubated overnight at 4 °C with a primary anti-cyclo-oxygenase (COX)-2 antibody (GTX100656, diluted 1:1000; GeneTex, Alton Pkwy, CA, USA) and a primary anti-cPLA2. After washing three times with 1x TBST, the membranes were probed with a secondary antibody (horseradish peroxidase (HRP)-conjugated anti-rabbit immunoglobulin G (IgG)) (GeneTex) for one hour and washed three times with TBST. Immunoreactivity was detected using an enhanced chemiluminescence (ECL) kit (GE Healthcare, Chicago, NY, USA) and visualized with the UVP system (Analytic Jena, Upland, CA, USA). Anti-GAPDH from Santa Cruz (Heidelberg, Germany) was used as an internal control. Image J software (1.52 k, Wayne Rasband, NIH) was used to quantify the intensity of the protein bands.

4.4. Animal Study

4.4.1. Ethics Statement

The animal experiment in the current study was conducted following the procedures approved by the Institutional Animal Care and Use Committee (IACUC) of Chung Yuan Christian University under project number 106011 (accessed on 10 August 2017).

4.4.2. Rat management

A total of 14 male Sprague Dawley (SD) rats were purchased from BioLASCO Taiwan Co., Ltd. The weight of the rats was around 250 to 300 g. The animals were divided into three groups, as indicated in Table 2. We performed the experiments according to the timeline in Figure 6.

Table 2. Rat groups and treatment administered.

Experimental Groups of Rats	LPS Injected 3 μL (5 μg/μL)	EGCG-Loaded Liposomes 2 μL (12.5 μM)	PBS 5 μL	Amount
Rats with Parkinson Disease syndrome	Yes	No	Yes	4
Rats with Parkinson Disease syndrome treated	Yes	Yes	No	5
Control group	No	No	Yes	5

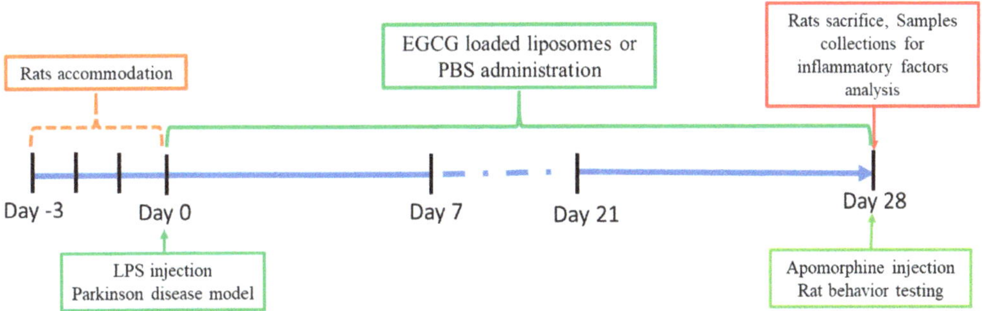

Figure 6. Experimental timeline of Parkinson's disease model in rats treated with EGCG-loaded liposomes (or PBS).

The LPS-induced PD rat model, which injected LPS unilaterally into the substantia nigra, which can elicit strong microglia-mediated neuroinflammation that is followed by the specific death of nigral dopaminergic neurons, is widely used to study the inflammatory process in the pathogenesis of PD [42,43], and the anti-inflammatory mechanism of some substances. In our experiment, LPS was injected into the substantia nigra of the right midbrain. The LPS injection led to neuroinflammation by destroying the dopamine neuropathway of the substantia nigra. The EGCG-loaded liposome was administered for four weeks to investigate its neuroprotective activity. The rotation behavior of rats and the expression of neuroinflammatory markers (TNF-α, IL-1β) and brain-derived neurotrophic factor (BDNF) have been evaluated.

4.4.3. Parkinsonian Syndrome Rat Model

The rats were anesthetized intraperitoneally with 0.06 mL/10 g of Zoletil 50 (50 mg/mL) dilute 5X (1 mL Zoletil in 0.1 mL Rompun 2% and 3.9 mL normal saline). The rats were then fixed in a stereotactic position. The rats' hair was shaved off, they were disinfected with povidone-iodine, and the skull was exposed. We injected into the right substantia nigra 3 μL LPS (5 μg/μL) (from *Escherichia coli* 0111:B4, Sigma, St. Louis, MO, USA). To achieve the nigrostriatal system's unilateral lesion, a microsyringe was used with a flow rate of 1 μL/min. The needle was held in the head for 7 min to prevent reflux of the drug. The skin was sutured for observation.

4.4.4. Animal Behavioral Test

Administration of apomorphine induced abnormal contralateral rotations in the PD model rats [44]. Amphetamine (2 mg/kg) was administered to the rats after four weeks. The rotation toward the side of injury was observed. For the control group, we injected 5 μL PBS in the unilateral brain substantia nigra area. The rat was placed into a 40 cm-diameter bowl, and the counterclockwise (contralateral) rotations were monitored. The image was recorded within 1.5 h to count the number of rotations, in which a circle is defined as a rotation of 270 ° of the rat's head. The damage was proportional to the number of rotations.

4.4.5. Animal Inflammatory Factor Analysis

Rats after the rotational test were anesthetized, sacrificed, and perfused. The brain's damaged nigrostriatal region was collected for RNA extraction with 1 mL TRIzol® (Invitrogen, Carlsbad, CA, USA) reagent according to the manufacturer's instruction. The RNA concentration was measured using a micro-volume spectrometer. Real-time reverse transcription-polymerase chain reaction (qRT-PCR) was performed by SuperScript III First-Strand Synthesis SuperMix and analyzed by 7300 Real-Time PCR System (Applied Biosystems, USA). cDNA was generated by reverse transcription of 1 μg RNA with reverse transcriptase (RT) and oligo(dT) primers (Table 3). The cDNA was diluted to 2.5 ng/μL, and cDNA, primers (Table 3), and FastStart Universal SYBR® Green Master (Rox) (Roche, Germany) were added to a qPCR 8-strip tube for the qPCR reaction.

Table 3. Primer sequences used for qRT-PCR analysis.

Gene	Forward Primers (5′→3′)	Reverse Primers (5′→3′)
TNF-α	TGA CTC GTG GGA TGA TGA CG	CTG GAG ACT GCC CAT TCT CG
IL-1β	CTC ACA CTC AGA TCA TCT TCT C	GGT ATG AAA TGG CAA ATC GG
BDNF	GGT CAC AGC GGC AGA TAA AAA G	TTC GGC ATT GCG AGT TCC AG
GAPDH	GA AGA GAG AGG CCC TCA G	TGT GAG GGA GAT GCT CAG TG

4.5. Statistical Analysis

All experimental data were expressed as mean ± standard deviation (mean ± SD). Data analysis was performed using Excel and GraphPad Prism software version 6.0 (GraphPad Software, La Jolla, CA, USA). The statistical analyses were T-test and one-way or two-way ANOVA with the component of Dunnett's multiple comparisons test. A *p*-value less than 0.05 was considered a statistically significant difference (SSD) between the experimental and control groups.

5. Conclusions

To improve the low oral bioavailability of catechins, EGCG was loaded in liposomes in this study. It is found that PS-containing liposomes were smaller and more stable. It has higher encapsulation efficiency, and the addition of vitamin E can protect EGCG from oxidation and improve encapsulation efficiency. In the in vitro study, expressions of TNF-α and NO production from BV-2 cells were all reduced after pretreatment of EGCG-loaded liposomes on LPS-induced BV-2 cells. Therefore, the EGCG-loaded liposomes have played an essential role in the neuroinflammatory response as an inhibitor. The beneficial effects have been to prevent cells from apoptosis in the neuroinflammatory reactions.

In in vivo study, the Parkinsonian syndrome in rats induced by LPS to the unilateral midbrain substantia nigra region has been improved. The neuro-inflammation mechanism, including the TNF-α secretion, has been inhibited by post-treatment of EGCG-loaded liposomes. We demonstrated that EGCG would be a more valuable candidate for treating neurodegenerative diseases by alleviating the inflammation in the brain. The subsequent investigation should focus on the inflammatory pathways by using a lysosome tracker to track the liposome entrance into the cell and the release of EGCG from the liposome.

Supplementary Materials: The following are available online at https://www.mdpi.com/1422-0067/22/6/3037/s1.

Author Contributions: C.-Y.C.: Investigation; Methodology; Validation. L.B.: Validation; Data curation; Formal analysis; Writing—review and editing. S.-T.T.: Data curation; Visualization; Roles/Writing—original draft; Writing—review and editing. T.-W.F.: Investigation; Methodology; Validation. X.-Y.W.: Data curation; Formal analysis; Investigation; Methodology; Software; Validation; Visualization. C.-W.C.: Data curation; Formal analysis; Investigation; Methodology; Software; Validation; Visualization. R.-S.Y.: Investigation; Resources. T.-Y.C.: Funding acquisition; Methodology; Project administration; Resources; Writing—review and editing. M.F.H.: Conceptualization; Funding acquisition; Methodology; Project administration; Resources; Supervision; Roles/Writing—original draft; Writing—review and editing. All authors have read and agreed to the published version of the manuscript.

Funding: This work was funded by the Ministry of Science and Technology of Taiwan, grant number MOST-108-2119-M-033-002 and the Changhua Christian Hospital, grant number 107CYCU-CCH. The APC was funded by MOST-109-2124-M-033-001.

Data Availability Statement: The data presented in this study are available in this article *Int. J. Mol. Sci.* **2021**, *22*, 3037 or Supplementary Materials here.

Conflicts of Interest: The authors declare no competing interests. The funding bodies played no role in the design of the study, collection, analysis, and interpretation of the data, or writing of the manuscript.

References

1. Lull, M.E.; Block, M.L. Microglial activation and chronic neurodegeneration. *Neurotherapeutics* **2010**, *7*, 354–365. [CrossRef]
2. Ekdahl, C.; Kokaia, Z.; Lindvall, O. Brain inflammation and adult neurogenesis: The dual role of microglia. *Neuroscience* **2009**, *158*, 1021–1029. [CrossRef] [PubMed]
3. Weissleder, R.; Nahrendorf, M.; Pittet, M.J. Imaging macrophages with nanoparticles. *Nat. Mater.* **2014**, *13*, 125–138. [CrossRef]
4. Papageorgiou, I.E.; Fetani, A.F.; Lewen, A.; Heinemann, U.; Kann, O. Widespread activation of microglial cells in the hippocampus of chronic epileptic rats correlates only partially with neurodegeneration. *Brain Struct. Funct.* **2015**, *220*, 2423–2439. [CrossRef] [PubMed]
5. Dutta, G.; Zhang, P.; Liu, B. The lipopolysaccharide Parkinson's disease animal model: Mechanistic studies and drug discovery. *Fundam. Clin. Pharmacol.* **2008**, *22*, 453–464. [CrossRef] [PubMed]
6. Otsuka, M.; Tsuchiya, S.; Aramaki, Y. Involvement of ERK, a MAP kinase, in the production of TGF-β by macrophages treated with liposomes composed of phosphatidylserine. *Biochem. Biophys. Res. Commun.* **2004**, *324*, 1400–1405. [CrossRef]

7. Li, R.; Huang, Y.-G.; Fang, D.; Le, W.-D. (-)-Epigallocatechin gallate inhibits lipopolysaccharide-induced microglial activation and protects against inflammation-mediated dopaminergic neuronal injury. *J. Neurosci. Res.* **2004**, *78*, 723–731. [CrossRef]
8. Liu, J.; Hong, Z.; Ding, J.; Liu, J.; Zhang, J.; Chen, S. Predominant release of lysosomal enzymes by newborn rat microglia after LPS treatment revealed by proteomic studies. *J. Proteome Res.* **2008**, *7*, 2033–2049. [CrossRef]
9. Cho, H.-S.; Kim, S.; Lee, S.-Y.; Park, J.A.; Kim, S.-J.; Chun, H.S. Protective effect of the green tea component, L-theanine on environmental toxins-induced neuronal cell death. *Neurotoxicology* **2008**, *29*, 656–662. [CrossRef]
10. Mariani, M.M.; Kielian, T. Microglia in infectious diseases of the central nervous system. *J. Neuroimmune Pharmacol.* **2009**, *4*, 448–461. [CrossRef]
11. Santiago, R.M.; Barbieiro, J.; Lima, M.M.; Dombrowski, P.A.; Andreatini, R.; Vital, M.A. Depressive-like behaviors alterations induced by intranigral MPTP, 6-OHDA, LPS and rotenone models of Parkinson's disease are predominantly associated with serotonin and dopamine. *Prog. Neuro-Psychopharmacol. Biol. Psychiatry* **2010**, *34*, 1104–1114. [CrossRef] [PubMed]
12. Liu, M.; Bing, G. Lipopolysaccharide animal models for Parkinson's disease. *Parkinson's Dis.* **2011**, *2011*, 327089. [CrossRef] [PubMed]
13. Wu, D.-C.; Teismann, P.; Tieu, K.; Vila, M.; Jackson-Lewis, V.; Ischiropoulos, H.; Przedborski, S. NADPH oxidase mediates oxidative stress in the 1-methyl-4-phenyl-1, 2, 3, 6-tetrahydropyridine model of Parkinson's disease. *Proc. Natl. Acad. Sci. USA* **2003**, *100*, 6145–6150. [CrossRef]
14. Rahman, I.; Biswas, S.K.; Kirkham, P.A. Regulation of inflammation and redox signaling by dietary polyphenols. *Biochem. Pharmacol.* **2006**, *72*, 1439–1452. [CrossRef] [PubMed]
15. Valko, M.; Rhodes, C.; Moncol, J.; Izakovic, M.; Mazur, M. Free radicals, metals and antioxidants in oxidative stress-induced cancer. *Chem. Biol. Interact.* **2006**, *160*, 1–40. [CrossRef]
16. Surh, Y.-J.; Chun, K.-S.; Cha, H.-H.; Han, S.S.; Keum, Y.-S.; Park, K.-K.; Lee, S.S. Molecular mechanisms underlying chemopreventive activities of anti-inflammatory phytochemicals: Down-regulation of COX-2 and iNOS through suppression of NF-κB activation. *Mutat. Res. Fundam. Mol. Mech. Mutagenesis* **2001**, *480*, 243–268. [CrossRef]
17. Renaud, J.; Nabavi, S.F.; Daglia, M.; Nabavi, S.M.; Martinoli, M.-G. Epigallocatechin-3-gallate, a promising molecule for Parkinson's disease? *Rejuvenation Res.* **2015**, *18*, 257–269. [CrossRef]
18. Yang, J.E.; Rhoo, K.Y.; Lee, S.; Lee, J.T.; Park, J.H.; Bhak, G.; Paik, S.R. EGCG-mediated Protection of the Membrane Disruption and Cytotoxicity Caused by the 'Active Oligomer'of α-Synuclein. *Sci. Rep.* **2017**, *7*, 1–10. [CrossRef]
19. Bieschke, J.; Russ, J.; Friedrich, R.P.; Ehrnhoefer, D.E.; Wobst, H.; Neugebauer, K.; Wanker, E.E. EGCG remodels mature α-synuclein and amyloid-β fibrils and reduces cellular toxicity. *Proc. Natl. Acad. Sci. USA* **2010**, *107*, 7710–7715. [CrossRef]
20. Yoshida, W.; Kobayashi, N.; Sasaki, Y.; Ikebukuro, K.; Sode, K. Partial peptide of α-synuclein modified with small-molecule inhibitors specifically inhibits amyloid fibrillation of α-synuclein. *Int. J. Mol. Sci.* **2013**, *14*, 2590–2600. [CrossRef] [PubMed]
21. Xu, Y.; Zhang, Y.; Quan, Z.; Wong, W.; Guo, J.; Zhang, R.; Yang, Q.; Dai, R.; McGeer, P.L.; Qing, H. Epigallocatechin gallate (EGCG) inhibits alpha-synuclein aggregation: A potential agent for Parkinson's disease. *Neurochem. Res.* **2016**, *41*, 2788–2796. [CrossRef] [PubMed]
22. Li, Y.; Chen, Z.; Lu, Z.; Yang, Q.; Liu, L.; Jiang, Z.; Zhang, L.; Zhang, X.; Qing, H. "Cell-addictive" dual-target traceable nanodrug for Parkinson's disease treatment via flotillins pathway. *Theranostics* **2018**, *8*, 5469. [CrossRef]
23. Xu, Q.; Langley, M.; Kanthasamy, A.G.; Reddy, M.B. Epigallocatechin gallate has a neurorescue effect in a mouse model of Parkinson disease. *J. Nutr.* **2017**, *147*, 1926–1931. [CrossRef] [PubMed]
24. Zhou, T.; Zhu, M.; Liang, Z. (-)-Epigallocatechin-3-gallate modulates peripheral immunity in the MPTP-induced mouse model of Parkinson's disease. *Mol. Med. Rep.* **2018**, *17*, 4883–4888. [CrossRef]
25. Park, E.; Chun, H.S. Green tea polyphenol Epigallocatechine gallate (EGCG) prevented LPS-induced BV-2 micoglial cell activation. *J. Life Sci.* **2016**, *26*, 640–645. [CrossRef]
26. Liu, J.-B.; Zhou, L.; Wang, Y.-Z.; Wang, X.; Zhou, Y.; Ho, W.-Z.; Li, J.-L. Neuroprotective Activity of ()-Epigallocatechin Gallate against Lipopolysaccharide-Mediated Cytotoxicity. *J. Immunol. Res.* **2016**, *2016*, 4962351. [CrossRef] [PubMed]
27. Seong, K.-J.; Lee, H.-G.; Kook, M.S.; Ko, H.-M.; Jung, J.-Y.; Kim, W.-J. Epigallocatechin-3-gallate rescues LPS-impaired adult hippocampal neurogenesis through suppressing the TLR4-NF-κB signaling pathway in mice. *Korean J. Physiol. Pharmacol.* **2016**, *20*, 41. [CrossRef] [PubMed]
28. Cheng, C.-Y.; Pho, Q.-H.; Wu, X.-Y.; Chin, T.-Y.; Chen, C.-M.; Fang, P.-H.; Lin, Y.-C.; Hsieh, M.-F. PLGA microspheres loaded with β-cyclodextrin complexes of epigallocatechin-3-gallate for the anti-inflammatory properties in activated microglial cells. *Polymers* **2018**, *10*, 519. [CrossRef]
29. Buszello, K.; Harnisch, S.; Müller, R.; Müller, B. The influence of alkali fatty acids on the properties and the stability of parenteral O/W emulsions modified with Solutol HS 15®. *Eur. J. Pharm. Biopharm.* **2000**, *49*, 143–149. [CrossRef]
30. Shashi, K.; Satinder, K.; Bharat, P. A complete review on: Liposomes. *Int. Res. J. Pharm.* **2012**, *3*, 10–16.
31. Wang, W.-Y.; Tan, M.-S.; Yu, J.-T.; Tan, L. Role of pro-inflammatory cytokines released from microglia in Alzheimer's disease. *Ann. Transl. Med.* **2015**, *3*. [CrossRef]
32. Smith, J.A.; Das, A.; Ray, S.K.; Banik, N.L. Role of pro-inflammatory cytokines released from microglia in neurodegenerative diseases. *Brain Res. Bull.* **2012**, *87*, 10–20. [CrossRef] [PubMed]
33. Hu, H.; Li, Z.; Zhu, X.; Lin, R.; Chen, L. Salidroside reduces cell mobility via NF-κB and MAPK signaling in LPS-induced BV2 microglial cells. *Evid. Based Complementary Altern. Med.* **2014**, *2014*, 383821. [CrossRef]

34. Tambuyzer, B.R.; Ponsaerts, P.; Nouwen, E.J. Microglia: Gatekeepers of central nervous system immunology. *J. Leukoc. Biol.* **2009**, *85*, 352–370. [CrossRef] [PubMed]
35. Lee, D.-S.; Jeong, G.-S. Butein provides neuroprotective and anti-neuroinflammatory effects through Nrf2/ARE-dependent haem oxygenase 1 expression by activating the PI3K/Akt pathway. *Br. J. Pharmacol.* **2016**, *173*, 2894–2909. [CrossRef]
36. Matsuno, R.; Aramaki, Y.; Tsuchiya, S. Involvement of TGF-β in inhibitory effects of negatively charged liposomes on nitric oxide production by macrophages stimulated with LPS. *Biochem. Biophys. Res. Commun.* **2001**, *281*, 614–620. [CrossRef] [PubMed]
37. Zhang, J.; Fujii, S.; Wu, Z.; Hashioka, S.; Tanaka, Y.; Shiratsuchi, A.; Nakanishi, Y.; Nakanishi, H. Involvement of COX-1 and up-regulated prostaglandin E synthases in phosphatidylserine liposome-induced prostaglandin E2 production by microglia. *J. Neuroimmunol.* **2006**, *172*, 112–120. [CrossRef]
38. Ramos, G.C.; Fernandes, D.; Charao, C.T.; Souza, D.G.; Teixeira, M.M.; Assreuy, J. Apoptotic mimicry: Phosphatidylserine liposomes reduce inflammation through activation of peroxisome proliferator-activated receptors (PPARs) in vivo. *Br. J. Pharmacol.* **2007**, *151*, 844–850. [CrossRef]
39. Abe, N.; Nishihara, T.; Yorozuya, T.; Tanaka, J. Microglia and Macrophages in the Pathological Central and Peripheral Nervous Systems. *Cells* **2020**, *9*, 2132. [CrossRef] [PubMed]
40. Chen, C.-H.; Hsieh, M.-F.; Ho, Y.-N.; Huang, C.-M.; Lee, J.-S.; Yang, C.-Y.; Chang, Y. Enhancement of catechin skin permeation via a newly fabricated mPEG-PCL-graft-2-hydroxycellulose membrane. *J. Membr. Sci.* **2011**, *371*, 134–140. [CrossRef]
41. Parthasarathy, S.; Bin Azizi, J.; Ramanathan, S.; Ismail, S.; Sasidharan, S.; Said, M.I.M.; Mansor, S.M. Evaluation of antioxidant and antibacterial activities of aqueous, methanolic and alkaloid extracts from Mitragyna speciosa (*Rubiaceae family)* leaves. *Molecules* **2009**, *14*, 3964–3974. [CrossRef] [PubMed]
42. Herrera, D.; Molina, A.; Buhlin, K.; Klinge, B. Periodontal diseases and association with atherosclerotic disease. *Periodontology 2000* **2020**, *83*, 66–89. [CrossRef] [PubMed]
43. Batista, C.R.A.; Gomes, G.F.; Candelario-Jalil, E.; Fiebich, B.L.; de Oliveira, A.C.P. Lipopolysaccharide-Induced Neuroinflammation as a Bridge to Understand Neurodegeneration. *Int. J. Mol. Sci.* **2019**, *20*, 2293. [CrossRef]
44. Creese, I.; Burt, D.R.; Snyder, S.H. Dopamine receptor binding enhancement accompanies lesion-induced behavioral supersensitivity. *Science* **1977**, *197*, 596–598. [CrossRef] [PubMed]

International Journal of
Molecular Sciences

Article

Neuroprotective Effect of Apolipoprotein D in Cuprizone-Induced Cell Line Models: A Potential Therapeutic Approach for Multiple Sclerosis and Demyelinating Diseases

Eva Martínez-Pinilla [1,2,3,*], Núria Rubio-Sardón [1], Rafael Peláez [4], Enrique García-Álvarez [1], Eva del Valle [1,2,3], Jorge Tolivia [1,2,3,*], Ignacio M. Larráyoz [4] and Ana Navarro [1,2,3]

1 Department of Morphology and Cell Biology, University of Oviedo, 33003 Oviedo, Spain; nuria199510@hotmail.com (N.R.-S.); garal.enrique@gmail.com (E.G.-Á.); valleeva@uniovi.es (E.d.V.); anavarro@uniovi.es (A.N.)
2 Instituto de Neurociencias del Principado de Asturias (INEUROPA), 33003 Oviedo, Spain
3 Instituto de Investigación Sanitaria del Principado de Asturias (ISPA), 33011 Oviedo, Spain
4 Biomarkers and Molecular Signaling Group, Neurodegeneration Area, Center for Biomedical Research of La Rioja (CIBIR), 26006 Logroño, Spain; rpelaez@riojasalud.es (R.P.); ilarrayoz@riojasalud.es (I.M.L.)
* Correspondence: martinezpinillaeva@gmail.com (E.M.-P.); jtolivia@uniovi.es (J.T.)

Citation: Martínez-Pinilla, E.; Rubio-Sardón, N.; Peláez, R.; García-Álvarez, E.; del Valle, E.; Tolivia, J.; Larráyoz, I.M.; Navarro, A. Neuroprotective Effect of Apolipoprotein D in Cuprizone-Induced Cell Line Models: A Potential Therapeutic Approach for Multiple Sclerosis and Demyelinating Diseases. *Int. J. Mol. Sci.* **2021**, *22*, 1260. https://doi.org/10.3390/ijms22031260

Academic Editor: Anne Vejux
Received: 20 January 2021
Accepted: 22 January 2021
Published: 27 January 2021

Publisher's Note: MDPI stays neutral with regard to jurisdictional claims in published maps and institutional affiliations.

Abstract: Apolipoprotein D (Apo D) overexpression is a general finding across neurodegenerative conditions so the role of this apolipoprotein in various neuropathologies such as multiple sclerosis (MS) has aroused a great interest in last years. However, its mode of action, as a promising compound for the development of neuroprotective drugs, is unknown. The aim of this work was to address the potential of Apo D to prevent the action of cuprizone (CPZ), a toxin widely used for developing MS models, in oligodendroglial and neuroblastoma cell lines. On one hand, immunocytochemical quantifications and gene expression measures showed that CPZ compromised neural mitochondrial metabolism but did not induce the expression of Apo D, except in extremely high doses in neurons. On the other hand, assays of neuroprotection demonstrated that antipsychotic drug, clozapine, induced an increase in Apo D synthesis only in the presence of CPZ, at the same time that prevented the loss of viability caused by the toxin. The effect of the exogenous addition of human Apo D, once internalized, was also able to directly revert the loss of cell viability caused by treatment with CPZ by a reactive oxygen species (ROS)-independent mechanism of action. Taken together, our results suggest that increasing Apo D levels, in an endo- or exogenous way, moderately prevents the neurotoxic effect of CPZ in a cell model that seems to replicate some features of MS which would open new avenues in the development of interventions to afford MS-related neuroprotection.

Keywords: breast cystic fluid; clozapine; endocytosis; glia; neurons; ROS

1. Introduction

Apolipoprotein D (Apo D) is a well-known lipocalin family member that plays a key role in the transport, metabolism and homeostasis of some lipids due to its ability to bind cholesterol, arachidonic acid, steroids, retinoic acid or anandamide, among other small hydrophobic ligands [1–3]. In the past decades, increasing evidence at biochemical and functional level suggested that Apo D acts as an antioxidant, being part of the body's defense system against oxidative stress, and also as an endogenous neuroprotective agent. Indeed, crystallographic analysis revealed that this 29 kDa glycoprotein comprises an eight-stranded antiparallel β-barrel flanked by a singled α-helix that encloses a fat specific ligand-binding pocket. Moreover, Apo D shows various exposed hydrophobic residues located in three of its extended loops which may contribute to Apo D association with lipids and seem to explain its potential as multiligand and multifunctional protein [4,5]. Most importantly, they have been linked to the ability of Apo D to bind and reduce oxidized lipids, and thereby inhibit radical-propagation of lipid hydroperoxides [6–8].

Studies in cell systems reported that different stress signal pathways may modulate Apo D transcription. In fact, stressful stimuli such as H_2O_2, Rose Bengal, kainic acid, ultraviolet (UV) light, paraquat or lipopolysaccharide, leading to extended growth arrest and apoptosis, increase Apo D expression in a dose and time-dependent manner [9,10]. The potential neuroprotective and prosurvival roles for Apo D have also been proven in animal models where the overexpression of this protein confers greater protection against oxidative stress and contributes significantly to the regulation of longevity; the experimental lack of Apo D causes opposite results [11,12].

In humans, Apo D is expressed in neural and peripheral tissues, detected in cerebrospinal fluid (CSF), in plasma as an important component of high-density plasma lipoproteins (HDL), and in breast cyst fluid (BCF) [1,2,13–15]. In nonpathological conditions of the central and peripheral nervous system (CNS and PNS, respectively), Apo D is widely expressed in neurons, glia (astrocytes, oligodendrocytes (OLGs), and Schwann cells), perivascular cells and pericytes [13,16–18], contributing to maintain neuronal homeostasis and myelin extracellular leaflet compaction [19,20]. Remarkably, Apo D is upregulated in neural cells and CSF during aging, and in brains affected by neurodegenerative diseases characterized by cellular stress and excitotoxicity such as multiple sclerosis (MS), Spongiform encephalopathy, Parkinson's disease (PD), Niemann–Pick disease, or Alzheimer's disease (AD) as well as psychiatric disorders (schizophrenia and bipolar disorder) [8,21,22].

MS is a devastating neurodegenerative disease that affects more than 2 million young adults worldwide, mainly women, with a complex and unknown etiology [23]. This demyelinating, autoimmune and inflammatory disease is manifested clinically in the form of multiple fully or partially reversible symptomatic episodes (reviewed in [24–26]), which reflect the progressive focal degeneration of OLGs and myelin membranes around axons in both white and grey matter areas throughout the brain and spinal cord [27–30]. Classically, OLG dysfunction has been linked to an exacerbated adaptive immune response, involving the recruitment of autoreactive T cells through a defective and permeable blood–brain barrier (BBB), and the activation of B cells [31–33]. The consequent inflammatory process activates microglia, astrocytes, and infiltrated macrophages that are able, in turn, to generate oxidative stress-related molecules as reactive oxygen species (ROS) and reactive nitrogen species (RNS) [34], which promote demyelination, compromise the neuro-axonal functional unit and contribute to the progressive tissue damage in MS [26,35,36]. However and contrary to what was thought, recent evidence shows that the biochemical alteration of myelin could be the initial event that triggers a secondary autoimmune response that results in the demyelinating inflammatory reaction taking place in the diseased brains, the so-called "inside-out" model of MS pathogenesis [37–39]. In the last two decades, extensive research has been carried out to find efficacious neuroprotective therapies in an attempt to alleviate symptoms and/or slow down or delay the progression of the MS [26,40–44]. Therefore, it is essential to know the root cause of the MS pathology in order to properly select the target for developing efficacious therapeutic interventions. For this purpose, a number of neurotoxin-induced in vivo and in vitro models of demyelination and MS-related neurodegeneration are used. Among all, neuronal and glial cell lines exposed to cuprizone (CPZ), a copper chelator that reversibly impacts on mitochondrial function, may be a convenient experimental approach instrumental in the advance of understanding of the functioning of the nervous system [45–48].

Previous studies showed that Apo D is upregulated in the CSF of MS patients [49,50], reactive astrocytes, and exhibits a characteristic expression pattern in MS lesions of the brain [51]. In this regard, levels of OLG-derived Apo D are lower in demyelinating plaques but appear to recover in areas of remyelination [51]. This study aims to assess the potential of Apo D (either by triggering its endogenous synthesis or by its exogenous addition), as well as its mechanism of action, to prevent the neurotoxic effect of CPZ in two cell models that mimic biochemical features of MS.

2. Results

2.1. Apo D Expression in HOG and SH-SY5Y Cells in Response to Cuprizone Treatment

Taking advantage of our experience in the CPZ-induced cell model, some previous results showing that CPZ was able to induce cytotoxic damage mediated by a mitochondrial dysfunction in the HOG and SH-SY5Y cell lines (data not shown), and the findings reported in this work (see next figures), we aimed to analyze the potential effect of CPZ on Apo D expression in these oligodendroglioma and neuroblastoma cell lines by qRT-PCR and immunocytochemistry. As shown in Figure 1, the analysis of Apo D gene expression (Figure 1a) and the immunosignal quantification (Figure 1b–d) revealed that CPZ induced changes in Apo D expression and, interestingly, only at the highest concentration (1000 µM) and at 48 h of treatment. In fact, a constant and almost invariable fluorescence signal was observed in control and treated cells (Figure 1b).

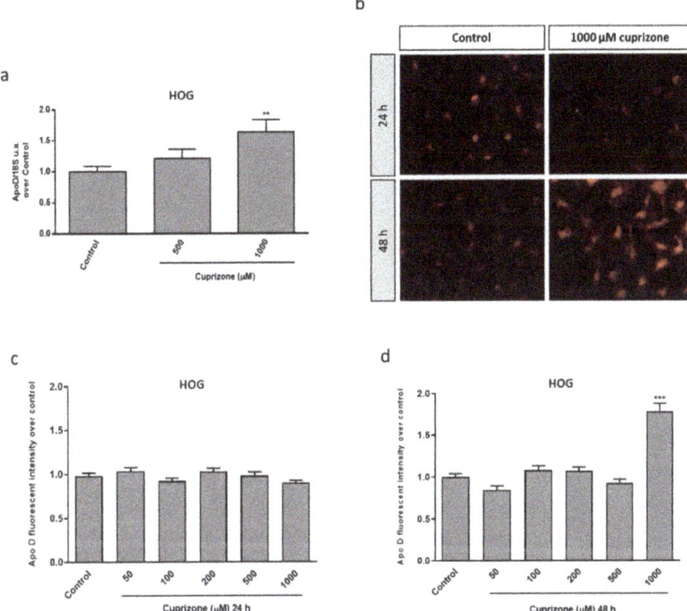

Figure 1. Relative Apo D gene expression in HOG cells treated with 0–1000 µM of CPZ following 24 h. Data represent the quotient between the gene and the expression of the housekeeping gene 18S rRNA. Bars represent the mean ± SEM of all measurements (n = 6–8) (**a**). Representative fluorescence microscopy images of Apo D levels in HOG cells treated or not with 1000 µM of CPZ during 24 and 48 h. 40× magnification (**b**). Densitometric quantification of Apo D immunocytochemical signal after 24 (**c**) and 48 h (**d**) of treatment with increasing concentrations of CPZ (50–1000 µM) in HOG cells (n = 6). Bars represent mean density per cell in a 40× field ± SEM (over control). Significant differences were analyzed by a one-way ANOVA followed by post-hoc Tukey's test. ** $p < 0.01$, *** $p < 0.001$ compared with control.

As expected in the case of SH-SY5Y neuroblastoma cells, which according to previous studies show a negligible expression of Apo D [52], we found that these cells exhibited a very scarce endogenous expression of Apo D only detected by immunocytochemistry, and that CPZ did not influence the apolipoprotein synthesis as observed in the images (Figure 2a) and the immunocytochemical quantification (Figure 2b,c).

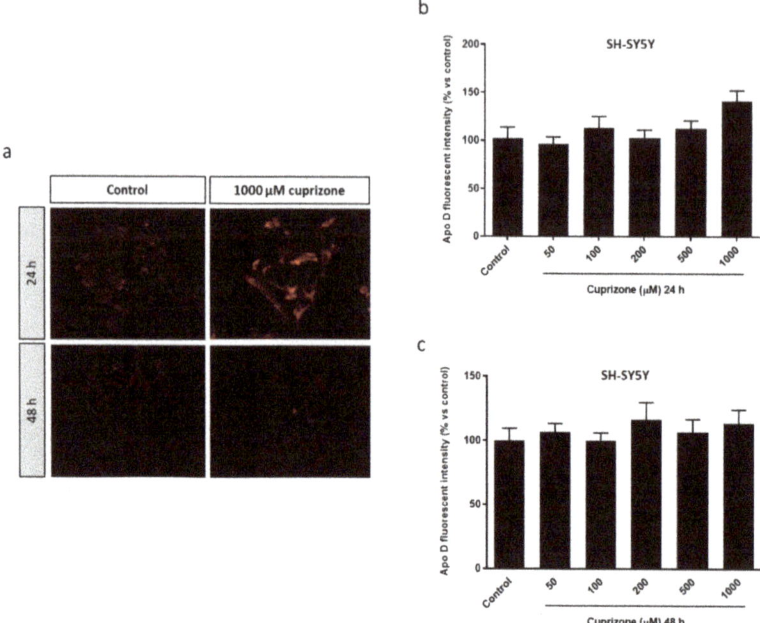

Figure 2. Representative fluorescence microscopy images of Apo D levels in SH-SY5Y cells treated or not with 1000 µM of CPZ during 24 and 48 h. 40× magnification (**a**). Densitometric quantification of Apo D immunocytochemical signal after 24 (**b**) and 48 h (**c**) of treatment with increasing concentrations of CPZ (50–1000 µM) in SH-SY5Y cells (n = 6). Bars represent mean density per cell in a 40× field ± SEM (% versus control).

2.2. *Clozapine Prevents Loss of Mitochondrial Functionality and Cell Viability in Oligodendroglial and Neuronal CPZ-Induced Models of MS*

The atypical antipsychotic drug, clozapine (CLO), widely used in the treatment of schizophrenia, among other psychiatric disorders, is considered as a therapeutic agent that seems to exert its beneficial effects by its ability to increase Apo D levels in the brain [53,54]. Therefore, we first evaluated the potential neuroprotective effect of CLO in the CPZ-induced cell models. For this purpose, a wide range of CLO concentrations, from 0.1 to 100 µM, was used to treat HOG or SH-SY5Y cells during 24 and 48 h in absence of CPZ. Once it was established that CLO did not cause loss of cell viability, except in extremely high doses and/or prolonged exposures (Figures A1 and A2), we assessed whether the addition of CLO could avoid the CPZ cytotoxicity. Of note, the two cell lines were differentially affected by CLO, being neurons more sensitive than glial cells to the same concentrations. Our findings demonstrated that CLO was able to prevent the mitochondrial dysfunction caused by the toxic in both HOG and SH-SY5Y cells. As shown in Figure 3, cell viability assessed by the MTT assay revealed that CLO (0.1–1 µM) prevented about 15–30% loss of cell viability when added 24 h before 500 µM of CPZ (Figure 3a,b). Similar results were obtained when cells were treated with CLO and CPZ at the same time. In contrast, this neuroprotective effect was not noticeable when cells were incubated with 500 µM of CPZ for 24 h and subsequently with increasing concentrations of CLO for, at least, another 24 h (data not shown).

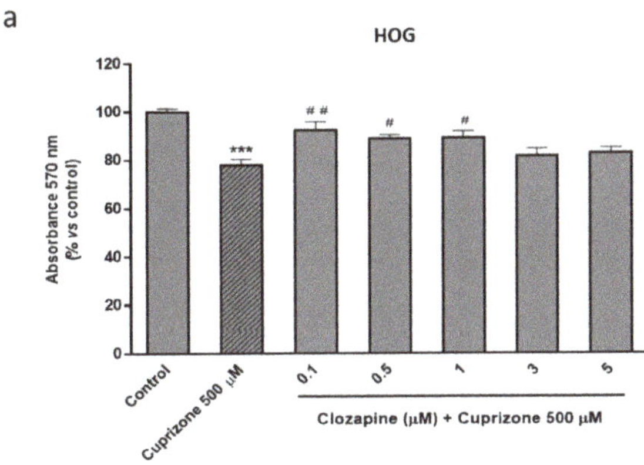

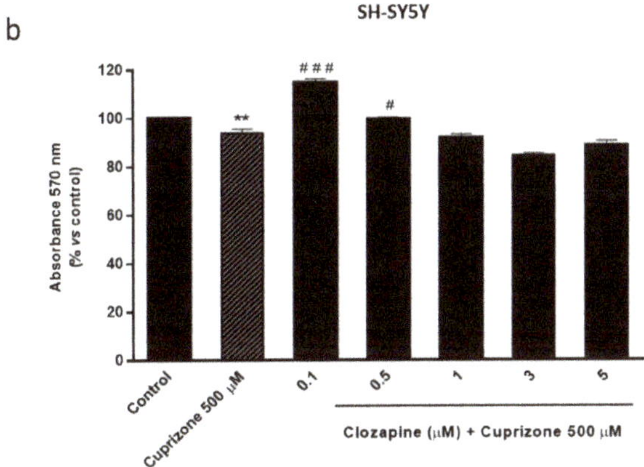

Figure 3. MTT assay in HOG (**a**) and SH-SY5Y cells (**b**) treated with increasing concentrations of CLO (0.1–5 μM) followed by 24 h with 500 μM of CPZ. Cell damage is represented as the percentage of viability versus control. Data are the mean ± SEM of five independent experiments. Significant differences were analyzed by a one-way ANOVA followed by post-hoc Tukey's test. ** $p < 0.01$, *** $p < 0.001$ compared with control; $^{\#}$ $p < 0.05$, $^{\#\#}$ $p < 0.01$, $^{\#\#\#}$ $p < 0.001$ compared with CPZ treatment.

2.3. Neuroprotective Doses of Clozapine Increase Apo D Expression in the CPZ-Induced Cell Models of MS

Then, and in order to check the possible link between the neuroprotective effect observed for CLO and the endogenous Apo D levels, the expression of this apolipoprotein was analyzed in HOG cells upon CLO treatment. qPCR and immunocytochemical analyses demonstrated that this antipsychotic drug did not produce changes in Apo D expression by itself, at least in the tested concentrations and times of treatment (Figure 4). However, CLO (0.1–3 μM) induced an increase in Apo D synthesis when it was coadministrated with CPZ in OLGs at the same concentrations that prevented the loss of viability caused by the toxin. As shown in Figure 5, the increase in Apo D signal was higher than the control values when added 24 h before 500 μM of CPZ.

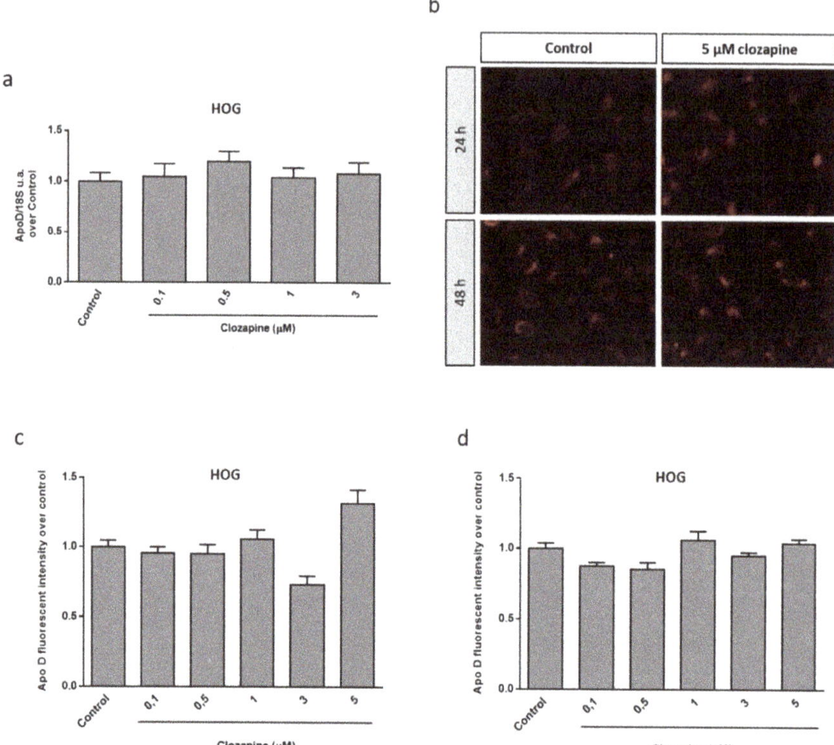

Figure 4. Relative Apo D gene expression in HOG cells treated or not with 5 µM of CLO during 24 h. Data represent the quotient between the gene and the expression of the housekeeping gene 18S rRNA. Bars represent the mean ± SEM of all measurements (n = 6–8) (**a**). Representative fluorescence microscopy images of Apo D expression in HOG cells treated or not with 3 µM of CLO during 24 and 48 h. 40× magnification (**b**). Densitometric quantification of Apo D immunocytochemical signal after 24 (**c**) and 48 h (**d**) of treatment with increasing concentrations of CLO (0.1–5 µM) in HOG cells (n = 6). Bars represent mean density per cell ± SEM (over control) in a 40× field.

Similar results, but with some nuances, were obtained in the neuroblastoma cell line. In fact, immunocytochemical assays revealed that CLO induced changes in Apo D expression in SH-SY5Y cells but only at the highest concentration (5 µM), at 24 and 48 h of treatment, as observed in the images (Figure 6a) and the immunocytochemical quantification (Figure 6b,c). When CLO was added 24 h before 500 µM of CPZ the Apo D immunosignal increased from 1.5 to 2-fold (compared to control) in the concentrations of the antipsychotic drug associated with the neuroprotective effects (Figure 7a,b). Interestingly, the treatment with 5 µM of CLO, which almost doubled Apo D levels in SH-SY5Y cells (Figure 7), was unable to prevent the cytotoxic effect of CPZ (Figure 3b).

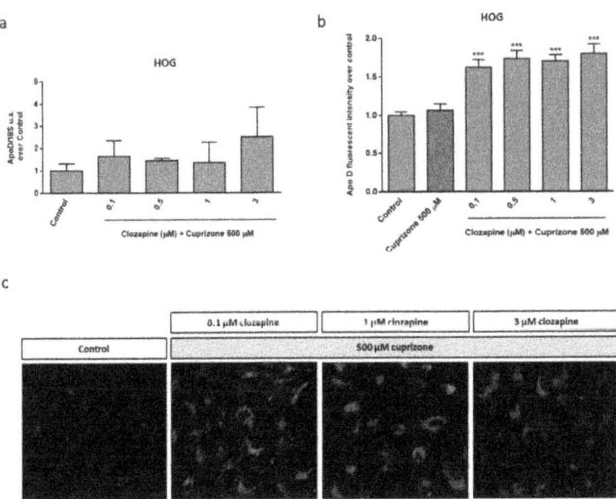

Figure 5. Relative Apo D gene expression in HOG cells treated with increasing concentrations of CLO (0.1–3 µM) during 24 h followed by 24 h with 500 µM of CPZ. Data represent the quotient between the gene and the expression of the housekeeping gene 18S rRNA. Bars represent the mean ± SEM of all measurements (n = 6–8) (**a**). Densitometric quantification of Apo D immunocytochemical signal after 24 h of treatment with increasing concentrations of CLO (0.1–3 µM) followed by 24 h with 500 µM of CPZ (n = 6). Bars represent mean density per cell in a 40× field ± SEM (over control) (**b**). Representative fluorescence microscopy images of Apo D expression in HOG cells treated with increasing concentrations of CLO (0.1–3 µM) followed by 24 h with 500 µM of CPZ. 40× magnification (**c**). Significant differences were analyzed by a one-way ANOVA followed by post-hoc Tukey's test. *** $p < 0.001$ compared with control.

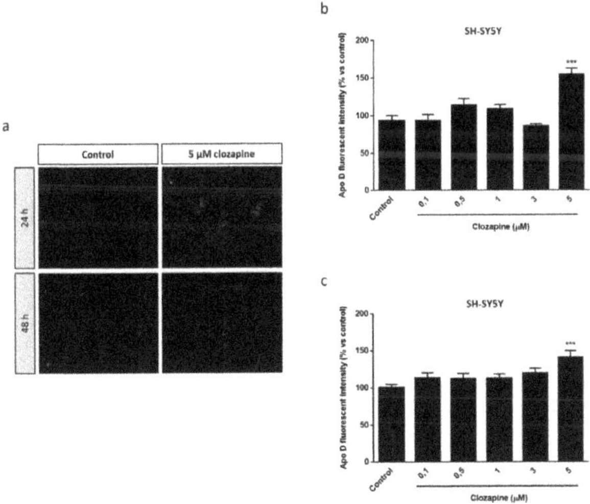

Figure 6. Representative fluorescence microscopy images of Apo D expression in SH-SY5Y cells treated or not with 5 µM of CLO during 24 and 48 h. 40× magnification (**a**). Densitometric quantification of Apo D immunocytochemical signal after 24 (**b**) and 48 h (**c**) of treatment with increasing concentrations of CLO (0.1–5 µM) in SH-SY5Y cells (n = 6). Bars represent mean density per cell ± SEM (% versus control) in a 40× field. Significant differences were analyzed by a one-way ANOVA followed by post-hoc Tukey's test. *** $p < 0.001$ compared with control.

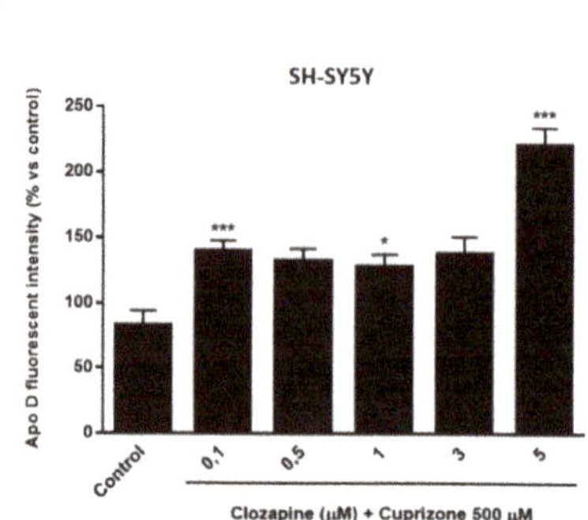

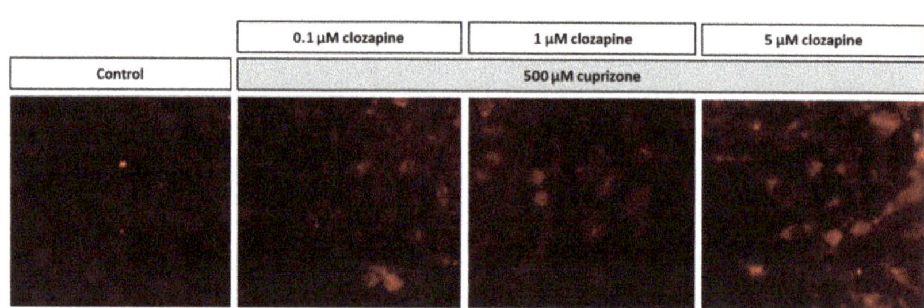

Figure 7. Densitometric quantification of Apo D immunocytochemical signal in SH-SY5Y cells after 24 h of treatment with increasing concentrations of CLO (0.1–5 μM) followed by 24 h with 500 μM of CPZ (n = 6). Bars represent mean density per cell in a 40× field ± SEM (% versus control). Significant differences were analyzed by a one-way ANOVA followed by post-hoc Tukey's test. * $p < 0.05$, *** $p < 0.001$ compared with control (**a**). Representative fluorescence microscopy images of Apo D expression in SH-SY5Y cells treated with increasing concentrations of CLO (0.1–5 μM) followed by 24 h with 500 μM of CPZ. 40× magnification (**b**).

2.4. Neuroprotective Effect of the Exogenously Added hApo D in Oligodendroglial and Neuronal CPZ-Induced Models of MS

The next step to assess the neuroprotective potential of Apo D was to check the impact of the exogenous addition of human Apo D (hApo D), hApo D purified from BCF or human recombinant Apo D (hrApo D), in the CPZ-based cell models of MS. On the one hand, we found that both apolipoproteins induced some improvement in mitochondrial oxidation and, consequently, an increase in OLGs (Figure 8) and neurons (Figure 9) viability under normal conditions. On the other hand, the analysis revealed that treatment with hApo D (5–100 nM) totally prevented the loss of viability caused by the addition of 500 μM CPZ for 24 h in the HOG cells (Figure 8c). Although to a lesser extent, similar results were found in cells pretreated with hrApo D (Figure 8d). Noteworthy, these findings were confirmed in the SH-SY5Y neuroblastoma cells that lack endogenous Apo D expression. Accordingly, both hApo D and hrApo D were able to prevent the toxic effect of CPZ after 24 h of treatment in neuroblastoma cells as well (Figure 9c,d).

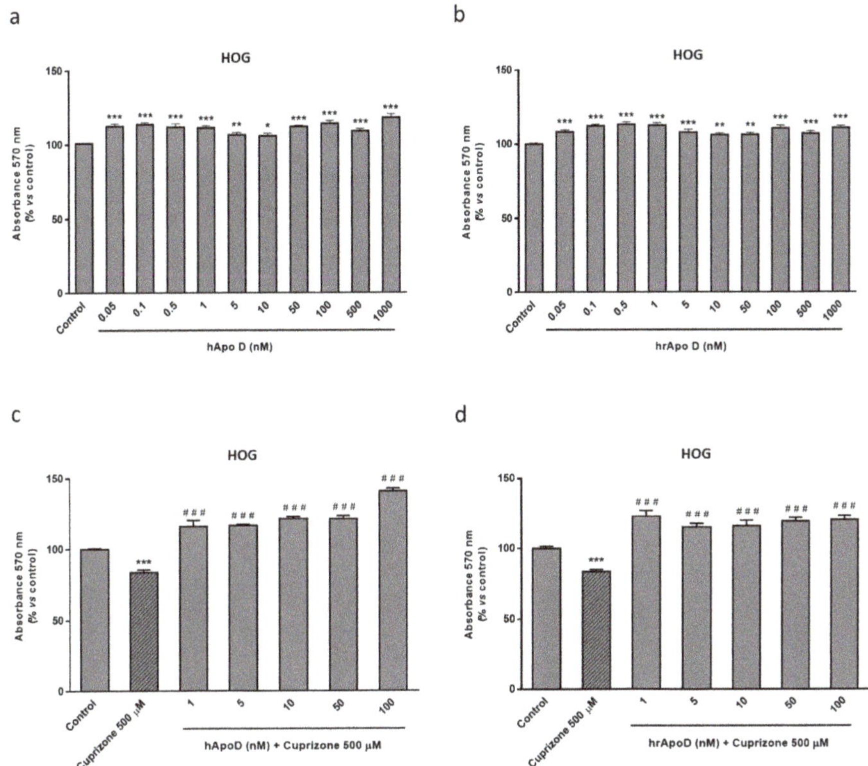

Figure 8. Top panel: MTT assays in HOG cells treated with increasing concentrations (0.05–1000 nM) of hApo D (**a**) or of hrApo D during 24 h (**b**). Bottom panel: MTT assays in HOG cells treated with increasing concentrations (1–100 nM) of hApo D (**c**) or of hrApo D (**d**) for 24 h followed by 24 h with 500 μM of CPZ. Cell damage is represented as the percentage of viability versus control. Data are the mean ± SEM of five independent experiments. Significant differences were analyzed by a one-way ANOVA followed by post-hoc Tukey's test. * $p < 0.05$, ** $p < 0.01$, *** $p < 0.001$ compared with control; ### $p < 0.001$ compared with CPZ treatment.

Finally, and in order to test whether Apo D exerts its antioxidant activity intra- or extracellularly by sequestering/blocking CPZ or oxidative stress-induced molecules, we subjected SH-SY5Y to different pharmacological inhibitors of endocytosis. First, we examined the above-described neuroprotective effect of Apo D against CPZ upon perturbation of clathrin-mediated endocytosis (CME), or upon alteration of actin-dependent phagocytosis and micropinocytosis by pretreatment of cells with chlorpromazine (5 μg/mL) and cytochalasin D (8 μg/mL), respectively. As shown in Figure 10, these conditions did not seem to influence the effect exerted by hApo D (50–100 nM), even they significantly enhance it, as demonstrated by the MTT assay. However, when cells were pretreated with dynasore (80 μM), a compound that blocks GTPase activity of dynamin and vesicle scission, hApo D was not able to prevent the significant decrease of about 20–25% in cell viability after 24 h of treatment with 500 μM CPZ (Figure 10). Nevertheless, it should be noted that dynasore drastically magnifies, in some way, the cytotoxic effect of CPZ.

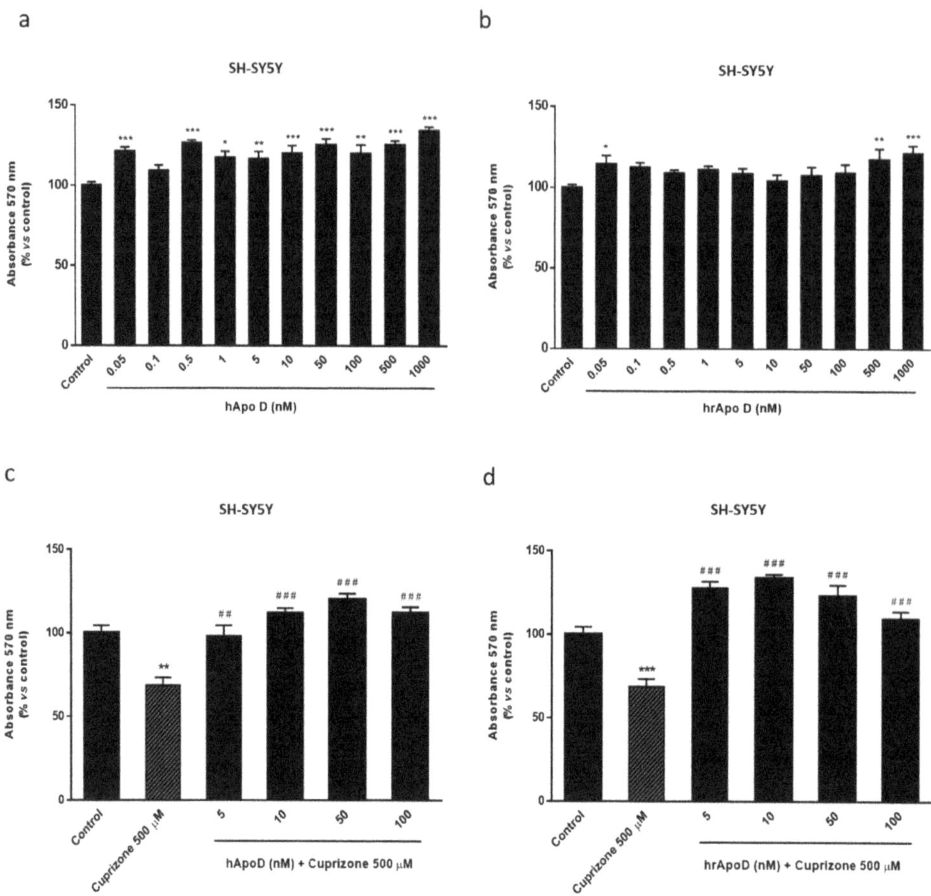

Figure 9. Top panel: MTT assays in SH-SY5Y cells treated with increasing concentrations (0.05–1000 nM) of hApo D (**a**) or of hrApo D (**b**) during 24 h. Bottom panel: MTT assays in SH-SY5Y cells treated with increasing concentrations (1–100 nM) of hApo D (**c**) or of hrApo D (**d**) for 24 h followed by 24 h with 500 μM of CPZ. Cell damage is represented as the percentage of viability versus control. Data are the mean ± SEM of five independent experiments. Significant differences were analyzed by a one-way ANOVA followed by post-hoc Tukey's test. * $p < 0.05$, ** $p < 0.01$, *** $p < 0.001$ compared with control; ## $p < 0.01$, ### $p < 0.001$ compared with CPZ treatment.

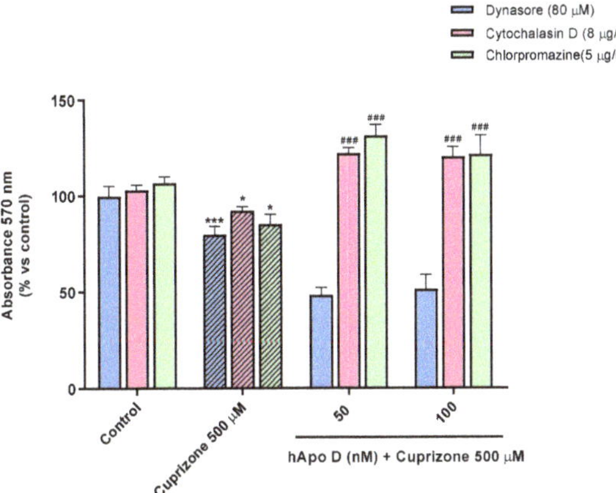

Figure 10. MTT assay in SH-SY5Y cells treated with 8 μg/mL cytochalasin D, 5 μg/mL chlorpromazine hydrochloride, or 80 μM dynasore prior to the addition of increasing concentrations (50–100 nM) of hApo D for 24 h followed by 24 h with 500 μM of CPZ. Cell damage is represented as the percentage of viability versus control. Data are the mean ± SEM of five independent experiments. Significant differences were analyzed by a one-way ANOVA followed by post-hoc Tukey's test. * $p < 0.05$, *** $p < 0.001$ compared with control; ### $p < 0.001$ compared with CPZ treatment.

2.5. Neuroprotective Effect of hApo D is not Related to a Decrease in CPZ-Induced ROS Levels

We previously demonstrated that CPZ affects mitochondrial function and aerobic cell respiration in neurons and glial cells which could lead to an increase in intracellular ROS production. In fact, the data summarized in Figure 11 show that treatment of HOG and SH-SY5Y cells with CPZ concentrations of 500 and 1000 μM significantly increased the levels of intracellular ROS. In the MTT assays Apo D totally prevented the loss of viability caused by CPZ so the next step was to measure ROS formation in these conditions. However, we did not find significant differences in the levels of intracellular ROS production between cells pretreated or not with hApo D (Figure 11a,b).

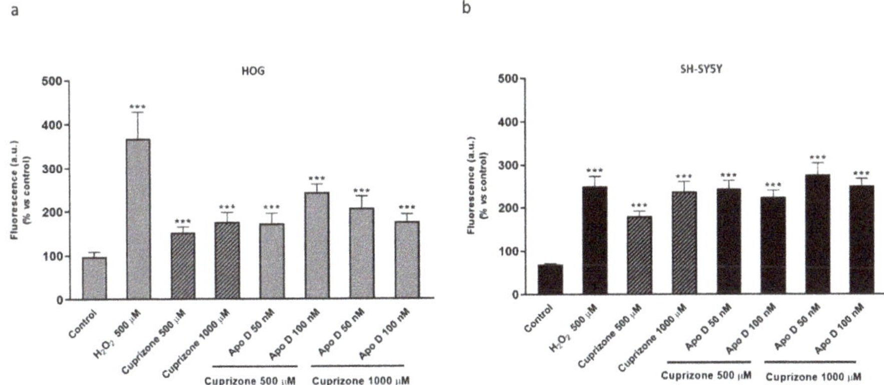

Figure 11. ROS production in HOG (**a**) and SH-SY5Y cells (**b**) treated with increasing concentrations (50–100 nM) of hApo D for 24 h followed by 24 h with 500 or 1000 μM of CPZ. Changes in ROS levels, measured with the oxidant-sensitive dye H_2DCFDA, are represented as the percentage of fluorescent DCF production versus control. H_2O_2 500 μM was used as positive control. Data are the mean ± SEM of five independent experiments. Significant differences were analyzed by a one-way ANOVA followed by post-hoc Tukey's test. *** $p < 0.001$ compared with control.

3. Discussion

Thirty years of research have provided significant insights to unravel the function of Apo D, which has helped to elucidate its antioxidant and anti-inflammatory role and a better understanding of the mechanisms whereby this specific apolipoprotein may exert its beneficial effects. To elucidate the function of Apo D requires the development of cellular models that allow studying the actions of this protein in a physiologically relevant but simple context. The data here presented aims to make progress in the knowledge of potential neuroprotective effect of Apo D in MS and other demyelinated diseases by both indirect and direct in vitro approximations.

A common trend in multiple pathological and nonpathological conditions of the nervous system, from neural development and aging to diverse neurodegenerative processes such as those observed in MS, is the Apo D upregulation with a seemingly neuroprotective purpose [12,20,51,55,56]. Valuable information has been gained concerning the Apo D expression in MS, i.e., it is increased in the CSF of MS patients and exhibits a characteristic pattern in the brain lesions [49–51]. However, mechanisms involved in the Apo D function in this pathology have been not fully elucidated until now. By taking advantage of the CPZ-induced model of MS, we aimed to analyze the expression of Apo D in HOG cells treated with CPZ. Our results showed that the changes induced by CPZ in Apo D expression are minimal despite the dose and time-dependent cytotoxic damage previously reported for CPZ in the same conditions. Unsurprisingly, a similar CPZ effect was demonstrated in the SH-SY5Y, a cell line that does not efficiently express Apo D at least in nonpathological conditions. In fact, we only found negligible levels of this apolipoprotein in the neuroblastoma cells by immunocytochemistry but not by other techniques, probably due to methodological differences. During the last decades, various authors, including us, have used in vitro assays to demonstrate that H_2O_2, amyloid beta-peptide, lipopolysaccharide, paraquat as well as other acute short-term oxidative stressors induce a time- and dose-dependent effect on Apo D expression [55–59]. At least in astrocytes, this effect seems to be regulated by the stress responsive JNK signaling pathway [9]. Here, we found that CPZ by itself does not promote a significant increase in Apo D levels in oligodendroglioma cells. Although the exact mechanism of action of CPZ is not completely understood, we previously demonstrated that the ion chelator impacts on functional state of mitochondria and aerobic cell respiration in neurons and glial cells. Now, we have also shown that these processes are accompanied by a significant increase in intracellular levels of ROS. Interestingly, the consequent compromise of mitochondrial function, cell metabolism, and the increase in oxidative stress are not immediately apparent in in vivo CPZ models. In this regard, the toxic/demyelinating effect induced by CPZ in mice does not peak until the third week of treatment [59,60], so it is reasonable to assume that CPZ may trigger the Apo D expression in the longer term.

Neuroprotection by Apo D may be afforded by either an indirect or a direct manner. On the one hand, it has been shown that CLO, an atypical antipsychotic drug used in the treatment of schizophrenia and bipolar syndrome, is able to increase Apo D levels in the brain [53,54]. However, the exact mechanism by which CLO regulates Apo D expression is still unknown. Our findings demonstrated that CLO induces an increase in Apo D synthesis in OLGs and neurons only in the presence of CPZ, at the same time that moderately prevents the loss of viability caused by the toxin, i.e., at neuroprotective doses of CLO. An important aspect, according to the findings obtained in SH-SY5Y cells, is that low concentrations of CLO would be the ones that may exert an Apo D-related neuroprotection against CPZ. In this way, we hypothesize that the great increase in Apo D synthesis induced by the treatment with 5 μM of CLO, unable to prevent the cytotoxic effect of CPZ, may be a consequence of some cell stress/damage caused by this dose of antipsychotic drug.

Based on our data, it seems that Apo D may contribute to the protective effect of the drug. This idea is sustained by some in vivo studies postulating that the effect of CLO in pathological situations would be related to the protective function of Apo D, thanks to its ability to (i) bind hydrophobic ligands, (ii) minimize their release, (iii) prevent their peroxidative degradation, and (iv) stabilize plasma membranes [53,61]. In particular, some authors propose that mechanism of action of CLO depends on the role of Apo D in arachidonic acid metabolism [6,62]. Therefore, targeting neural cells to increase Apo D and prevent further death would be one promising choice for MS modifying approaches.

On the other hand, this study was designed to test the effect of Apo D when it is added exogenously. The results obtained, using either a purified or a recombinant version of the hApo D, were interesting as these compounds afforded some neuroprotection against the CPZ insult. In fact, our analysis revealed that the exogenous addition of hApo D, purified from BCF or produced in a mammalian expression system, induces an increase in cell viability/proliferation in normal conditions. At the same time, this apolipoprotein also completely prevents the mitochondrial damage and loss of viability caused by the treatment with moderate to high doses of CPZ in oligodendroglial cells and, more importantly, in a neuroblastoma cell line that lacks endogenous Apo D expression [52]. Moreover, and in order to check the possibility that Apo D exerts, in this case, its protective activity in an extracellular way, a chemical perturbation of endocytosis in SH-SY5Y cells was carried out. Our data showed that Apo D neuroprotection is largely independent of CME, phagocytosis and macropinocytosis, but it is significantly reduced by inhibitors of dynamin, i.e., dynamin-dependent mechanisms that are consistent with clathrin-independent endocytosis (CIE) modes (caveolae- and RhoA-dependent) [63]. The above-mentioned results may have several implications. First, these results suggest that OLGs and neurons would be able to capture and internalize hApo D from the medium, triggering an increase in the cell metabolic activity and/or proliferation rate. Apo D uptake by some cells is not an unknown phenomenon. For instance, it is clearly demonstrated that astrocytes and OLGs synthesize and secrete this protein [16,18,64] which is captured by certain neurons in some specific situations [9,64–66]. Although technically challenging, pioneering studies in last years stated that Apo D may enter cells as a clathrin-independent cargo mediated by a specific cell surface receptor, basignin [67]. The recent discovery that Apo D is located inside the endosome-lysosome-autophagosomal compartment [66] and the results here presented support this hypothesis. Second, the effect of exogenous Apo D, once internalized, turns neuroprotective in pathological situations, which is consistent across studies. For example, Najyb et al. (2017) demonstrated that hApo D internalization and accumulation in primary hippocampal neurons were accentuated by kainate treatment. In addition, these authors reported that hApo D could act by decreasing abnormally increased cholesterol levels in damaged neurons [68]. In this line, He et al. (2009) showed that hApo D purified from BCF was able to bind arachidonic acid and cholesterol, attenuating the increase in oxidants and proinflammatory derivatives as F(2)-isoprostanes and 7-ketocholesterol in similar pathological conditions [69]. These neuroprotective and antioxidant roles of Apo D may be closely associated with its capacity of reducing radical-propagating lipid hydroperoxides by three methionine (Met) residues (Met49, Met93, and Met157), highly conserved in mammals [70]. Alternatively, Apo D has an extra cysteine, Cys116, with a thiol group that can be implicated in a direct antioxidant activity [70,71]. Despite this, and according with our findings, the protective mechanism of Apo D against oxidative damage induced by CPZ may directly target mitochondrial function but would not act through the production levels of ROS.

Noteworthy, although Apo D has been generally described as a monomeric protein [4,72], it dimerizes when reducing peroxidized lipids [7,15]. Thus, small-angle X-ray scattering analysis revealed that this apolipoprotein is mainly present as a tetramer in BCF or an oligomer in CSF [15]. As a general rule, heteromers are currently considered as novel molecular entities with new ligand and signaling characteristics, and probably different antioxidant properties which could explain the greater neuroprotective effect of hApo D, compared with hrApo D, demonstrated here.

In summary, valuable information has been gained in this work concerning neuroprotective effect of Apo D against CPZ, a neurotoxin used to produce models of MS. Although the development of simpler models, as the ones shown in this work, constitutes a way to provide reliable answers in some pathological situations, these results must be validated on more physiological models such as primary cultures in order to check whether increasing either endogenously and/or exogenously the levels of Apo D could be a feasible intervention as part of medical therapy in neurodegenerative diseases. In this regard, the origin and the native structure of this protein must be taken into account in order to design the most effective approach.

4. Materials and Methods

4.1. Cell Lines

HOG cell line, established from a surgically removed human oligodendroglioma by Dr. A. T. Campagnoni (University of California, UCLA, Berkeley, CA, USA) [73] was kindly provided by Dr. J. A. López-Guerrero (Universidad Autónoma de Madrid, Madrid, Spain) [74]. Cells were grown in DMEM, low glucose, pyruvate, HEPES (22320-022, Invitrogen, Paisley, Scotland, UK), 100 units/mL penicillin/streptomycin (17-602E, Invitrogen, Paisley, Scotland, UK), and 10% (*v*/*v*) heat inactivated fetal bovine serum (FBS) (10270-106, Invitrogen, Paisley, Scotland, UK).

Human neuroblastoma SH-SY5Y cell line was obtained from Sigma (ref 94030304, Sigma-Aldrich, St. Louis, MO, USA) and was grown in DMEM supplemented with 2 mM L-glutamine (61965-059, Invitrogen, Paisley, Scotland, UK), 100 units/mL penicillin/streptomycin (17-602E, Invitrogen, Paisley, Scotland, UK), 1% nonessential amino acids (11140-035, Invitrogen, Paisley, Scotland, UK), and 10% (*v*/*v*) heat inactivated FBS (10270-106, Invitrogen, Paisley, Scotland, UK).

Cells were maintained at 37 °C in a humidified atmosphere of 5% CO_2 and were passaged when they were 80–90% confluent, i.e., approximately twice a week for no more than 20 passages.

4.2. Human Apo D Purification

Human Apo D (hApo D) was purified from BCF samples provide by the Pathology Unit of the Hospital Universitario Central de Asturias (HUCA). First, a cell fractionation was performed by differential centrifugation. Then, Amicon® Ultra-15, 100 kDa, centrifugal filter units (Z740211, Sigma-Aldrich, St. Louis, MO, USA) were used. The solution containing the protein was flow through two consecutively ion-exchange chromatographic columns (HiTrap® Q Fast Flow, GE Healthcare, Chicago, IL, USA) with 25 mM Tris pH 8.0, followed by a size-exclusion chromatography (HiLoad® 16/60 Superdex® 200 prep grade, GE Healthcare, Chicago, IL, USA) in 50 mM Tris pH 8.0, 75 mM NaCl. Elution fractions with the protein of interest can be further concentrated using an appropriate 30 kDa cut-off Amicon® centrifuge filter (Z717185, Sigma-Aldrich, St. Louis, MO, USA). The presence of hApo D in these fractions was checked by Western blot, the amount of this apolipoprotein (concentration in the fraction) was quantified and its functionality was tested.

4.3. Cell Treatments

For CPZ treatment, a stock solution of 30 mM CPZ (C9012-25G, Sigma-Aldrich, St. Louis, MO, USA) was prepared freshly. For this, CPZ powder was dissolved in 50% ethanol/medium and shaken at 225 rpm at 60 °C for 15–20 min until its complete dissolution. Working solutions were prepared by diluting the stock in the specific medium for each cell type in a series of sequential solutions to reduce the ethanol concentration [72,73]. After 24–48 h of plating (30−40% cellular confluence), cellular toxicity was induced by the addition of CPZ in growing concentrations (50–1000 μM; see corresponding figure legends), for 24, 48, or 72 h. In order to prove that results are only attributable to CPZ not ethanol, the vehicle effect was also tested for each sequential solution in the cell models. As it can be observed in the graphs (Figure A3), we demonstrated that even the highest concentration of ethanol used to dissolve CPZ did not negatively affect, in a statistically significant way, cell viability of HOG and SH-SY5Y cells after 24 and 48 h of treatment.

To investigate the effect of the antipsychotic drug CLO (C6305-100G, Sigma-Aldrich, St. Louis, MO, USA) in the Apo D expression, cells were treated with different concentrations (0.1–5 nM; see corresponding figure legends) for 24 h, before fresh addition of CPZ. For the exogenous addition of Apo D, hApo D purified from BCF or hrApo D derived from human cells (P05090, Novoprotein, Summit, NJ, USA) were added (0.05–1000 nM; see corresponding figure legends) to cell cultures 24 h prior to CPZ.

For inhibition of endocytic mechanisms, SH-SY5Y cells were treated with different chemical inhibitors, cytochalasin D (8 μg/mL; C2618, Sigma-Aldrich, St. Louis, MO, USA), chlorpromazine hydrochloride (5 μg/mL; C8138, Sigma-Aldrich, St. Louis, MO, USA) and dynasore (80 μM; 324410, Sigma-Aldrich, St. Louis, MO, USA). Stock solutions were prepared in dimethyl sulfoxide (DMSO) and diluted in serum-free medium supplemented with 30 mM HEPES on the day of the experiment. The final DMSO concentration added was kept <0.1%. Cells were washed in serum-free medium and treated with the respective inhibitors (see corresponding figure legends) for 30 min before Apo D addition. H_2O_2 was used as positive control.

Drug concentrations and times of treatments were based on the bibliography and on our previous experience [75,76].

4.4. MTT Assay

Cell viability was studied by 3-(4,5-dimethylthiazol-2-yl)-2,5-diphenyltetrazolium bromide (MTT) reduction assay, a method based on the activity of mitochondrial NAD dependent oxidoreductases as indicator of the functional state of mitochondria. For this, 3000–5000 cells/well were seeded in 96-well plates and grown in 100μL/well of complete medium. Once treatments were completed, 10 μL of MTT (5mg/mL in phosphate buffered saline (PBS; 10010-023, Gibco, Invitrogen, Paisley, Scotland, UK)) (M5655; Sigma-Aldrich, St. Louis, MO, USA) were added to each well. Four hours later, 100 μL of lysis solution (20% sodium dodecyl sulfate (SDS); 50% dimethylformamide; pH 4) were added to the culture and incubated overnight at 37 °C. Absorbance at 570 nm was measured using a Multiskan EX Microplate Reader (ThermoFisher Scientific, Waltham, MA, USA). Values from blank wells, containing only medium, were subtracted from the values of the samples. Cell viability was expressed as the percentage of the controls.

4.5. Determination of ROS

The intracellular level of ROS, as an important biomarker for oxidative stress, was estimated with the dye 2′7′-dichlorodihydrofluorescein diacetate (H_2DCFDA; D399, Molecular Probes, Invitrogen, Paisley, Scotland, UK), a nonpolar compound that easily penetrates into the cell where it is hydrolyzed to the nonpermeant H_2DCF. This nonfluorescent compound becomes oxidized by various ROS to highly fluorescent 2′,7′-dichlorofluorescein (DCF). For determination, 3000–5000 cells/well were seeded in 96-well plates and grown in 100 μL/well of complete medium. Once treatments were completed, medium was removed and prewarmed PBS containing the probe (final working concentration of 10 μM dye) was added to the cells. After incubation for 60 min at 37 °C, the dye was removed and cells were returned to prewarmed growth medium. Then, fluorescence was measured in a microplate fluorimeter FLX-800 (Bio-Tek Instruments, Inc., Winooski, VT, USA) at an excitation wavelength of 485 nm and an emission wavelength of 528 nm.

4.6. Immunocytochemistry

Cells were seeded over glass coverslips (10 mm diameter) in 6-well plates at a density of 50,000 cells/well in a final volume of 2 mL of medium. Once the treatments were concluded, cells were washed three times with PBS and fixed in bouin solution for 15 min. After fixation, cells were washed three times and then permeabilized by incubation with 1% Triton X-100 at room temperature for 15 min. Nonspecific binding was blocked by incubation with bovine serum 30 min at room temperature. Incubation with anti-human Apo D antibody 1:2000 (provided by Dr. Carlos López-Otín, department of Biochemistry and Molecular Biology, University of Oviedo; see [58,77,78]) was carried out overnight in a humid chamber at 4 °C. After three washes in PBS, coverslips were incubated 30 min at room temperature using a biotinylated horse universal antibody (Universal quick, PK-8800, Vector Laboratories, Inc., Burlingame, CA, USA) diluted 1:50. After that, cells were incubated with streptavidin Alexa Fluor® 550 conjugate (1:500; S2138, Invitrogen, Paisley, Scotland, UK). Finally, cells were washed in distilled water, dehydrated, cleared in eucalyptol and mounted with Fluoromount. The fluorescence was visualized in a Nikon Eclipse E400 microscope equipped with a Nikon G2-A and recorded by a digital camera (Nikon DN-100). The resulting immunocytochemical signal was selected with Photoshop and quantified with ImageJ 1.57 software (NIH, Bethesda, MD, USA) [79]. Images were acquired under the same conditions of illumination, diaphragm and condenser adjustments, exposure time, and background correction. For control purposes, representative cell cultures were processed in the same way with a nonimmune serum or with specifically absorbed sera instead of the primary antibody. Under these conditions no specific immunostaining was observed.

4.7. RNA Purification

Cells were recovered from culture dishes using scrapers and TRIzol (15596018, Invitrogen, Paisley, Scotland, UK), then total RNA was purified using the RNeasy mini-kit (74104, Qiagen, Valencia, CA, USA) with a DNAse digestion step performed (79204, Qiagen, Valencia, CA, USA) following the manufacturer's instructions.

4.8. Quantitative Real-Time PCR

Random primers and the SuperScript III kit (11752050, Invitrogen, Paisley, Scotland, UK) were used to reverse-transcribe 1 μg of total RNA into first-strand cDNA in a total volume of 20 μL according to the manufacturer's instructions. SYBR Green PCR Master Mix (a25778, Applied Biosystems, Carlsbad, CA, USA) was mixed with cDNA for quantitative real time polymerase chain reaction (qRT-PCR) using 0.3 μM forward and reverse oligonucleotide primers (Table 1). 7300 Real Time PCR System (Applied Biosystems, Carlsbad, CA, USA) was used for quantitative measure of gene expression. Cycling conditions were an initial denaturation at 95 °C for 10 min, followed by 40 cycles of 95 °C for 15 s 60 °C for 1 min. At the end, a dissociation curve was implemented from 60 to 95 °C to validate amplicon specificity. Relative quantification of gene expression was calculated by interpolation into a standard curve. All values were divided by the expression of the house keeping gene 18S rRNA.

Table 1. Primers used for qRT-PCR in this study.

Oligonucleotide		Sequence
R2-ApoD-F	——	TGCATCCAGGCCAACTACTC
R2-ApoD-Rev	——	GGGTGGCTTCACCTTCGATT
18S-Fw	——	ATGCTCTTAGCTGAGTGTCCCG
18S-Rev	——	ATTCCTAGCTGCGGTATCCAGG

The annealing temperature was 60 °C for all primers. 18S rRNA was used as a housekeeping gene.

4.9. Data Analysis

The data in the graphs are presented as the mean ± S.E.M, from at least five independent experiments. The normality of population and the homogeneity of variance were evaluated by the test of Kolmogorov–Smirnov with the correction of Lilliefors and the test of Levene, respectively. Then one- or two-way ANOVA tests followed by post hoc Tukey's test for multiple comparisons were used to compare the values. Statistical analysis was carried out with SPSS 18.0 software (IBM, Armonk, NY, USA). Significant differences were considered when $p < 0.05$.

Author Contributions: Conceptualization was agreed by J.T., I.M.L. and A.N., who also participated in the design of the project and analyzed the results. E.M.-P., N.R.-S. and E.G.-Á. performed the majority of the experiments; E.d.V. performed some of the immunocytochemistry assays and participated in data analysis; R.P. performed the qRT-PCR, participated in data analysis and in providing data for final figures; E.M-P. and J.T. did many of the imaging assays in the confocal microscope, took images and participated in data analysis; E.M.-P., A.N. and I.M.L. wrote the first draft of the manuscript that was further edited by all coauthors, who agreed with submission. All authors have read and agreed to the published version of the manuscript.

Funding: This research was funded by Instituto de Salud Carlos III through the project (PI15/00601) (co-funded by European Regional Development Fund/European Social Fund "Investing in your future"). This work was also funded in part by a grant (PI19/01805) from the Instituto de Salud Carlos III, co-funded by European Regional Development Fund (ERDF) "A way to build Europe" and by Fundación Rioja Salud. I.M.L. is supported by a Miguel Servet contract (CPII20/00029) from the Instituto de Salud Carlos III, co-funded by European Social fund (ESF) "Investing in your future".

Institutional Review Board Statement: Not applicable.

Informed Consent Statement: Not applicable.

Data Availability Statement: The data presented in this study are available on request from the corresponding author. The data are not publicly available due to privacy restrictions.

Conflicts of Interest: The authors declare no conflict of interest.

Abbreviations

AD	Alzheimer's disease
Apo D	Apolipoprotein D
BBB	Blood–brain barrier
BCF	Breast cyst fluid
CIE	Clathrin-independent endocytosis
CLO	Clozapine
CME	Clathrin-mediated endocytosis
CNS	Central nervous system
CPZ	Cuprizone
CSF	Cerebrospinal fluid
DCF	2′,7′-dichlorofluorescein
DMSO	Dimethyl sulfoxide
FBS	Fetal bovine serum
H_2DCFDA	2′7′-dichlorodihydrofluorescein diacetate
hApo D	Human Apo D
HDL	High-density plasma lipoproteins
hrApo D	Human recombinant Apo D
HUCA	Hospital Universitario Central de Asturias
MS	Multiple sclerosis
MTT	3-(4,5-dimethylthiazol-2-yl)-2,5-diphenyltetrazolium bromide
OLGs	Oligodendrocytes
PBS	Phosphate buffered saline
PD	Parkinson's disease
PNS	Peripheral nervous system
RNS	Reactive nitrogen species
ROS	Reactive oxygen species
SDS	Sodium dodecyl sulfate
UV	Ultraviolet

Appendix A

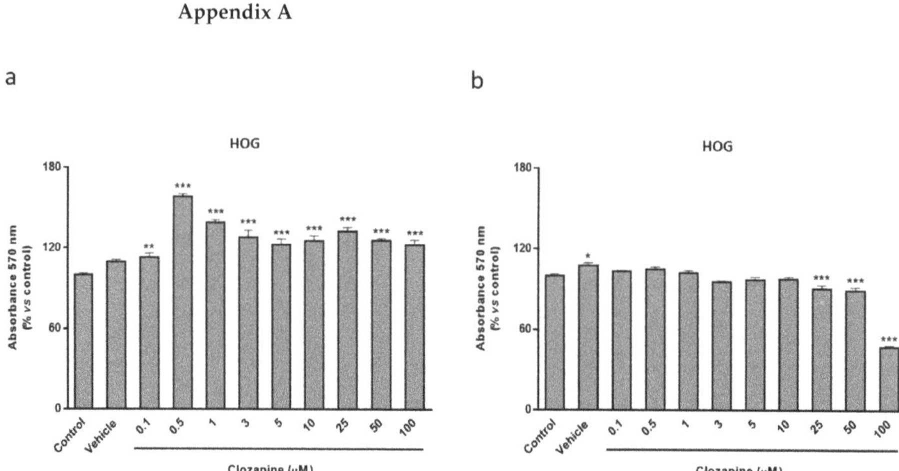

Figure A1. MTT assay in HOG cells treated with increasing concentrations of CLO (0.1–100 μM) during 24 h (**a**) and 48 h (**b**). Cell damage is represented as the percentage of viability versus control. Data are the mean ± SEM of five independent experiments. Significant differences were analyzed by a one-way ANOVA followed by post-hoc Tukey's test. * $p < 0.05$, ** $p < 0.01$, *** $p < 0.001$ compared with control.

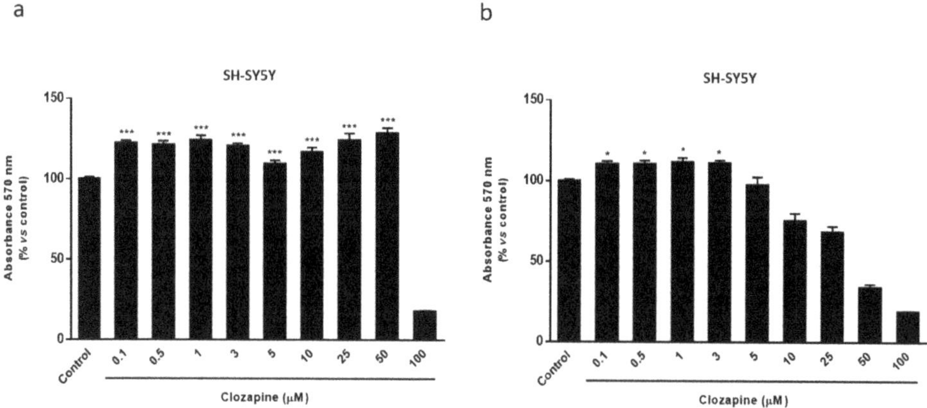

Figure A2. MTT assay in SH-SY5Y cells treated with increasing concentrations of CLO (0.1–100 μM) during 24 h (**a**) and 48 h (**b**). Cell damage is represented as the percentage of viability versus control. Data are the mean ± SEM of five independent experiments. Significant differences were analyzed by a one-way ANOVA followed by post-hoc Tukey's test. * $p < 0.05$, *** $p < 0.001$ compared with control.

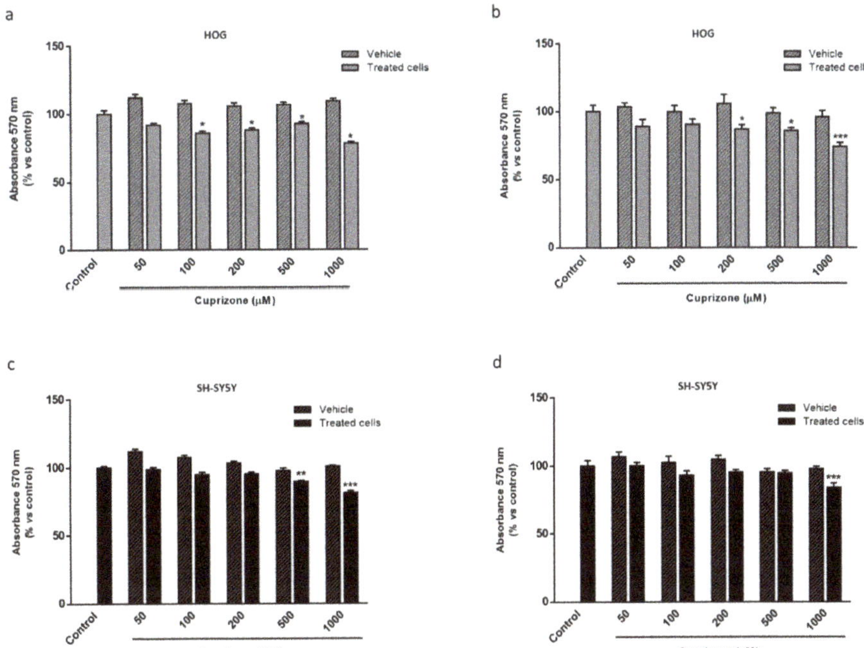

Figure A3. MTT assay in HOG (**a**,**b**) and SH-SY5Y (**c**,**d**) cells treated with increasing concentrations of CPZ (50–1000 μM) and their respective vehicles for 24 h (**a**,**c**) or 48 h (**b**,**d**). Cell damage is represented as the percentage of viability versus control. Data are the mean ± S.E.M of five independent experiments. Significant differences were analyzed by a one-way ANOVA followed by post-hoc Tukey's test. * $p < 0.05$, ** $p < 0.01$, *** $p < 0.001$ compared to control.

References

1. Rassart, E.; Bedirian, A.; Do Carmo, S.; Guinard, O.; Sirois, J.; Terrisse, L.; Milne, R. Apolipoprotein D. *Biochim. Biophys. Acta Protein Struct. Mol. Enzymol.* **2000**, *1482*, 185–198. [CrossRef]
2. Rassart, E.; Desmarais, F.; Najyb, O.; Bergeron, K.F.; Mounier, C. Apolipoprotein D. *Gene* **2020**, *756*, 144874. [CrossRef] [PubMed]
3. Fernández-Ruiz, J.; Gómez-Ruiz, M.; García, C.; Hernández, M.; Ramos, J.A. Modeling Neurodegenerative Disorders for Developing Cannabinoid-Based Neuroprotective Therapies. In *Methods in Enzymology*; Elsevier: Amsterdam, The Netherlands, 2017; Volume 593, pp. 175–198.
4. Nasreen, A.; Vogt, M.; Kim, H.J.; Eichinger, A.; Skerra, A. Solubility engineering and crystallization of human apolipoprotein D. *Protein Sci.* **2006**, *15*, 190–199. [CrossRef]
5. Eichinger, A.; Nasreen, A.; Kim, H.J.; Skerra, A. Structural insight into the dual ligand specificity and mode of high density lipoprotein association of apolipoprotein D. *J. Biol. Chem.* **2007**, *282*, 31068–31075. [CrossRef] [PubMed]
6. Thomas, E.A.; George, R.C.; Gregor Sutcliffe, J. Apolipoprotein D modulates arachidonic acid signaling in cultured cells: Implications for psychiatric disorders. *Prostaglandins Leukot. Essent. Fat. Acids* **2003**, *69*, 421–427. [CrossRef] [PubMed]
7. Bhatia, S.; Knoch, B.; Wong, J.; Kim, W.S.; Else, P.L.; Oakley, A.J.; Garner, B. Selective reduction of hydroperoxyeicosatetraenoic acids to their hydroxy derivatives by apolipoprotein D: Implications for lipid antioxidant activity and Alzheimer's disease. *Biochem. J.* **2012**, *442*, 713–721. [CrossRef]
8. Waldner, A.; Dassati, S.; Redl, B.; Smania, N.; Gandolfi, M. Apolipoprotein D Concentration in Human Plasma during Aging and in Parkinson's Disease: A Cross-Sectional Study. *Parkinsons. Dis.* **2018**, *2018*. [CrossRef]
9. Bajo-Grañeras, R.; Ganfornina, M.D.; Martín-Tejedor, E.; Sanchez, D. Apolipoprotein D mediates autocrine protection of astrocytes and controls their reactivity level, contributing to the functional maintenance of paraquat-challenged dopaminergic systems. *Glia* **2011**, *59*, 1551–1566. [CrossRef]
10. Do Carmo, S.; Levros, L.C.; Rassart, E. Modulation of apolipoprotein D expression and translocation under specific stress conditions. *Biochim. Biophys. Acta Mol. Cell Res.* **2007**, *1773*, 954–969. [CrossRef]
11. Ganfornina, M.D.; Do Carmo, S.; Lora, J.M.; Torres-Schumann, S.; Vogel, M.; Allhorn, M.; González, C.; Bastiani, M.J.; Rassart, E.; Sanchez, D. Apolipoprotein D is involved in the mechanisms regulating protection from oxidative stress. *Aging Cell* **2008**, *7*, 506–515. [CrossRef]

12. Sanchez, D.; Bajo-Grañeras, R.; Del Caño-Espinel, M.; Garcia-Centeno, R.; Garcia-Mateo, N.; Pascua-Maestro, R.; Ganfornina, M.D. Aging without apolipoprotein d: Molecular and cellular modifications in the hippocampus and cortex. *Exp. Gerontol.* **2015**, *67*, 19–47. [CrossRef] [PubMed]
13. Li, H.; Ruberu, K.; Karl, T.; Garner, B. Cerebral apolipoprotein-D Is hypoglycosylated compared to peripheral tissues and is variably expressed in mouse and human brain regions. *PLoS ONE* **2016**, *11*, e0148238. [CrossRef] [PubMed]
14. Scheen, L.; Brandt, L.; Bodén, R.; Tiihonen, J.; Andersen, M.; Kieler, H.; Reutfors, J. Predictors for initiation of pharmacological prophylaxis in patients with newly diagnosed bipolar disorder-A nationwide cohort study. *J. Affect. Disord.* **2015**, *172*, 204–210. [CrossRef] [PubMed]
15. Kielkopf, C.S.; Low, J.K.K.; Mok, Y.-F.; Bhatia, S.; Palasovski, T.; Oakley, A.J.; Whitten, A.E.; Garner, B.; Brown, S.H.J. Identification of a novel tetrameric structure for human apolipoprotein-D. *J. Struct. Biol.* **2018**, *203*, 205–218. [CrossRef]
16. Navarro, A.; Tolivia, J.; Astudillo, A.; Del Valle, E. Pattern of apolipoprotein D immunoreactivity in human brain. *Neurosci. Lett.* **1998**, *254*, 17–20. [CrossRef]
17. Navarro, A.; Del Valle, E.; Juárez, A.; Martinez, E.; Ordóñez, C.; Astudillo, A.; Tolivia, J. Apolipoprotein D synthesis progressively increases in frontal cortex during human lifespan. *Age (Omaha)* **2010**, *32*, 85–96. [CrossRef]
18. Hu, C.Y.; Ong, W.Y.; Sundaram, R.K.; Chan, C.; Patel, S.C. Immunocytochemical localization of apolipoprotein D in oligodendrocyte precursor-like cells, perivascular cells, and pericytes in the human cerebral cortex. *J. Neurocytol.* **2001**, *30*, 209–218. [CrossRef]
19. García-Mateo, N.; Pascua-Maestro, R.; Pérez-Castellanos, A.; Lillo, C.; Sanchez, D.; Ganfornina, M.D. Myelin extracellular leaflet compaction requires apolipoprotein D membrane management to optimize lysosomal-dependent recycling and glycocalyx removal. *Glia* **2018**, *66*, 670–687. [CrossRef]
20. Bhatia, S.; Kim, W.S.; Shepherd, C.E.; Halliday, G.M. Apolipoprotein D Upregulation in Alzheimer's Disease but Not Frontotemporal Dementia. *J. Mol. Neurosci.* **2018**, *67*, 125. [CrossRef]
21. Muffat, J.; Walker, D.W. Apolipoprotein D: An overview of its role in aging and age-related diseases. *Cell Cycle* **2010**, *9*, 269–273. [CrossRef]
22. Dassati, S.; Waldner, A.; Schweigreiter, R. Apolipoprotein D takes center stage in the stress response of the aging and degenerative brain. *Neurobiol. Aging* **2014**, *35*, 1632–1642. [CrossRef] [PubMed]
23. Feigin, V.L.; Nichols, E.; Alam, T.; Bannick, M.S.; Beghi, E.; Blake, N.; Culpepper, W.J.; Dorsey, E.R.; Elbaz, A.; Ellenbogen, R.G.; et al. Global, regional, and national burden of neurological disorders, 1990–2016: A systematic analysis for the Global Burden of Disease Study 2016. *Lancet Neurol.* **2019**. [CrossRef]
24. Noseworthy, J.H.; Lucchinetti, C.; Rodriguez, M.; Weinshenker, B.G. Multiple sclerosis. *N. Engl. J. Med.* **2000**, *343*, 938–952. [CrossRef] [PubMed]
25. Gelfand, J.M. Multiple sclerosis: Diagnosis, differential diagnosis, and clinical presentation. In *Handbook of Clinical Neurology*; Elsevier: Amsterdam, The Netherlands, 2014; Volume 122, pp. 269–290.
26. Correale, J.; Gaitán, M.I.; Ysrraelit, M.C.; Fiol, M.P. Progressive multiple sclerosis: From pathogenic mechanisms to treatment. *Brain* **2017**, *140*, 527–546. [CrossRef] [PubMed]
27. Haines, J.D.; Inglese, M.; Casaccia, P. Axonal damage in multiple sclerosis. *Mt. Sinai J. Med.* **2011**, *78*, 231–243. [CrossRef] [PubMed]
28. McLaughlin, J.E. Practical Review of Neuropathology.: Gregory N, Fuller J, Goodman C. 2001, Lippincott Williams & Wilkins, $69.95. ISBN 0 7817 2778 2. *J. Clin. Pathol.* **2011**, *55*, 158. [CrossRef]
29. Bö, L.; Geurts, J.J.G.; Mörk, S.J.; Van Der Valk, P. Grey matter pathology in multiple sclerosis. In *Acta Neurologica Scandinavica*; Wiley-Blackwell: Hoboken, NJ, USA, 2006; Volume 113, pp. 48–50.
30. Zindler, E.; Zipp, F. Neuronal injury in chronic CNS inflammation. *Best Pract. Res. Clin. Anaesthesiol.* **2010**, *24*, 551–562. [CrossRef]
31. McQualter, J.L.; Bernard, C.C.A. Multiple sclerosis: A battle between destruction and repair. *J. Neurochem.* **2007**, *100*, 295–306. [CrossRef]
32. McGuire, C.; Beyaert, R.; van Loo, G. Death receptor signalling in central nervous system inflammation and demyelination. *Trends Neurosci.* **2011**, *34*, 619–628. [CrossRef]
33. Disanto, G.; Morahan, J.M.; Barnett, M.H.; Giovannoni, G.; Ramagopalan, S.V. The evidence for a role of B cells in multiple sclerosis. *Neurology* **2012**, *78*, 823–832. [CrossRef]
34. Ohl, K.; Tenbrock, K.; Kipp, M. Oxidative stress in multiple sclerosis: Central and peripheral mode of action. *Exp. Neurol.* **2016**, *277*, 58–67. [CrossRef] [PubMed]
35. Dendrou, C.A.; Fugger, L.; Friese, M.A. Immunopathology of multiple sclerosis. *Nat. Rev. Immunol.* **2015**, *15*, 545–558. [CrossRef] [PubMed]
36. Reich, D.S.; Lucchinetti, C.F.; Calabresi, P.A. Multiple sclerosis. *N. Engl. J. Med.* **2018**, *378*, 169–180. [CrossRef] [PubMed]
37. Stys, P.K.; Zamponi, G.W.; Van Minnen, J.; Geurts, J.J.G. Will the real multiple sclerosis please stand up? *Nat. Rev. Neurosci.* **2012**, *13*, 507–514. [CrossRef]
38. Caprariello, A.V.; Rogers, J.A.; Morgan, M.L.; Hoghooghi, V.; Plemel, J.R.; Koebel, A.; Tsutsui, S.; Dunn, J.F.; Kotra, L.P.; Ousman, S.S.; et al. Biochemically altered myelin triggers autoimmune demyelination. *Proc. Natl. Acad. Sci. USA* **2018**, *115*, 5528–5533. [CrossRef] [PubMed]

39. Titus, H.E.; Chen, Y.; Podojil, J.R.; Robinson, A.P.; Balabanov, R.; Popko, B.; Miller, S.D. Pre-clinical and Clinical Implications of "Inside-Out" vs. "Outside-In" Paradigms in Multiple Sclerosis Etiopathogenesis. *Front. Cell. Neurosci.* **2020**, *14*. [CrossRef] [PubMed]
40. Ontaneda, D.; Hyland, M.; Cohen, J.A. Multiple Sclerosis: New Insights in Pathogenesis and Novel Therapeutics. *Annu. Rev. Med.* **2012**, *63*, 389–404. [CrossRef]
41. Leibowitz, S.M.; Yan, J. NF-κB pathways in the pathogenesis of multiple sclerosis and the therapeutic implications. *Front. Mol. Neurosci.* **2016**, *9*, 84. [CrossRef]
42. Vandenbark, A.A.; Culbertson, N.E.; Bartholomew, R.M.; Huan, J.; Agotsch, M.; LaTocha, D.; Yadav, V.; Mass, M.; Whitham, R.; Lovera, J.; et al. Therapeutic vaccination with a trivalent T-cell receptor (TCR) peptide vaccine restores deficient FoxP3 expression and TCR recognition in subjects with multiple sclerosis. *Immunology* **2008**, *123*, 66–78. [CrossRef]
43. Goldberg, P.; Fleming, M.C.; Picard, E.H. Multiple sclerosis: Decreased relapse rate through dietary supplementation with calcium, magnesium and vitamin D. *Med. Hypotheses* **1986**, *21*, 193–200. [CrossRef]
44. Kappos, L.; Bar-Or, A.; Cree, B.A.C.; Fox, R.J.; Giovannoni, G.; Gold, R.; Vermersch, P.; Arnold, D.L.; Arnould, S.; Scherz, T.; et al. Siponimod versus placebo in secondary progressive multiple sclerosis (EXPAND): A double-blind, randomised, phase 3 study. *Lancet* **2018**, *391*, 1263–1273. [CrossRef]
45. Denic, A.; Johnson, A.J.; Bieber, A.J.; Warrington, A.E.; Rodriguez, M.; Pirko, I. The relevance of animal models in multiple sclerosis research. *Pathophysiology* **2011**, *18*, 21–29. [CrossRef] [PubMed]
46. Vega-Riquer, J.M.; Mendez-Victoriano, G.; Morales-Luckie, R.A.; Gonzalez-Perez, O. Five Decades of Cuprizone, an Updated Model to Replicate Demyelinating Diseases. *Curr. Neuropharmacol.* **2019**, *17*, 129–141. [CrossRef] [PubMed]
47. Nyamoya, S.; Schweiger, F.; Kipp, M.; Hochstrasser, T. Cuprizone as a model of myelin and axonal damage. *Drug Discov. Today Dis. Model.* **2017**, *25–26*, 63–68. [CrossRef]
48. Procaccini, C.; De Rosa, V.; Pucino, V.; Formisano, L.; Matarese, G. Animal models of Multiple Sclerosis. *Eur. J. Pharmacol.* **2015**, *759*, 182–191. [CrossRef]
49. Stoop, M.P.; Singh, V.; Dekker, L.J.; Titulaer, M.K.; Stingl, C.; Burgers, P.C.; Sillevis Smitt, P.A.E.; Hintzen, R.Q.; Luider, T.M. Proteomics comparison of cerebrospinal fluid of relapsing remitting and primary progressive multiple sclerosis. *PLoS ONE* **2010**, *5*, e12442. [CrossRef]
50. Reindl, M.; Knipping, G.; Wicher, I.; Dilitz, E.; Egg, R.; Deisenhammer, F.; Berger, T. Increased intrathecal production of apolipoprotein D in multiple sclerosis. *J. Neuroimmunol.* **2001**, *119*, 327–332. [CrossRef]
51. Navarro, A.; Rioseras, B.; Del Valle, E.; Martínez-Pinilla, E.; Astudillo, A.; Tolivia, J. Expression Pattern of Myelin-Related Apolipoprotein D in Human Multiple Sclerosis Lesions. *Front. Aging Neurosci.* **2018**, *10*, 254. [CrossRef]
52. Pascua-Maestro, R.; González, E.; Lillo, C.; Ganfornina, M.D.; Falcón-Pérez, J.M.; Sanchez, D. Extracellular vesicles secreted by astroglial cells transport apolipoprotein D to neurons and mediate neuronal survival upon oxidative stress. *Front. Cell. Neurosci.* **2019**, *12*, 526. [CrossRef]
53. Thomas, E.A.; Danielson, P.E.; Austin Nelson, P.; Pribyl, T.M.; Hilbush, B.S.; Hasel, K.W.; Gregor Sutcliffe, J. Clozapine increases apolipoprotein D expression in rodent brain: Towards a mechanism for neuroleptic pharmacotherapy. *J. Neurochem.* **2001**, *76*, 789–796. [CrossRef]
54. Mahadik, S.P.; Khan, M.M.; Evans, D.R.; Parikh, V.V. Elevated plasma level of apolipoprotein D in schizophrenia and its treatment and outcome. *Schizophr. Res.* **2002**, *58*, 55–62. [CrossRef]
55. Do Carmo, S.; Séguin, D.; Milne, R.; Rassart, E. Modulation of apolipoprotein D and apolipoprotein E mRNA expression by growth arrest and identification of key elements in the promoter. *J. Biol. Chem.* **2002**, *277*, 5514–5523. [CrossRef] [PubMed]
56. Sarjeant, J.M.; Lawrie, A.; Kinnear, C.; Yablonsky, S.; Leung, W.; Massaeli, H.; Prichett, W.; Veinot, J.P.; Rassart, E.; Rabinovitch, M. Apolipoprotein D Inhibits Platelet-Derived Growth Factor-BB–Induced Vascular Smooth Muscle Cell Proliferated by Preventing Translocation of Phosphorylated Extracellular Signal Regulated Kinase 1/2 to the Nucleus. *Arterioscler. Thromb. Vasc. Biol.* **2003**, *23*, 2172–2177. [CrossRef] [PubMed]
57. Martínez, E.; Navarro, A.; Ordóez, C.; Del Valle, E.; Tolivia, J. Amyloid-$\beta_{25\text{-}35}$ induces apolipoprotein D synthesis and growth arrest in HT22 hippocampal cells. *J. Alzheimer's Dis.* **2012**, *30*, 233–244. [CrossRef] [PubMed]
58. Martínez, E.; Navarro, A.; Ordóñez, C.; Del Valle, E.; Tolivia, J. Oxidative stress induces apolipoprotein d overexpression in hippocampus during aging and alzheimer's disease. *J. Alzheimer's Dis.* **2013**, *36*, 129–144. [CrossRef] [PubMed]
59. Pasquini, L.A.; Calatayud, C.A.; Bertone Uña, A.L.; Millet, V.; Pasquini, J.M.; Soto, E.F. The neurotoxic effect of cuprizone on oligodendrocytes depends on the presence of pro-inflammatory cytokines secreted by microglia. *Neurochem. Res.* **2007**, *32*, 279–292. [CrossRef] [PubMed]
60. Benetti, F.; Ventura, M.; Salmini, B.; Ceola, S.; Carbonera, D.; Mammi, S.; Zitolo, A.; D'Angelo, P.; Urso, E.; Maffia, M.; et al. Cuprizone neurotoxicity, copper deficiency and neurodegeneration. *Neurotoxicology* **2010**, *31*, 509–517. [CrossRef]
61. Yao, J.K.; Thomas, E.A.; Reddy, R.D.; Keshavan, M.S. Association of plasma apolipoproteins D with RBC membrane arachidonic acid levels in schizophrenia. *Schizophr. Res.* **2005**, *72*, 259–266. [CrossRef]
62. Thomas, E.A.; Yao, J.K. Clozapine specifically alters the arachidonic acid pathway in mice lacking apolipoprotein D. *Schizophr. Res.* **2007**, *89*, 147–153. [CrossRef]
63. Dutta, D.; Donaldson, J.G. Search for inhibitors of endocytosis. *Cell. Logist.* **2012**, *2*, 203–208. [CrossRef]

64. Patel, S.C.; Asotra, K.; Patel, Y.C.; McConathy, W.J.; Patel, R.C.; Suresh, S. Astrocytes synthesize and secrete the lipophilic ligand carrier apolipoprotein D. *Neuroreport* **1995**, *6*, 653–657. [CrossRef] [PubMed]
65. Ong, W.Y.; Lau, C.P.; Leong, S.K.; Kumar, U.; Suresh, S.; Patel, S.C. Apolipoprotein d gene expression in the rat brain and light and electron microscopic immunocytochemistry of apolipoprotein D expression in the cerebellum of neonatal, immature and adult rats. *Neuroscience* **1999**, *90*, 913–922. [CrossRef]
66. Pascua-Maestro, R.; Diez-Hermano, S.; Lillo, C.; Ganfornina, M.D.; Sanchez, D. Protecting cells by protecting their vulnerable lysosomes: Identification of a new mechanism for preserving lysosomal functional integrity upon oxidative stress. *PLoS Genet.* **2017**, *13*, e1006603. [CrossRef] [PubMed]
67. Najyb, O.; Brissette, L.; Rassart, E. Apolipoprotein D internalization is a basigin-dependent mechanism. *J. Biol. Chem.* **2015**, *290*, 16077–16087. [CrossRef] [PubMed]
68. Najyb, O.; Do Carmo, S.; Alikashani, A.; Rassart, E. Apolipoprotein D Overexpression Protects Against Kainate-Induced Neurotoxicity in Mice. *Mol. Neurobiol.* **2017**, *54*, 3948–3963. [CrossRef]
69. He, X.; Jittiwat, J.; Kim, J.H.; Jenner, A.M.; Farooqui, A.A.; Patel, S.C.; Ong, W.Y. Apolipoprotein D modulates F2-isoprostane and 7-ketocholesterol formation and has a neuroprotective effect on organotypic hippocampal cultures after kainate-induced excitotoxic injury. *Neurosci. Lett.* **2009**, *455*, 183–186. [CrossRef]
70. Zhang, Y.; Cong, Y.; Wang, S.; Zhang, S. Antioxidant activities of recombinant amphioxus (Branchiostoma belcheri) apolipoprotein D. *Mol. Biol. Rep.* **2011**, *38*, 1847–1851. [CrossRef]
71. Oakley, A.J.; Bhatia, S.; Ecroyd, H.; Garner, B. Molecular dynamics analysis of apolipoprotein-D - lipid hydroperoxide interactions: Mechanism for selective oxidation of Met-93. *PLoS ONE* **2012**, *7*, e34057. [CrossRef]
72. Åkerstrom, B.; Flower, D.R.; Salier, J.P. Lipocalins: Unity in diversity. *Biochim. Biophys. Acta Protein Struct. Mol. Enzymol.* **2000**, *1482*, 1–8. [CrossRef]
73. Post, G.R.; Dawson, G. Characterization of a cell line derived from a human oligodendroglioma. *Mol. Chem. Neuropathol.* **1992**, *16*, 303–317. [CrossRef]
74. Bello-Morales, R.; Crespillo, A.J.; García, B.; Dorado, L.Á.; Martín, B.; Tabarés, E.; Krummenacher, C.; De Castro, F.; López-Guerrero, J.A. The effect of cellular differentiation on HSV-1 infection of oligodendrocytic cells. *PLoS ONE* **2014**, *9*, 303–317. [CrossRef] [PubMed]
75. Cammer, W. The neurotoxicant, cuprizone, retards the differentiation of oligodendrocytes in vitro. *J. Neurol. Sci.* **1999**, *168*, 116–120. [CrossRef]
76. Bénardais, K.; Kotsiari, A.; Škuljec, J.; Koutsoudaki, P.N.; Gudi, V.; Singh, V.; Vulinović, F.; Skripuletz, T.; Stangel, M. Cuprizone [bis(cyclohexylidenehydrazide)] is selectively toxic for mature oligodendrocytes. *Neurotox. Res.* **2013**, *24*, 244–250. [CrossRef] [PubMed]
77. Díez-Itza, I.; Vizoso, F.; Merino, A.M.; Sánchez, L.M.; Tolivia, J.; Fernández, J.; Ruibal, A.; López-Otín, C. Expression and prognostic significance of apolipoprotein D in breast cancer. *Am. J. Pathol.* **1994**, *144*, 310–320.
78. Navarro, A.; Del Valle, E.; Tolivia, J. Differential expression of apolipoprotein D in human astroglial and oligodendroglial cells. *J. Histochem. Cytochem.* **2004**, *52*, 1031–1036. [CrossRef]
79. Tolivia, J.; Navarro, A.; Del Valle, E.; Perez, C.; Ordoñez, C.; Martínez, E. Application of photoshop and scion image analysis to quantification of signals in histochemistry, immunocytochemistry and hybridocytochemistry. *Anal. Quant. Cytol. Histol.* **2006**, *28*, 43–53.

International Journal of
Molecular Sciences

MDPI

Article

Long-Term Effects of Repetitive Mild Traumatic Injury on the Visual System in Wild-Type and TDP-43 Transgenic Mice

Kristina Pilipović [1], Jelena Rajič Bumber [1], Petra Dolenec [1], Nika Gržeta [1], Tamara Janković [1], Jasna Križ [2] and Gordana Župan [1,*]

1 Department of Basic and Clinical Pharmacology and Toxicology, Faculty of Medicine, University of Rijeka, Braće Branchetta 20, 51 000 Rijeka, Croatia; kristina.pilipovic@medri.uniri.hr (K.P.); jelena.rajic@medri.uniri.hr (J.R.B.); petra.dolenec@medri.uniri.hr (P.D.); nika.grzeta@medri.uniri.hr (N.G.); tamara.jankovic@medri.uniri.hr (T.J.)

2 Department of Psychiatry and Neuroscience, Faculty of Medicine, University Laval, Québec City, QC G1V 0A6, Canada; jasna.kriz@fmed.ulaval.ca

* Correspondence: gordana.zupan@medri.uniri.hr

Abstract: Little is known about the impairments and pathological changes in the visual system in mild brain trauma, especially repetitive mild traumatic brain injury (mTBI). The goal of this study was to examine and compare the effects of repeated head impacts on the neurodegeneration, axonal integrity, and glial activity in the optic tract (OT), as well as on neuronal preservation, glial responses, and synaptic organization in the lateral geniculate nucleus (LGN) and superior colliculus (SC), in wild-type mice and transgenic animals with overexpression of human TDP-43 mutant protein (TDP-43^{G348C}) at 6 months after repeated closed head traumas. Animals were also assessed in the Barnes maze (BM) task. Neurodegeneration, axonal injury, and gliosis were detected in the OT of the injured animals of both genotypes. In the traumatized mice, myelination of surviving axons was mostly preserved, and the expression of neurofilament light chain was unaffected. Repetitive mTBI did not induce changes in the LGN and the SC, nor did it affect the performance of the BM task in the traumatized wild-type and TDP-43 transgenic mice. Differences in neuropathological and behavioral assessments between the injured wild-type and TDP-43^{G348C} mice were not revealed. Results of the current study suggest that repetitive mTBI was associated with chronic damage and inflammation in the OT in wild-type and TDP-43^{G348C} mice, which were not accompanied with behavioral problems and were not affected by the TDP-43 genotype, while the LGN and the SC remained preserved in the used experimental conditions.

Keywords: brain injuries; traumatic; diffuse axonal injury; geniculate bodies; mice; nerve degeneration; neurofilament proteins; neuroglia; optic tract; superior colliculi; synaptophysin; TDP-43 proteinopathies

Citation: Pilipović, K.; Rajič Bumber, J.; Dolenec, P.; Gržeta, N.; Janković, T.; Križ, J.; Župan, G. Long-Term Effects of Repetitive Mild Traumatic Injury on the Visual System in Wild-Type and TDP-43 Transgenic Mice. *Int. J. Mol. Sci.* **2021**, *22*, 6584. https://doi.org/10.3390/ijms22126584

Academic Editor: Anne Vejux

Received: 11 May 2021
Accepted: 17 June 2021
Published: 19 June 2021

1. Introduction

Repetitive mild traumatic brain injury (mTBI) represents a current and growing serious medical and economic problem worldwide. It is particularly common in athletes engaged in contact sports, such as soccer, ice hockey, American football, boxing, wrestling, and mixed martial arts [1,2], as well as in victims of domestic spousal violence or child abuse [3] and military personnel [4,5]. The true prevalence of repetitive mTBI is not known because the symptoms of a single mTBI or concussion frequently resolve without medical care, pass spontaneously, and stay unrecognized, unreported, or undiagnosed. For example, in most patients, especially adult athletes, some of the post-concussion symptoms, such as dizziness, disorientation, confusion, or headache, subside within 10 days [6] and, in some cases, within several months following the first head trauma without specific interventions [7]. Increasing evidence suggests that, in humans or experimental animals with prior mTBI history, the susceptibility to brain damage induced by a future TBI is augmented and that repetitive injuries have cumulative effects, enhancing a risk for long-term and later-life cognitive, behavioral, and psychiatric disturbances, as well as the development of neurodegeneration [8–10].

Repetitive mTBI has long been recognized as a risk factor for chronic traumatic encephalopathy (CTE), a condition characterized by generalized cerebral atrophy associated with widespread deposits of phosphorylated tau protein occurring as neurofibrillary tangles, diffuse beta-amyloid deposits, neuroinflammation, axonal pathology through the brain, and, in the majority of cases, by transactivation response element (TAR) DNA/RNA-binding protein 43 (TDP-43) immunoreactive intraneuronal and intraglial inclusions [10–13].

TDP-43 is predominantly a nuclear protein with the primary amino-acid structure similar to the members of the heterogeneous ribonucleoprotein family that shuttles between nucleus and cytoplasm [14]. Its intracellular functions in physiological conditions are insufficiently characterized, but it is becoming increasingly evident that TDP-43 is involved in specific pre-mRNA splicing and transcription events, in the regulation of mRNA stability, transport, translation, and degradation, and in chromatin condensation [14–16].

Recent evidence suggests that TDP-43 proteinopathy has been identified not only in CTE, but also in most cases of amyotrophic lateral sclerosis (ALS) [17], in a subset of the frontotemporal lobar degeneration (FTLD) with tau-negative ubiquitin-positive TDP-43-positive inclusions [18,19], and in specific disorders such as Alzheimer's disease [20], Lewy body disease [21], hippocampal sclerosis [22], and corticobasal degeneration [20], suggesting its important role in the pathogenesis of neurodegeneration [23].

While TDP-43 dysregulation and accompanying neuropathological changes were documented in humans with previous history of repetitive mTBI [11], to our knowledge, they were investigated in only four published experimental studies. Elevated TDP-43 expression levels in the whole-cell lysates from the injured mouse cortical and hippocampal tissue [24,25], as well as the protein changes in the rat brain following blast TBI [26], were described. We detected transitory TDP-43 cytoplasmatic translocation and overexpression of the protein and its pathological forms in the frontal cortex within the first week following repetitive mTBI in mice [27]. Neurodegeneration and gliosis in the optic tracts (OT) of injured wild-type mice and animals with overexpression of human mutant TDP-43 protein (TDP-43^{G348C}), a model of ALS/FTLD [28], were also demonstrated [27]. In addition, the level of damage in the OT was significantly increased in TDP-43 transgenic animals compared with wild-type mice at the end of the first week after the last injury [27]. TDP-43^{G348C} mice used in the mentioned study were 9–11 week old animals at the beginning of the study and did not show any neurodegenerative and behavioral impairments before head traumas. Here, we expanded our previous research to investigate the changes in the OT, as well as in the lateral geniculate nucleus (LGN) of the thalamus and the superior colliculus (SC), the brain structures that receive input from axons traveling in the OT, in wild-type and TDP-43^{G348C} animals at 6 months after the last brain trauma. We were interested in the level of neurodegeneration and glial activity in the OT and the mentioned nuclei, the presence of axonal injury and demyelination in the OT, and the possible synaptic changes included in visual information processing from retinal ganglion cells toward the nuclei. Chronic pathological changes in the OT, the LGN, and the SC of transgenic TDP-43 animals have not yet been studied. Furthermore, to the best of our knowledge, the preservation, glial responses, and synaptic organization in the mentioned nuclei of the visual pathway in wild-type mice in a model characterized with unconstrained head and body movements following the brain traumas have not been previously investigated. Moreover, because of different visual impairments detected in patients after mTBI [29], spatial learning and memory testing, which require preserved visual information processing, was conducted in mice of both genotypes. Such behavior in TDP-43 transgenic mice after repetitive mTBI has not been previously examined. Therefore, considering the results of our previous study and the fact that behavioral and the pathological brain changes are present for months after the initial injuries [12,27,30–33], this research hypothesized that the damage, gliosis, and synaptic reorganization in the OT and the investigated nuclei, as well as behavioral impairments induced by repeated mTBI, would be detected 6 months after the last injury and that they would be more pronounced in TDP-43^{G348C} mice.

2. Results

2.1. Repetitive mTBI Induced Neurodegeneration, Axonal Injury, and Gliosis in the Optic Tract in Wild-Type and TDP-43^{G348C} Mice at 6 Months Following the Last Head Impact

In the first part of the study, we investigated whether repetitive mTBI causes the OT pathology 6 months after the last hit in wild-type and TDP-43 transgenic mice. We were especially focused on neuronal and axonal degeneration, demyelination, and glial activity in the mentioned brain structure after repeated brain impacts in the animals of both genotypes.

Fluoro-Jade C was used as the marker of neurodegeneration [34,35]. This dye was found to stain degenerating nerve cell bodies and distal dendrites, axons, and terminals [36]. Figure 1A shows representative microphotographs of Fluoro-Jade C-stained sections of the OT in the sham animals and the traumatized mice of both genotypes. It is evident that Fluoro-Jade C-positive staining was detectable in the OT of the traumatized mice, both wild-type and TDP-43 transgenic, but not in the sham animals. Quantitative analysis of Fluoro-Jade C staining intensity confirmed these observations (Figure 1B). It was demonstrated that Fluoro-Jade C intensity levels were significantly higher in traumatized wild-type and TDP-43^{G348C} animals than in the sham animals of the corresponding control groups ($p = 0.012$; $p = 0.012$). No significant differences were observed between the levels of Fluoro-Jade C-positive staining in the OT of wild-type and transgenic TDP-43 injured mice ($p = 0.676$) (Figure 1B).

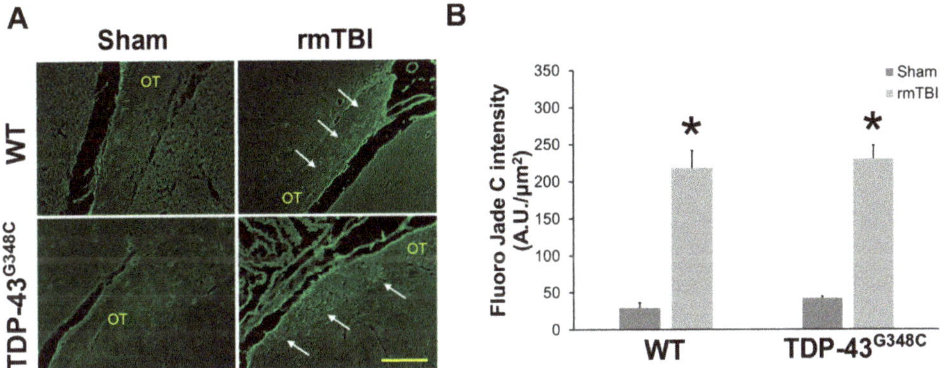

Figure 1. Neurodegeneration in the optic tract (OT) of wild-type (WT) and TDP-43^{G348C} mice at 6 months after repetitive mild traumatic brain injury (rmTBI). (**A**) Representative microphotographs of the OT stained with the Fluoro-Jade C fluorescent dye. Arrows point to Fluoro-Jade C-positive staining. Scale bar: 100 µm. (**B**) The histogram shows the intensity levels of the fluorescent staining (AU/µm^2) in the OT of WT and TDP-43^{G348C} mice of the control groups (Sham) and animals with rmTBI. Results are expressed as means ± SEM ($N = 5$). * $p < 0.05$, significantly different from the related Sham.

Silver staining is a method that has been used for visualization and localization of degenerating axons [37]. Representative microphotographs of silver-stained sections of the OT in mice of all experimental groups are shown in Figure 2. There was an increased silver uptake and staining in the OT of traumatized mice of both genotypes demonstrating evidence of axonal abnormalities compared with the sham-treated animals (Figure 2A). Furthermore, spheroids, a sign of axonal swelling, were observed in the axons of the OT in the injured wild-type and TDP-43 transgenic mice (Figure 2B). Results shown in Figure 2 suggest axonal injury in the OT at 6 months following repetitive mTBI in mice of both genotypes.

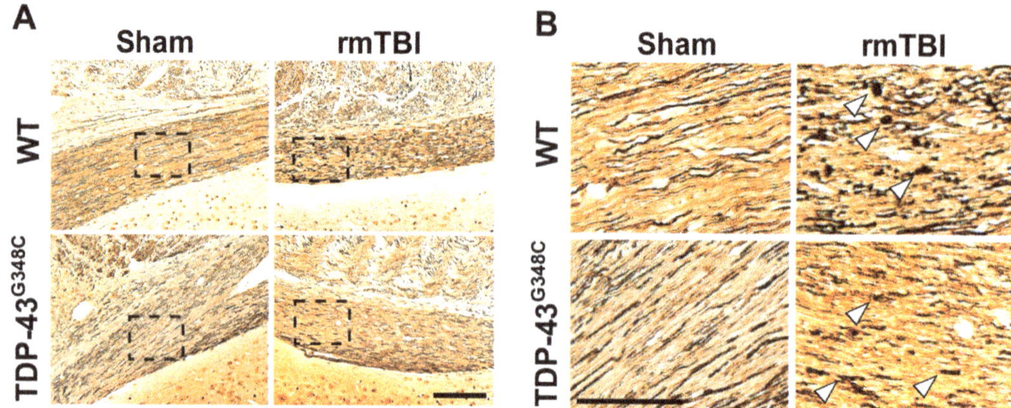

Figure 2. Axonal degeneration in the optic tract (OT) of wild-type (WT) and TDP-43^{G348C} mice at 6 months after repetitive mild traumatic brain injury (rmTBI). (**A**) Representative microphotographs of the OT stained with the silver staining in the mice of the control group (Sham) and the animals with rmTBI. Scale bar: 200 μm. (**B**) Microphotographs are higher-magnification images of the areas in the boxes of the corresponding panels. Arrowheads point to the spheroids of degenerating axons. Scale bar: 50 μm.

In order to analyze the integrity of myelinated neuronal fibers in the OT, we used conventional histological methods, i.e., staining with luxol fast blue (LFB) and immunohistochemical staining with anti-myelin basic protein (MBP) antibody.

Representative photomicrographs of the OT sections stained with LFB in the traumatized wild-type and TDP-43^{G348C} mice, as well as in the sham animals, are shown in Figure 3A. It is evident that myelin density and integrity of myelinated fibers were approximately equal in sham-treated wild-type and transgenic TDP-43 animals, while reduced myelin in some parts of the OT, characterized by porous and weaker LFB staining, was detectable in traumatized mice of both genotypes (Figure 3A). Quantitative analysis demonstrated that the staining densities in the OT of the injured wild-type and TDP-43^{G348C} mice were slightly decreased in comparison to the levels in the related sham animals, but a statistically significant difference was not detected (p = 0.097). In addition, a significant difference in the levels of the LFB staining between the traumatized wild-type and TDP-43^{G348C} transgenic mice was also not revealed (Figure 3B).

Representative photomicrographs of the coronal OT sections that were stained with anti-MBP antibody and the quantitative analysis of the MBP optical density for the mice of all experimental groups are shown in Figure 3C,D. There were no differences in the MBP immunoreactivity (Figure 3C) and optical density (Figure 3D) between the traumatized groups and their related control groups for both genotypes or between the injured wild-type and TDP-43 transgenic animals (p = 0.234) at 6 months after the last head trauma. The results obtained by LFB and MBP staining suggest that the myelination of surviving axons in the OT was mostly preserved at the investigated time point after the last brain trauma.

The integrity of surviving axons of retinal ganglion cells following repetitive mTBI was also examined using immunohistochemistry for neurofilament light (NfL) chain that is a major constituent of the neuronal cytoskeleton. Approximately equal NfL-positive staining was revealed in the axons of the OT in the traumatized and the sham mice of both genotypes (Figure 4A), while a significant difference in the NfL optical densities between the experimental groups was not revealed (p = 0.254) (Figure 4B). These results suggest that the NfL expression in the axons of the OT was unchanged at 6 months following repetitive mTBI.

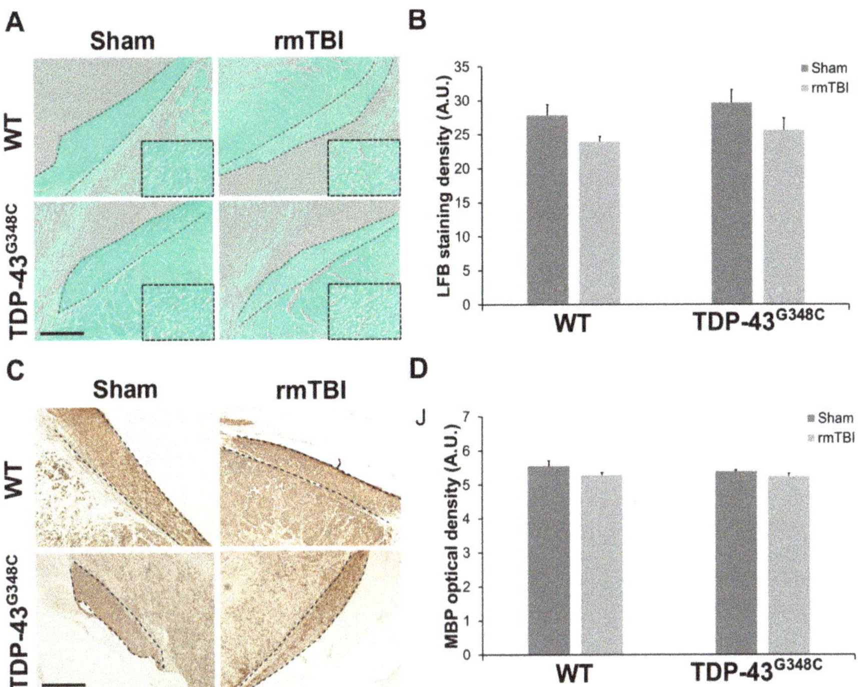

Figure 3. The integrity of myelinated neuronal fibers in the optic tract (OT) of wild-type (WT) and TDP-43^{G348C} mice at 6 months after repetitive mild traumatic brain injury (rmTBI). (**A**) Representative microphotographs of the OT stained with luxol fast blue (LFB). Higher magnification of the boxed regions reveals an area with reduced myelin and characterized by porous and weaker LFB staining. Dashed lines indicate the OT. Scale bar: 200 μm. (**B**) The histogram shows the LFB staining density (AU) in WT and TDP-43^{G348C} mice with rmTBI and related control groups (Sham). Results are expressed as means ± SEM (*N* = 4–6). (**C**) Representative microphotographs of the OT sections immunostained with anti-myelin basic protein (MBP). Dashed lines indicate the OT. Scale bar: 200 μm. (**D**) The histogram shows the MBP optical density (AU) in WT and TDP-43^{G348C} mice with rmTBI and related control groups (Sham). Results are expressed as means ± SEM (*N* = 4–5).

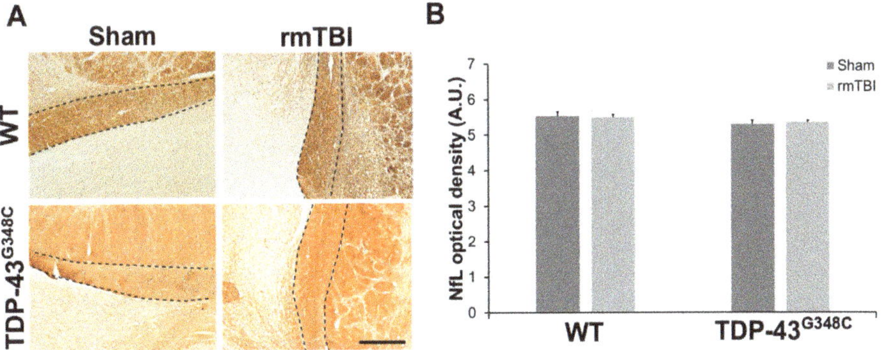

Figure 4. Neurofilament light chain (NfL) expression in the optic tract (OT) of wild-type (WT) and TDP-43^{G348C} mice at 6 months after repetitive mild traumatic brain injury (rmTBI). (**A**) Representative microphotographs of the OT stained with anti-neurofilament light chain protein. Dashed lines indicate the OT. Scale bar: 200 μm (**B**) The histogram shows NfL optical density (AU) in the axons of the OT in WT and TDP-43^{G348C} mice with rmTBI and related control groups (Sham). Results are expressed as means ± SEM (*N* = 3–5).

The activity of the glial cells in the OT was examined using microglial marker ionized calcium-binding adaptor molecule 1 (Iba1) and the astrocytic marker glial fibrillary acidic protein (GFAP) (Figure 5).

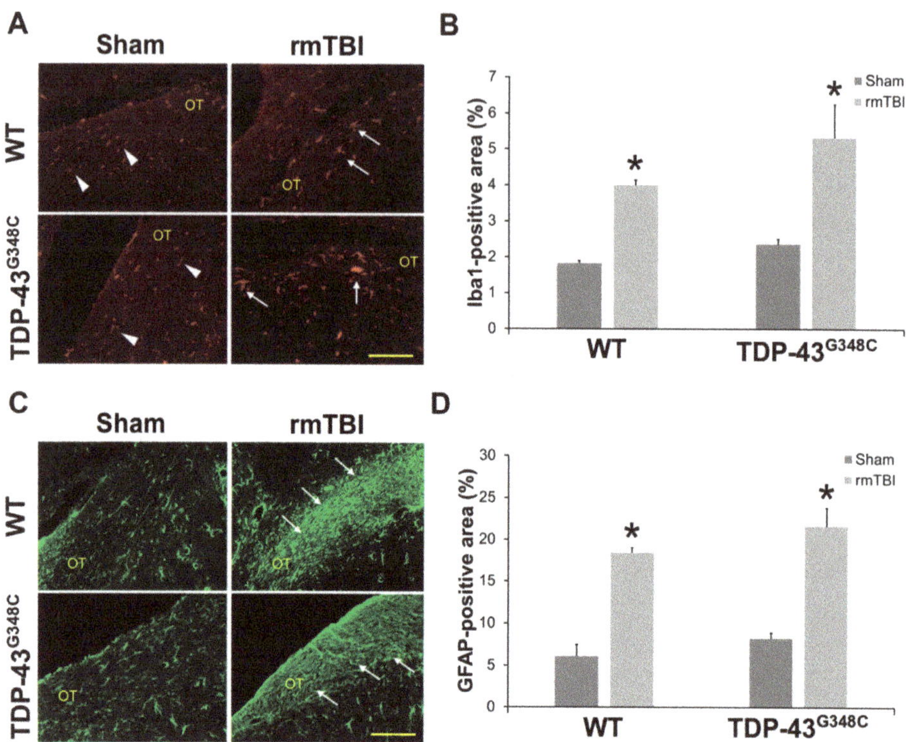

Figure 5. Gliosis at 6 months after repetitive mild traumatic brain injury (rmTBI) in the optic tract (OT) of wild-type (WT) and TDP-43^{G348C} mice. (**A**) Representative microphotographs of the OT immunostained with anti-ionized calcium-binding adaptor molecule 1 (Iba1). Arrowheads point to Iba1-positive cells with "resting" microglial morphology, and arrows point to Iba1-positive cells with activated microglial morphology. Scale bar: 100 µm. (**B**) The histogram shows the Iba1-positive area (%) in the OT of WT and TDP-43^{G348C} mice with rmTBI and related control groups (Sham). Results are expressed as means ± SEM ($N = 5$). * $p < 0.05$, significantly different from the related Sham. (**C**) Representative microphotographs of the OT immunostained with anti-glial fibrillary acidic protein (GFAP). Arrows point to the GFAP-positive cells with hypertrophic morphology. Scale bar: 100 µm. (**D**) The histogram shows the GFAP-positive area (%) in the OT of WT and TDP-43^{G348C} mice with rmTBI and related control groups (Sham). Results are expressed as means ± SEM ($N = 5$). * $p < 0.05$, significantly different from the related Sham.

Figure 5A shows representative microphotographs of the Iba1 immunostained OT sections in mice of all the experimental groups. "Resting" microglia, i.e., the microglial cells with thin Iba1-immunoreactive processes, were detected in the sham-treated mice of both control groups. Moreover, in all the traumatized animals, activated microglia, i.e., microglial cells with hypertrophic and large cell bodies, thick processes, and with amoeboid and migrating morphology, were noticed (Figure 5A). The higher magnifications images of "resting" and "activated" microglia are shown in Figure S1 (Supplementary Materials).

Quantitative analysis of the percentages of the Iba1 immunoreactive areas in the OT demonstrated statistically significant higher values in the injured wild-type and TDP-43^{G348C} animals compared with related sham mice ($p = 0.012$; $p = 0.012$) (Figure 5B). However, a statistically significant difference in the percentages of the Iba1 immunoreactive areas between the traumatized wild-type and TDP-43 transgenic mice was not found (p = 0.296) (Figure 5B).

Morphological changes that would suggest astroglial activation were not detected in the OT of the sham injured wild-type and TDP-43^{G348C} mice at 6 months after the last head trauma (Figure 5C). Contrarily, the GFAP immunoreactivities were more pronounced in the OT of traumatized wild-type and TDP-43 transgenic mice than in sham-treated animals, suggesting astrocytic hypertrophy after repetitive mTBI (Figure 5C). Furthermore, the percentages of the OT areas covered by hypertrophic astrocytes in the traumatized animals of both genotypes were significantly higher compared with the values noticed in the non-injured mice (p = 0.012; p = 0.012) (Figure 5D). In addition, the values of GFAP-positive areas in the injured TDP-43^{G348C} animals did not significantly differ from the values in wild-type mice (p = 0.210) (Figure 5D).

2.2. *Repetitive mTBI Did Not Cause Neurodegeneration, Glial Activation, and Synaptic Reorganization in the Lateral Geniculate Nucleus and the Superior Colliculus in Wild-Type and TDP-43^{G348C} Mice at 6 Months Following the Last Head Impact*

The presence of neurodegenerative changes in the LGN and the SC, the regions that receive direct innervation from the retinal ganglion cells via the OT, were analyzed using Fluoro-Jade C and cresyl-violet staining. From the representative microphotographs of Fluoro-Jade C-stained sections of the LGN, shown in Figure S2A (Supplementary Materials), it is evident that the staining used was not detected in any of the experimental animals, suggesting no neurodegeneration in this structure at 6 months after the last mTBI. Moreover, individual Fluoro-Jade C-positive staining was detectable in the superficial SC of the injured mice of both genotypes (Figure S2D, Supplementary Materials). Quantitative analysis demonstrated a slight increase in the Fluoro-Jade C intensity levels in the SC of the traumatized wild-type and TDP-43^{G348C} animals compared with sham control animals. However, a statistically significant difference between the experimental groups was not revealed (p = 0.235) (Figure S2E, Supplementary Materials).

Additionally, cresyl-violet staining revealed approximately equal neuronal cell density in the LGN (Figure S2B, Supplementary Materials) and the SC (Figure S2F, Supplementary Materials) of all the experimental groups of mice, which was confirmed by subsequent quantitative analysis (p = 0.113, p = 0.617) (Figure S2C,G, Supplementary Materials).

Repetitive mTBI did not cause changes in the activity of the glial cells in the LGN and the SC of wild-type and TDP-43 transgenic mice in our experimental conditions (Figure S3, Supplementary Materials). Specifically, in the investigated nuclei of the traumatized and sham-treated mice of both genotypes, the "resting" but not activated microglia was detected (Figure S3A,D, Supplementary Materials). Furthermore, a statistically significant difference in the number of Iba1-positive cells between the groups was not revealed for the LGN (p = 0.200) or for the SC (p = 0.446). Moreover, the signs of astrocytosis in the examined nuclei were not detected in any of the experimental animals (Figure S3C,F, Supplementary Materials).

To detect if repetitive mTBI affects synaptic density in wild-type and TDP-43^{G348C} animals, anti-synaptophysin (SYP) immunostaining was performed. Although it seemed to be more pronounced in the LGN of the sham and injured TDP-43^{G348C} animals compared with wild-type mice (Figure S4A, Supplementary Materials), a significant difference between the experimental groups in the SYP staining intensities was not obtained (p = 0.069) (Figure S4B, Supplementary Materials). Moreover, significant differences in the SYP expression (Figure S4C, Supplementary Materials) and density intensities (Figure S4D, Supplementary Materials) between the groups were not observed in the SC (p = 0.100).

2.3. *Repetitive mTBI Did Not Affect the Barnes Maze Task in Wild-Type and TDP-43^{G348C} Mice at 6 Months Following Repetitive mTBI*

To explore the long-term effects of repetitive mTBI on the function of the visual system, we trained experimental animals in the Barnes maze. Figure 6 shows the latency time to reach the target hole and the time spent inside the target quadrant during the tests and the retests for all the experimental groups of mice. There was no significant difference in the latency time to reach the target hole between the experimental groups of mice during the tests ($p = 0.148$) and the retests ($p = 0.763$) (Figure 6A,B). Additionally, no statistically significant differences in the time spent in the target quadrant between the experimental animals were obtained for the tests ($p = 0.527$) and the retests ($p = 0.381$) (Figure 6C,D).

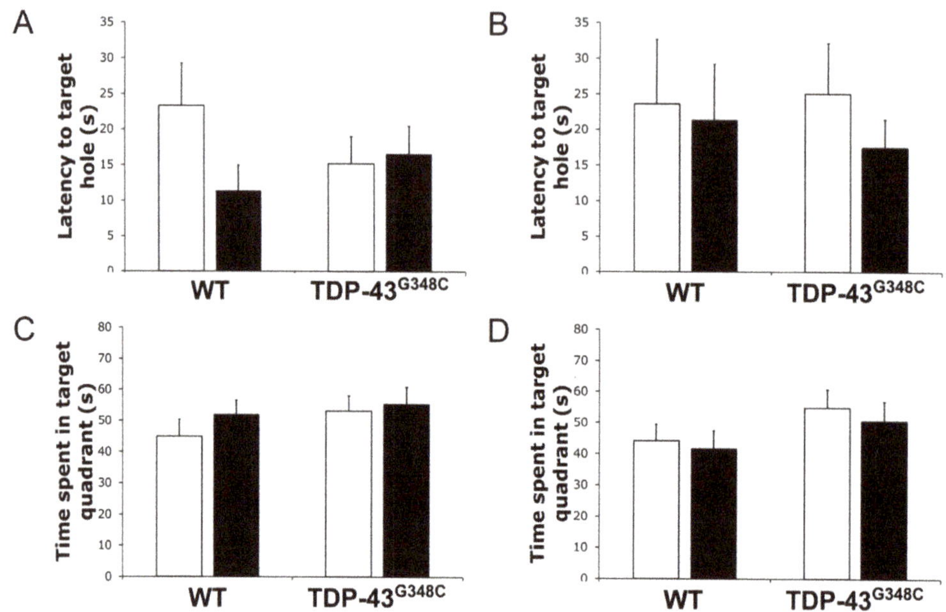

Figure 6. Barnes maze task performance in wild-type (WT) and TDP-43^{G348C} mice at 6 months after repetitive mild traumatic brain injury (rmTBI). The histogram shows the latency time (s) to reach the target hole during (**A**) the tests and (**B**) the retests or the time (s) spent in the target quadrant during (**C**) the tests and (**D**) the retests for mice with rmTBI (■) and related control groups (□). Results are expressed as means ± SEM (N = 12–14).

3. Discussion

Different visual problems, such as blurred vision, visual acuity loss, and visual field defects, have been reported in patients with previous history of moderate to severe TBI [38], but much less is known about these impairments and pathological changes in the visual system following mild and especially repetitive mTBI [39]. Studies exploring neuropathological changes in the visual system in animal models of repeated head traumas are scant [27,40–43]. The aims of this study were focused on the chronic effects of repetitive mTBI on some parts of the visual system in wild-type and TDP-43^{G348C} mice. As in our previous study [27], we used the abovementioned mice with overexpression of human familial ALS-linked mutant TDP-43 protein and a predisposition to the pathological accumulation of its aggregates, cytotoxic cleavage fragments, axonopathy, and neuroinflammation in the brain and the spinal cord [28]. The mentioned pathological and biochemical changes are age-related [28]. Thus, the TDP-43 ^{G348C} mice did not develop TDP-43-positive aggregates at an early age of 9–11 weeks, as they were at the beginning of the study. The TDP-43-positive aggregates can be detected in these mice starting at 10 months of age [28]. The rationale behind using the young TDP-43 transgenic mice was to explore whether a subtle (2.3–3-fold) overexpression of TDP-43 represents an additional risk factor in the context of mild repetitive TBI, as well as to observe whether it would predispose and/or trigger more intense neurodegeneration, as observed in the model of stroke [44]. Furthermore, there is growing evidence that exposure to repetitive mTBI is associated with an increased risk for ALS/FTLD, especially in a subset of vulnerable individuals with genetic predisposition [12,45–48]. Furthermore, since the results of our previous study suggested that genetically acquired TDP-43 dysregulation might predispose the OT to more intense acute and subacute damage following repetitive mTBI [27], we wanted to further explore whether TDP-43 proteinopathy is associated with marked long-term posttraumatic changes in the mentioned brain structure, as well as in the investigated nuclei of the visual pathway. We used a clinically relevant model of repetitive TBI that includes the elements of acceleration/deceleration and rotational injuries of the freely moving mouse head and body resulting in diffuse brain damage. This method produced mild injuries with no skull fractures, intracranial bleeding, respiratory arrest, or seizures, and the mice quickly recovered and demonstrated normal behavior following head impacts.

3.1. Repetitive mTBI Induced Neurodegeneration, Axonal Injury, and Gliosis in the Optic Tract in Wild-Type and TDP-43^{G348C} Mice at 6 Months Following the Last Head Impact

In our previous research, we demonstrated that the OT was the only damaged brain structure in the injured wild-type and TDP-43 transgenic mice already on the first day following the last head impact, lasting up to the end of the first week, suggesting early vulnerability of this structure to the investigated type of injury [27]. In the current study, we histologically detected neurodegeneration using Fluoro-Jade C staining and degenerating, argyrophylic, and swollen axons using neurosilver staining in the OT of the traumatized animals of both genotypes at 6 months after repetitive mTBI, indicating chronic axonal posttraumatic damage. Our results are in agreement with some previous studies in which the destruction of this structure in rodents was reported in different models by using various injury paradigms and time points of 1, 3, 7, or 60 days, 3 or 10 weeks, and 8 or 12 months following repeated mTBI [27,31,33,42,49–52]. Taking all these results together, including the fact that repetitive mTBI increases the sensitivity of the brain to each subsequent trauma, it is plausible that the time for the recovery between individual impacts was insufficient in the experimental models and protocols used in the mentioned investigations, resulting in neurodegeneration and axonal degeneration months after the final injury. Increased sensitivity and vulnerability of the OT to the damage induced by repeated mTBI may be a consequence of its position below the brain, as well as of its anatomic characteristics. Specifically, it consists of very long myelinated axons which are susceptible to compression in the optic canal during the direct injury, as well as to tension and torsion during acceleration and deceleration forces caused by the hits [53]. Moreover, the blood supply of the optic nerve arrives from pial arteries. Its swelling, induced by repeated head traumas, may cause localized ischemic injuries that may additionally contribute to the OT damage [31]. Moreover, neurodegenerative changes in the proximal part of the visual system, including the optic nerves and the optic chiasma, and decreased cellularity in the ganglion cell layer of the retina were previously described in mice at different time points following repetitive mTBI [40–42,50].

The current study is the first in which the neural and axonal degeneration in the OT of TDP-43 transgenic mice was investigated at a chronic time point following repetitive mTBI. Neurodegeneration was found in the injured TDP-43^{G348C} animals compared with the related sham mice. Contrarily, significant differences in Fluoro-Jade C intensity were not observed between traumatized transgenic TDP-43 mice compared with wild-type animals, suggesting that human genetic TDP-43 background did not affect chronic damage of this structure.

In order to detect other chronic effects of repeated head impacts on the surviving axons of the OT, we investigated the levels of their myelination and the expression of the cytoskeleton NfL protein. Myelin preservation was determined by LFB and MBP stains and their quantification. Reduced LFB staining was evident in some parts of the OT in the injured mice of both genotypes. Furthermore, there was no significant difference in the staining density between the traumatized wild-type and transgenic TDP-43 mice compared with their related sham or between injured wild-type and TDP-43^{G348C} animals in this brain structure at 6 months after head traumas. Previously, no changes in LFB-positive staining were observed in the brains of mice at 6 months following the first injury in the model of repetitive mTBI induced by electromagnetic controlled impact device [32], in which local areas of reduced myelination were described in the optic nerve at 3 and 13 weeks after the last head trauma [40,41]. To our knowledge, the level of myelination was not previously investigated in the OT in the models of repeated mTBI in TDP-43 transgenic animals.

In our experiments, no changes were observed in the MBP immunoreactivity and optical density in the OT of the injured wild-type and TDP-43 transgenic mice compared with related controls or in traumatized wild-type compared with TDP-43^{G348C} animals. Taken together, the results of this study obtained by the LFB and MBP staining demonstrated that repetitive mTBI did not significantly affect myelination of the surviving axons of the retinal ganglion cells at 6 months after the final head impact and that the transgenic genotype did not influence it. In addition, Gangolli et al. [54] did not detect Myelin Black Gold staining in the OT 1 year following injury induced by CHIMERA in mice. In the same experimental model, MBP immunoreactivity was not altered in the OT 7 days after the final injury [51].

NfL chain is abundantly expressed in the long and large-caliber myelinated white-matter axons, and it is considered a promising candidate biomarker of axonal injury in different diseases of the central nervous system [55], including repetitive mTBI [56–58]. Increased exosomal and plasma levels of NfL chain have been detected in humans even years following repeated head traumas, suggesting chronic, long-term axonal dysregulation and degeneration induced by sustained brain injuries [57].

To our knowledge, our study is the first in which NfL chain staining was investigated in the OT at a later time point following repetitive mTBI. We did not detect statistically significant differences in the levels of this protein's optical densities between the injured wild-type group and related sham 6 months after the final head trauma. Similar results were obtained by Cheng et al. [33] and Vonder Haar et al. [52], who also reported no differences in the Nf medium and heavy chains or the NfL chain staining in the OT between traumatized and sham mice at chronic post-injury time points in the CHIMERA model. Taking into account the results of our and other mentioned animal studies, it can be suggested that the changes in the NfL chain protein were not evident at later time points following repetitive mTBI. This could be due to previous death of the affected neurons and preserved cytoskeletal NfL architecture in surviving axons. Contrarily, axonal swellings and varicosities in the OT of the traumatized mice were observed 2 days after the injury induced by the CHIMERA method, but they disappeared by the seventh day following the last impact [59]. Taking all abovementioned results regarding the changes of the NfL following repetitive mTBI, it seems that this protein can be used as a brain marker of early axonal damage in animal models, in contrast to the human studies in which it has been detected in the blood 1 h to years after repeated head traumas [56–58].

Among other roles, under physiological conditions, TDP-43 binds and stabilizes NfL mRNA, regulating its transcription, metabolism, and axonal transport [60–62]. Contrarily, TDP-43 dysregulation, observed, e.g., in FTLD, is associated with NfL alterations and white-matter pathology [63]. In a recent study, Kumar et al. [64] found that cytoplasmic TDP-43 accumulation in mice expressing ALS-linked human TDP-43^{A315T} mutant caused marked suppression of mRNA translation for NfL, Nf medium, and α-internexin, resulting in a decrease in the levels of these proteins at 12 months of their age when they exhibited TDP-43 proteinopathy in cortical neurons. In our research, we were interested if repetitive mTBI affects NfL chain staining intensity in the OT of mice presenting with cytoplasmic TDP-43 aggregates in the spinal cord starting at approximately 10 months of age and increased pathological TDP fragment in the brain and spinal cord at 10 months of age [28]. We did not detect any significant changes in NfL chain staining in the OT of injured TDP-43^{G348C} mice compared to the related sham group or between traumatized wild-type and TDP-43 transgenic animals at 6 months following repeated head injury, suggesting that repetitive mTBI and the investigated genotype did not affect the structure and, consequently, function of this neuronal cytoskeletal protein in the used experimental conditions. Because the age of our experimental animals was approximately 8.5 months at the time of the experiments, it remains to be investigated whether there are age-dependent changes in NfL chain expression in mice of the tested TDP-43 genotype.

Neuroinflammation is one of the most important processes developing after head trauma that may have beneficial or detrimental effects in the acute TBI [65]; however, if it is chronic, it usually contributes to the brain damage [66–68]. In the current study, pronounced microglial and astrocytic response to the repetitive mTBI was found in the OT of the injured wild-type and TDP-43 transgenic mice related to their sham groups, suggesting chronic neuroinflammation of this structure as a result of synergistic exacerbating effects of repeated head traumas that provoked an increase in inflammatory responses during short time periods between each injury. In our previous study, using the same closed head weight drop method, significant microgliosis and astrocytosis were also demonstrated in traumatized wild-type and TDP-43^{G348C} mice, in the acute and subacute posttraumatic periods [27]. Our previous and current results regarding gliosis in the OT following repetitive mTBI are in agreement with those obtained in animal studies in which the rodents were subjected to impacts induced by other methods and in which various head trauma protocols and different posttraumatic time points from 1 to 365 days after the injury were used [31,33,43,49–52,59,69,70]. Our study found no differences in the microglial and astrocytic hyperactivities between injured transgenic TDP-43 and wild-type mice, suggesting that the investigated genetic background did not affect inflammatory parameters used in this research. Taking all the abovementioned results together, it may be suggested that the OT is particularly susceptible and vulnerable to neuroinflammation induced by repetitive mTBI.

3.2. Repetitive mTBI Did Not Cause Neurodegeneration, Changes in the Responses of Glial Cells, and Synaptic Reorganization in the Lateral Geniculate Nucleus and the Superior Colliculus in Wild-Type and TDP-43^{G348C} Mice at 6 Months Following the Last Head Impact

To the best of our knowledge, we are the first group to explore possible chronic damage, glial activity, and synaptic organization in the LGN and the SC following repetitive mTBI in mice of both genotypes.

The LGN, situated in the thalamus, was reported to receive external visual information mostly by the axons of the retinal ganglion cells, conducting them to the visual cortex [71]. The SC has a laminar structure, and its three superficial layers are primarily visual sensory in nature [72]. In mice, the SC also receives the projections, but from at least 70% [73] and possibly even approximately 88% of the retinal ganglion cells [74].

In the current study, no signs of neurodegeneration in the LGN and only a few scattered Fluoro-Jade C-positive signals in the superficial SC were detected in the injured mice of both genotypes. Additionally, when using cresyl-violet, no differences were revealed in the neuronal density between the injured wild-type and TDP-43 transgenic mice compared with their related sham or between traumatized wild-type and TDP-43^{G348C} mice. Considering the abovementioned results, we suggest that repetitive mTBI did not induce chronic damage of the target nuclei and that the investigated genotype did not influence it. Furthermore, reactive microgliosis or astrocytosis were not detected. Previously, microglial infiltration and activation in the SC were found 7 days after the last head trauma [50]. In the same research, the injury of the LGN was mentioned [50]. Both results were obtained using a method of repeated head traumas different from ours.

The central nervous system has the capacity of neuroplasticity and recovery following different insults that include various processes, ranging from molecular, cellular, and synaptic to global [75]. Some of these processes are associated with changes in the expression of different specific proteins, synaptogenesis markers, and synapse remodeling, such as SYP [76–79]. In our study, no differences in the SYP immunostaining intensities between the experimental groups were detected, suggesting that repetitive mTBI or the genotype did not induce synaptic perturbations in the LGN or the SC at 6 months following repetitive mTBI. To our knowledge, synaptogenesis markers have not been previously studied in the LGN and the SC following repeated head traumas.

3.3. Repetitive mTBI Did Not Affect Barnes Maze Task in Wild-Type and TDP-43^{G348C} Mice at 6 Months Following Repetitive mTBI

In the current research, we evaluated the effects of repetitive mTBI on the possibility of successful performance in the Barnes maze task. The Barnes maze task is usually used as a spatial learning/memory test in rodents taking advantage of their innate behavior to run away from brightly illuminated to dark areas [80]; however, in our study, it was applied primarily to test the functional status of the visual system following repeated head traumas. No differences in the time to escape to the target hole or in the time spent in the target quadrant were observed between the experimental groups, suggesting that neuronal damage, axonal damage, and neuroinflammation detected in the OT at 6 months after the last head trauma, as well as genetic TDP-43 background, did not influence the performance in the Barnes maze task. In addition, the Barnes maze task, as a cognitive test, was performed in some other studies in rodents of different ages and genotypes in which various repetitive mTBI methods, severity of the injuries, protocols, tested parameters, and posttraumatic periods were used, which is the reason why the obtained results are not consistent and hardly reciprocally comparable [33,59,69,81–83]. Previously, it was shown that TDP-43^{G348C} animals exhibited a significant reduction in the time spent in the target quadrant and increased primary errors in the Barnes maze test as compared with age-matched wild-type mice at 10 months of age [28]. In our study, all mice, including these transgenic traumatized and control animals, performed the task equally at approximately 8.5 months of age. Further studies are needed in order to explore age-dependent behavior of TDP-43^{G348C} mice in the Barnes maze task.

In summary, current study results suggest that repetitive mTBI induced damage of the OT, but did not affect the nuclei that transmit information from the retinal ganglion cells to the visual cortex, as well as mouse behavior that includes preserved vision, at 6 months after the last head impact. In addition, genetic TDP-43 background did not influence the assessed neuropathology and behavior in experimental animals. This study could improve our knowledge and understanding of chronic neuropathological changes in the visual system following repetitive mTBI and the role of TDP-43 proteinopathy in these processes.

4. Materials and Methods

4.1. Animals and Treatment

This study was performed on wild-type C57BL/6J and transgenic TDP-43^{G348C} male mice of C57BL/6J background. At the beginning of the experiments, the mice were 9–11 weeks old. Transgenic TDP-43^{G348C} mice were obtained from the University Laval, Quebec, Canada, and the colony was raised in the Laboratory for Mice Breeding and Engineering Rijeka, Faculty of Medicine Rijeka, University of Rijeka, Croatia. All the experimental procedures were performed according to the Faculty's Ethical Committee approval and in accordance with the Croatian laws and rules (NN 135/06; NN 37/13; NN 125/13; NN 39/17), as well as the guidelines of the European Community Council Directive (86/609/EEC). Mice were maintained in the animal facility of the Faculty's Department of Basic and Clinical Pharmacology and Toxicology in temperature- and humidity-controlled holding rooms, with an alternating 12 hour light/dark cycle. Fresh water and standard rodent chew were available to animals ad libitum.

Mild brain traumas were induced using the closed head weight drop method previously described by Kane et al. [84]. In brief, mice were anesthetized with 3.5% isoflurane in a nitrous oxide/oxygen (2:1) mixture in an induction chamber and rapidly positioned on aluminum foil placed over a Plexiglas box, lined with a sponge. The box was situated beneath the vertical metal tube of the apparatus. A steel weight (1.2 cm diameter, mass 97 g), set above the mouse head and between the ears, was pulled rapidly upward to 1 m height and released. Following the impact, the mice fell down through the foil and onto the surface of the sponge, all while rotating their bodies by 180° horizontally. After each mild brain trauma, the mice were returned to their holding cages to recover. In our experiments, mice rapidly recovered and showed normal interactions with other animals without demonstrating signs of pain or indisposition nor resistance to manipulations after mTBI. We did not observe respiratory arrest or seizures in any of the tested animals. For the control group, sham treated animals were only briefly anesthetized without receiving impacts. Sham procedures or mild brain traumas were repeated twice daily, in intervals of 6 h, for five consecutive days, i.e., a total of 10 impacts. Mice were euthanized at 6 months after the final impact or sham procedure.

4.2. Polymerase Chain Reaction

In the transgenic animals, the presence of TDP-43^{G348C} transgene was determined by polymerase chain reaction (PCR) as previously described [28]. GoTaq® G2 Green Master Mix (Promega Corporation, Madison, WI, USA) and the primers CTCTTTGTGGAGAGGAC and TTATTACCCGATGGGCA (Metabion international AG, Planegg, Germany) were used for the reaction.

4.3. Tissue Preparation

For histochemistry purposes, the animals were anesthetized with xylazine and ketamine mixture and transcardially perfused, first with phosphate-buffered saline (PBS) and then prefixed with 4% paraformaldehyde in PBS. Their brains were dissected and post-fixed for 20–22 h in the same fixative solution at room temperature and then embedded in paraffin. Brain sections were cut to 3 µm thickness. For the analyses of the OTs, coronal sections ranging from bregma +1.18 to −2.30 [85] were used. To detect changes in the LGN, we analyzed coronal sections cut at approximately −2.46 from bregma, and the SC nuclei were investigated at around −3.52 from bregma [85].

4.4. Fluoro-Jade C Staining

The slides were deparaffinized in xylene, rehydrated in ethanol and water, and then treated for 10 min with a 0.06% potassium permanganate solution. Sections were rinsed twice with distilled water (dH_2O) for 1 min and incubated in 0.0001% Fluoro-Jade C (Chemicon, Millipore, Billerica, MA, USA) staining solution for 20 min in the dark. After that, they were washed in dH_2O thrice per minute and dried on a hot plate on 50 °C for 20 min. Sections were dehydrated in xylene two times for 10 min, mounted in *Entellan*® (Merck Millipore, Billerica, MA, USA), and coverslipped. Stained sections were examined by epifluorescence microscopy using the appropriate light filter cube (Olympus BX 51 microscope with Olympus DP 70 digital camera, Olympus, Tokyo, Japan).

Quantification of Fluoro-Jade C intensity in the OT was done on microphotographs taken at ×400 final magnification; for each animal, two images were used for the analyses. Within each microphotograph, three ROIs of 0.0057 mm^2 were analyzed. By subtracting the background fluorescent intensity from those ROIs, we could determine only degenerating axons within that field.

4.5. Bielschowsky Silver Staining

Bielschowsky silver staining is a method that can be used to detect nerve fibers and stain axons, neurofibrils, and senile plaques in the central nervous system. Brain sections were deparaffinized, dehydrated, and then immersed for 15 min in the solution with 20% silver nitrate, preheated at 37 °C. Sections were then washed in distilled water, after which they were submerged in silver ammonia solution for 15 min, all at 37 °C. Next, slides were developed by placing them in 50 mL of distilled water with eight drops of both ammonium hydroxide and developer stock (8% *v*/*v* formaldehyde, 0.5% *w*/*v* citric acid, 0.1% *v*/*v* nitric acid) for 2 min maximum, i.e., until their color changed to the desired intensity of brown. Following the washing step in distilled water, sections were immersed in 5% sodium thiosulfate for 5 min at room temperature, washed with tap water for 5 min, dehydrated, cleared, and mounted with Entellan®.

Microphotographs of the OTs stained with Bielschowsky silver stain were taken at ×400 magnification using an Olympus BX 51 microscope with an Olympus DP 70 digital camera (Olympus, Tokyo, Japan).

4.6. Luxol Fast Blue Staining

LFB staining of the OT was performed to determine the degree of myelination. Following deparaffinization and rehydration, brain sections were stained with LFB solution at 56 °C overnight. The next day, slides were rinsed with 95% ethanol and distilled water, after which they were differentiated in the lithium carbonate solution for 10–15 s, and then immersed briefly in the 70% ethanol three times. Following that, slides were washed in distilled water, dehydrated through rising ethanol concentrations, cleared in xylene, and mounted with Entellan®.

LFB-stained sections were photographed at 200× magnification using an Olympus BX 51 microscope with an Olympus DP 70 digital camera (Olympus, Japan).

Myelin densities on LFB stained photographs were quantified by using the ImageJ software (NIH, Bethesda, Md, USA), according to the protocol described by Underhill et al. [86] with modifications suggested by Khodanovich et al. [87]. Briefly, mean intensities of the red channel in a region of interest (ROI), i.e., in the OTs, were measured from RGB images as a quantity characterizing the complementary blue channel saturation. The background mean intensity of the red channel was also measured on each photograph in ROIs outside the brain tissue, which served for the calculation of the background correction factor. LFB optical density (in %) was calculated for each ROI according to the following formula: LFB density = 100 × (1 − (red channel intensity/background intensity)).

4.7. Immunofluorescence/Immunohistochemistry

To investigate the expression of the proteins of interest, we used immunofluorescent labeling in combination with DAPI nuclear counterstaining or immunohistological staining visualized with 3,3′-diaminobenzidine (DAB) chromogen (Dako).

After deparaffinization and rehydration of slides, antigen retrieval was achieved by heat-induced epitope retrieval procedure in the citric acid buffer (10 mM, pH 6.0). Nonspecific binding sites were blocked with Tris-buffered saline (TBS) containing 5% bovine serum albumin (BSA) and 0.025% Triton X-100. Slides were incubated overnight at 4 °C with primary antibodies, as listed in Table 1.

Table 1. List of antibodies used for the immunofluorescent (IF) and immunohistological (IHC) analyses.

Primary Antibody	Dilution	Manufacturer (Reference Number)
Rabbit anti-Iba1	1:1000 (IF)	Wako Chemicals, Richmond, VA, USA (019-19741)
Mouse anti-GFAP	1:200 (IF)	Cell Signaling Technology, Beverly, MA, USA (#3670)
Rabbit anti-MBP	1:2500 (IHC)	Abcam, Cambridge, UK (ab218011)
Rabbit anti-NfL	1:100 (IHC)	Cell Signaling Technology, Beverly, MA, USA (#2837)
Mouse anti-SYP	1:200 (IF)	Santa Cruz Biotechnology, Santa Cruz, CA, USA (sc-17750)
Secondary Antibody	**Dilution**	**Manufacturer (Reference Number)**
Goat anti-rabbit Alexa Fluor 594	1:200 (IF)	Abcam, Cambridge, UK (ab6901)
Rabbit anti-mouse Alexa Fluor 594	1:200 (IF)	Cell Signaling Technology (#4408)
Biotinylated goat anti-rabbit	1:200 (IHC)	Invitrogen, Carlsbad, CA, USA (65-6140)

Abbreviations: Iba1, ionized calcium-binding adapter molecule 1; GFAP, glial fibrillary acidic protein; MBP, myelin basic protein; NfL, neurofilament light chain; SYP, synaptophysin.

For the sections immunolabeled and visualized by using the DAB chromogen, the brain slices were incubated with an appropriate biotinylated secondary antibody (Table 1), diluted in the antibody solution buffer for 1 h at RT, followed by the streptavidin–HRP conjugate for 30 min at RT. Following the application of DAB, reaction with HRP produced a brown precipitate. The slides were then dehydrated, immersed in xylene, and mounted. For the immunofluorescence labeling, appropriate fluorochrome-conjugated secondary antibodies were applied for 1 h at RT. Cell nuclei were counterstained with DAPI, and the slides were mounted in anti-fade mounting medium.

Immunolabeled sections were examined by light or epifluorescence microscopy (Olympus BX 51 microscope with Olympus DP 70 digital camera, Olympus, Tokyo, Japan).

Quantification of neurodegeneration and the glial response in the OTs was done on Fluoro-Jade C-stained and Iba1- or GFAP-immunolabeled coronal sections cut in the range of −1.34 to−2.30 from bregma [85] using ImageJ software (NIH, Bethesda, Md, USA). Quantification of microgliosis and astrocytosis was made by measuring the percentage (%) of the Iba1- or GFAP-immunoreactive areas. Microphotographs of two sections from each animal, at ×400 magnification, were transformed to 8 bit images and auto-thresholded (0 being white and 255 being black), which enabled differentiating positive immunoreactions from the background and calculating the immunoreactive area fraction. A region of interest (ROI) was drawn around the OT. The area fractions were averaged for each animal and each experimental group.

In the OTs, quantification of the DAB staining intensity was also done with ImageJ software. Briefly, mean gray values were collected from the ROIs selected in the images of the OTs, and optical density (OD) was calculated with the following formula: OD = log(max gray intensity/mean gray intensity).

In the LGN and the SC, we evaluated the intensity of SYP immunofluorescent staining. Conditions of the microscopy and photography were maintained constant throughout the experiment, and immunoreactivity was quantified by measuring the integrated optical density. In the mentioned nuclei, we also evaluated the microglial response by counting the number of Iba1-stained cells in the immunofluorescently labeled sections.

4.8. Cresyl-Violet

Cresyl-violet (Nissl) staining was used to detect the effects of repetitive mTBI on the number of neurons in the investigated nuclei of the visual system. Deparaffinized and rehydrated slides were stained with 0.1% cresyl-violet acetate (Sigma Aldrich, St Louis, MO, USA) by incubation for 10 min at room temperature. Differentiation of the brain tissue sections was done by immersing the slides in 95% alcohol with glacial acetic acid. Finally, brain sections were dehydrated in alcohol, cleared in xylene, and mounted with Entellan®.

Microphotographs of the cresyl-violet stained sections were taken at approximately −2.46 from bregma for the LGN and at −3.52 from bregma for the SC [85], at ×400 magnification, using an Olympus BX 51 microscope equipped with an Olympus DP 70 digital camera (Olympus, Tokyo, Japan). With the help of the ImageJ software, neuronal density estimation in the investigated nuclei was carried out by a blind investigator using the random simple counting method. Two to four random ROIs were taken from at least two serial cuts of the selected areas and used to estimate the number of the neurons. Only cells with visible nuclei were counted.

4.9. Barnes Maze Task

To test the behavior of experimental animals following repetitive mTBI that requires preserved vision, the Barnes maze task was performed. A homemade maze was situated in an experimental room with distinct visual cues. The apparatus consisted of an elevated circular platform 92 cm in diameter with 20 escape holes, 5 cm in diameter, spaced evenly around the perimeter, and an escape box placed underneath target hole. The maze was divided into quadrants consisting of five holes each, and the target hole was located in the center of the target quadrant. Visual cues enabled mice proper space orientation to learn and reach the target hole and quadrant. On the habituation day, on posttraumatic day 167, the mouse was placed under the black start chamber in the center of the platform for 10 s. After that, the box was removed and the animal was trained to find and enter the escape box. The mouse was allowed for 1 min inside the escape box and then returned to the holding cage. After this procedure, the animal was subjected to four daily described training trials that lasted 3 min each, with an intertrial interval of 15 min, on days 167 to 170 after the last mTBI. The test, during which escape box was removed from the maze, was conducted 24 h after the last training day and lasted 1.5 min per mouse. Seven days after the test, mice were retested. Using video tracking software (ANY-maze, Stoelting Europe, Dublin, Ireland), we recorded the time taken by the individual mouse to reach the target hole, as well as the time that each mouse spent in the target quadrant during the test and retest.

4.10. Laboratory Data and Statistical Analyses

Data were collected using the Microsoft Excel 2016 (Microsoft Corp., Redmond, WA, USA) and, when necessary, corrected for between-session variation as described previously [88]. All the statistical analyses were performed in the Statistica software version 13.5 (StatSoft Inc., Tulsa, OK, USA). According to the normality of the results, we used the nonparametric Kruskal–Wallis test followed by Mann–Whitney U test for all analyses, except the analyses of the time spent in the target quadrant for the Barnes maze task performance, for which parametric one-way analysis of variance test was utilized. Results are expressed as means ± standard error of the mean. In all comparisons, $p < 0.05$ was considered to indicate statistical significance.

Supplementary Materials: The following are available online at https://www.mdpi.com/article/10.3390/ijms22126584/s1: Figure S1. Representative microphotographs of the optic tracts in wild-type (WT) and TDP-43^{G348C} mice at 6 months after repetitive mild traumatic brain injury (rmTBI) or sham procedure (Sham), immunostained with anti-ionized calcium-binding adaptor molecule 1 (Iba1), showing different morphological forms of microglial cells; Figure S2. Fluoro-Jade C- and cresyl-violet-stained sections of the lateral geniculate nucleus (LGN) and the superior colliculus (SC) in wild-type (WT) and TDP-43^{G348C} mice at 6 months after repetitive mild traumatic brain injury (rmTBI); Figure S3. The activity of the glial cells in the lateral geniculate nucleus (LGN) and the superior colliculus (SC) in wild-type (WT) and TDP-43^{G348C} mice at 6 months after repetitive mild traumatic brain injury (rmTBI); Figure S4. Synaptic plasticity in the lateral geniculate nucleus (LGN) and the superior colliculus (SC) in wild-type (WT) and TDP-43^{G348C} mice at 6 months after repetitive mild traumatic brain injury (rmTBI).

Author Contributions: Conceptualization, G.Ž., K.P., and J.K.; resources, G.Ž. and J.K.; investigation, K.P., J.R.B., P.D., N.G., and T.J.; formal analysis, K.P., P.D., and J.R.B.; visualization, K.P., P.D., and J.R.B.; writing—original draft preparation, G.Ž. and K.P.; writing—review and editing, J.K., P.D., J.R.B., T.J., and N.G.; project administration, G.Ž.; supervision, G.Ž.; funding acquisition, G.Ž. All authors have read and agreed to the published version of the manuscript.

Funding: This work was fully funded by the Croatian Science Foundation under the project number IP-2016-06-4602 to G.Ž.

Institutional Review Board Statement: The study was conducted according to the guidelines of the European Community Council Directive (86/609/EEC) and the Croatian laws and rules (NN 135/06; NN 37/13; NN 125/13; NN 39/17), and approved by the Faculty Ethical Committee approval, 2 March 2017 (project "Mild repetitive traumatic brain injury: a model-system to study TDP-43-mediated neuropathology and neuroinflammation").

Informed Consent Statement: Not applicable.

Data Availability Statement: The data that support the findings of this study are available within the article and supplemental data or upon request from the corresponding author.

Acknowledgments: The authors thank Ljerka Delač, Marina Jakovac, Maja Rukavina, and Tanja Mešanović from the Department of Pharmacology for their much-appreciated technical assistance.

The authors are grateful to Bojan Polić and Sali Slavić Stupac from the Department of Histology and Miro Samsa from the Laboratory for Mice Breeding and Engineering Rijeka, all from the Faculty of Medicine, University of Rijeka, Rijeka, Croatia, for the breeding and maintenance of mice. Preliminary results regarding chronic neurodegeneration and glial response in the optic tract of wild and transgenic TDP-43 mice were presented at the FENS 2020 Virtual Forum of Neuroscience Glasgow, UK, 11–15 July 2020, and are available at https://www.bib.irb.hr/1084542 (18 June 2021).

Conflicts of Interest: The authors declare no conflict of interests.

References

1. Daneshvar, D.H.; Nowinski, C.J.; McKee, A.C.; Cantu, R.C. The Epidemiology of Sport-Related Concussion. *Clin. Sports Med.* **2011**, *30*, 1–17. [CrossRef]
2. McKee, A.C.; Daneshvar, D.H.; Alvarez, V.E.; Stein, T.D. The Neuropathology of Sport. *Acta Neuropathol.* **2014**, *127*, 29–51. [CrossRef]
3. Lancon, J.A.; Haines, D.E.; Parent, A.D. Anatomy of the Shaken Baby Syndrome. *Anat. Rec.* **1998**, *253*, 13–18. [CrossRef]
4. Bryan, C.J.; Clemans, T.A. Repetitive Traumatic Brain Injury, Psychological Symptoms, and Suicide Risk in a Clinical Sample of Deployed Military Personnel. *JAMA Psychiatry* **2013**, *70*, 686–691. [CrossRef]
5. Peskind, E.R.; Brody, D.; Cernak, I.; McKee, A.; Ruff, R.L. Military- and Sports-Related Mild Traumatic Brain Injury: Clinical Presentation, Management, and Long-Term Consequences. *J. Clin. Psychiatry* **2013**, *74*, 180–188. [CrossRef]
6. McCrory, P.; Meeuwisse, W.H.; Aubry, M.; Cantu, B.; Dvorák, J.; Echemendia, R.J.; Engebretsen, L.; Johnston, K.; Kutcher, J.S.; Raftery, M.; et al. Consensus Statement on Concussion in Sport: The 4th International Conference on Concussion in Sport Held in Zurich, November 2012. *Br. J. Sports Med.* **2013**, *47*, 250–258. [CrossRef] [PubMed]
7. Levin, H.S.; Diaz-Arrastia, R.R. Diagnosis, Prognosis, and Clinical Management of Mild Traumatic Brain Injury. *Lancet Neurol.* **2015**, *14*, 506–517. [CrossRef]
8. Laurer, H.L.; Bareyre, F.M.; Lee, V.M.; Trojanowski, J.Q.; Longhi, L.; Hoover, R.; Saatman, K.E.; Raghupathi, R.; Hoshino, S.; Grady, M.S.; et al. Mild Head Injury Increasing the Brain's Vulnerability to a Second Concussive Impact. *J. Neurosurg.* **2001**, *95*, 859–870. [CrossRef]
9. Belanger, H.G.; Spiegel, E.; Vanderploeg, R.D. Neuropsychological Performance Following a History of Multiple Self-Reported Concussions: A Meta-Analysis. *J. Int. Neuropsychol. Soc.* **2010**, *16*, 262–267. [CrossRef] [PubMed]
10. McKee, A.C.; Cantu, R.C.; Nowinski, C.J.; Hedley-Whyte, E.T.; Gavett, B.E.; Budson, A.E.; Santini, V.E.; Lee, H.-S.; Kubilus, C.A.; Stern, R.A. Chronic Traumatic Encephalopathy in Athletes: Progressive Tauopathy after Repetitive Head Injury. *J. Neuropathol. Exp. Neurol.* **2009**, *68*, 709–735. [CrossRef] [PubMed]
11. McKee, A.C.; Gavett, B.E.; Stern, R.A.; Nowinski, C.J.; Cantu, R.C.; Kowall, N.W.; Perl, D.P.; Hedley-Whyte, E.T.; Price, B.; Sullivan, C.; et al. TDP-43 Proteinopathy and Motor Neuron Disease in Chronic Traumatic Encephalopathy. *J. Neuropathol. Exp. Neurol.* **2010**, *69*, 918–929. [CrossRef]
12. McKee, A.C.; Daneshvar, D.H. The Neuropathology of Traumatic Brain Injury. *Handb. Clin. Neurol.* **2015**, *127*, 45–66. [CrossRef]
13. McKee, A.C.; Stein, T.D.; Kiernan, P.T.; Alvarez, V.E. The Neuropathology of Chronic Traumatic Encephalopathy. *Brain Pathol. Zurich Switz.* **2015**, *25*, 350–364. [CrossRef] [PubMed]
14. Gendron, T.F.; Josephs, K.A.; Petrucelli, L. Review: Transactive Response DNA-Binding Protein 43 (TDP-43): Mechanisms of Neurodegeneration. *Neuropathol. Appl. Neurobiol.* **2010**, *36*, 97–112. [CrossRef] [PubMed]
15. Ratti, A.; Buratti, E. Physiological Functions and Pathobiology of TDP-43 and FUS/TLS Proteins. *J. Neurochem.* **2016**, *138* (Suppl. 1), 95–111. [CrossRef] [PubMed]
16. Liu, E.Y.; Russ, J.; Cali, C.P.; Phan, J.M.; Amlie-Wolf, A.; Lee, E.B. Loss of Nuclear TDP-43 Is Associated with Decondensation of LINE Retrotransposons. *Cell Rep.* **2019**, *27*, 1409–1421. [CrossRef]
17. Mackenzie, I.R.A.; Bigio, E.H.; Ince, P.G.; Geser, F.; Neumann, M.; Cairns, N.J.; Kwong, L.K.; Forman, M.S.; Ravits, J.; Stewart, H.; et al. Pathological TDP-43 Distinguishes Sporadic Amyotrophic Lateral Sclerosis from Amyotrophic Lateral Sclerosis with SOD1 Mutations. *Ann. Neurol.* **2007**, *61*, 427–434. [CrossRef]
18. Ling, S.-C.; Polymenidou, M.; Cleveland, D.W. Converging Mechanisms in ALS and FTD: Disrupted RNA and Protein Homeostasis. *Neuron* **2013**, *79*, 416–438. [CrossRef]
19. Arai, T.; Hasegawa, M.; Akiyama, H.; Ikeda, K.; Nonaka, T.; Mori, H.; Mann, D.; Tsuchiya, K.; Yoshida, M.; Hashizume, Y.; et al. TDP-43 Is a Component of Ubiquitin-Positive Tau-Negative Inclusions in Frontotemporal Lobar Degeneration and Amyotrophic Lateral Sclerosis. *Biochem. Biophys. Res. Commun.* **2006**, *351*, 602–611. [CrossRef]
20. Uryu, K.; Nakashima-Yasuda, H.; Forman, M.S.; Kwong, L.K.; Clark, C.M.; Grossman, M.; Miller, B.L.; Kretzschmar, H.A.; Lee, V.M.-Y.; Trojanowski, J.Q.; et al. Concomitant TAR-DNA-Binding Protein 43 Pathology Is Present in Alzheimer Disease and Corticobasal Degeneration but Not in Other Tauopathies. *J. Neuropathol. Exp. Neurol.* **2008**, *67*, 555–564. [CrossRef]
21. Higashi, S.; Iseki, E.; Yamamoto, R.; Minegishi, M.; Hino, H.; Fujisawa, K.; Togo, T.; Katsuse, O.; Uchikado, H.; Furukawa, Y.; et al. Concurrence of TDP-43, Tau and Alpha-Synuclein Pathology in Brains of Alzheimer's Disease and Dementia with Lewy Bodies. *Brain Res.* **2007**, *1184*, 284–294. [CrossRef] [PubMed]

22. Amador-Ortiz, C.; Lin, W.-L.; Ahmed, Z.; Personett, D.; Davies, P.; Duara, R.; Graff-Radford, N.R.; Hutton, M.L.; Dickson, D.W. TDP-43 Immunoreactivity in Hippocampal Sclerosis and Alzheimer's Disease. *Ann. Neurol.* **2007**, *61*, 435–445. [CrossRef] [PubMed]
23. Hatanpaa, K.J.; Bigio, E.H.; Cairns, N.J.; Womack, K.B.; Weintraub, S.; Morris, J.C.; Foong, C.; Xiao, G.; Hladik, C.; Mantanona, T.Y.; et al. TAR DNA-Binding Protein 43 Immunohistochemistry Reveals Extensive Neuritic Pathology in FTLD-U: A Midwest-Southwest Consortium for FTLD Study. *J. Neuropathol. Exp. Neurol.* **2008**, *67*, 271–279. [CrossRef] [PubMed]
24. Zhang, J.; Teng, Z.; Song, Y.; Hu, M.; Chen, C. Inhibition of Monoacylglycerol Lipase Prevents Chronic Traumatic Encephalopathy-like Neuropathology in a Mouse Model of Repetitive Mild Closed Head Injury. *J. Cereb. Blood Flow Metab. Off. J. Int. Soc. Cereb. Blood Flow Metab.* **2015**, *35*, 443–453. [CrossRef]
25. Saykally, J.N.; Ratliff, W.A.; Keeley, K.L.; Pick, C.G.; Mervis, R.F.; Citron, B.A. Repetitive Mild Closed Head Injury Alters Protein Expression and Dendritic Complexity in a Mouse Model. *J. Neurotrauma* **2018**, *35*, 139–148. [CrossRef]
26. Heyburn, L.; Abutarboush, R.; Goodrich, S.; Urioste, R.; Batuure, A.; Statz, J.; Wilder, D.; Ahlers, S.T.; Long, J.B.; Sajja, V.S.S.S. Repeated Low-Level Blast Overpressure Leads to Endovascular Disruption and Alterations in TDP-43 and Piezo2 in a Rat Model of Blast TBI. *Front. Neurol.* **2019**, *10*, 766. [CrossRef]
27. Rajič Bumber, J.; Pilipović, K.; Janković, T.; Dolenec, P.; Gržeta, N.; Križ, J.; Župan, G. Repetitive Traumatic Brain Injury Is Associated With TDP-43 Alterations, Neurodegeneration, and Glial Activation in Mice. *J. Neuropathol. Exp. Neurol.* **2021**, *80*, 2–14. [CrossRef] [PubMed]
28. Swarup, V.; Phaneuf, D.; Bareil, C.; Robertson, J.; Rouleau, G.A.; Kriz, J.; Julien, J.P. Pathological Hallmarks of Amyotrophic Lateral Sclerosis/Frontotemporal Lobar Degeneration in Transgenic Mice Produced with TDP-43 Genomic Fragments. *Brain* **2011**, *134*, 2610–2626. [CrossRef] [PubMed]
29. Ventura, R.E.; Balcer, L.J.; Galetta, S.L.; Rucker, J.C. Ocular Motor Assessment in Concussion: Current Status and Future Directions. *J. Neurol. Sci.* **2016**, *361*, 79–86. [CrossRef]
30. Bailes, J.E.; Petraglia, A.L.; Omalu, B.I.; Nauman, E.; Talavage, T. Role of Subconcussion in Repetitive Mild Traumatic Brain Injury. *J. Neurosurg.* **2013**, *119*, 1235–1245. [CrossRef] [PubMed]
31. Winston, C.N.; Noël, A.; Neustadtl, A.; Parsadanian, M.; Barton, D.J.; Chellappa, D.; Wilkins, T.E.; Alikhani, A.D.; Zapple, D.N.; Villapol, S.; et al. Dendritic Spine Loss and Chronic White Matter Inflammation in a Mouse Model of Highly Repetitive Head Trauma. *Am. J. Pathol.* **2016**, *186*, 552–567. [CrossRef]
32. Ojo, J.O.; Mouzon, B.; Algamal, M.; Leary, P.; Lynch, C.; Abdullah, L.; Evans, J.; Mullan, M.; Bachmeier, C.; Stewart, W.; et al. Chronic Repetitive Mild Traumatic Brain Injury Results in Reduced Cerebral Blood Flow, Axonal Injury, Gliosis, and Increased T-Tau and Tau Oligomers. *J. Neuropathol. Exp. Neurol.* **2016**, *75*, 636–655. [CrossRef] [PubMed]
33. Cheng, W.H.; Martens, K.M.; Bashir, A.; Cheung, H.; Stukas, S.; Gibbs, E.; Namjoshi, D.R.; Button, E.B.; Wilkinson, A.; Barron, C.J.; et al. CHIMERA Repetitive Mild Traumatic Brain Injury Induces Chronic Behavioural and Neuropathological Phenotypes in Wild-Type and APP/PS1 Mice. *Alzheimers Res. Ther.* **2019**, *11*, 6. [CrossRef]
34. Schmued, L.C.; Albertson, C.; Slikker, W. Fluoro-Jade: A Novel Fluorochrome for the Sensitive and Reliable Histochemical Localization of Neuronal Degeneration. *Brain Res.* **1997**, *751*, 37–46. [CrossRef]
35. Chidlow, G.; Wood, J.P.M.; Sarvestani, G.; Manavis, J.; Casson, R.J. Evaluation of Fluoro-Jade C as a Marker of Degenerating Neurons in the Rat Retina and Optic Nerve. *Exp. Eye Res.* **2009**, *88*, 426–437. [CrossRef] [PubMed]
36. Schmued, L.C.; Stowers, C.C.; Scallet, A.C.; Xu, L. Fluoro-Jade C Results in Ultra High Resolution and Contrast Labeling of Degenerating Neurons. *Brain Res.* **2005**, *1035*, 24–31. [CrossRef]
37. Tenkova, T.I.; Goldberg, M.P. A Modified Silver Technique (de Olmos Stain) for Assessment of Neuronal and Axonal Degeneration. *Methods Mol. Biol. Clifton NJ* **2007**, *399*, 31–39. [CrossRef]
38. Greenwald, B.D.; Kapoor, N.; Singh, A.D. Visual Impairments in the First Year after Traumatic Brain Injury. *Brain Inj.* **2012**, *26*, 1338–1359. [CrossRef]
39. Sen, N. An Insight into the Vision Impairment Following Traumatic Brain Injury. *Neurochem. Int.* **2017**, *111*, 103–107. [CrossRef]
40. Tzekov, R.; Quezada, A.; Gautier, M.; Biggins, D.; Frances, C.; Mouzon, B.; Jamison, J.; Mullan, M.; Crawford, F. Repetitive Mild Traumatic Brain Injury Causes Optic Nerve and Retinal Damage in a Mouse Model. *J. Neuropathol. Exp. Neurol.* **2014**, *73*, 345–361. [CrossRef]
41. Tzekov, R.; Dawson, C.; Orlando, M.; Mouzon, B.; Reed, J.; Evans, J.; Crynen, G.; Mullan, M.; Crawford, F. Sub-Chronic Neuropathological and Biochemical Changes in Mouse Visual System after Repetitive Mild Traumatic Brain Injury. *PLoS ONE* **2016**, *11*, e0153608. [CrossRef] [PubMed]
42. Das, M.; Tang, X.; Han, J.Y.; Mayilsamy, K.; Foran, E.; Biswal, M.R.; Tzekov, R.; Mohapatra, S.S.; Mohapatra, S. CCL20-CCR6 Axis Modulated Traumatic Brain Injury-Induced Visual Pathologies. *J. Neuroinflamm.* **2019**, *16*, 115. [CrossRef]
43. Desai, A.; Chen, H.; Kim, H.-Y. Multiple Mild Traumatic Brain Injuries Lead to Visual Dysfunction in a Mouse Model. *J. Neurotrauma* **2020**, *37*, 286–294. [CrossRef] [PubMed]
44. Thammisetty, S.S.; Pedragosa, J.; Weng, Y.C.; Calon, F.; Planas, A.; Kriz, J. Age-Related Deregulation of TDP-43 after Stroke Enhances NF-KB-Mediated Inflammation and Neuronal Damage. *J. Neuroinflamm.* **2018**, *15*, 312. [CrossRef]
45. Chen, H.; Richard, M.; Sandler, D.P.; Umbach, D.M.; Kamel, F. Head Injury and Amyotrophic Lateral Sclerosis. *Am. J. Epidemiol.* **2007**, *166*, 810–816. [CrossRef]

46. Costanza, A.; Weber, K.; Gandy, S.; Bouras, C.; Hof, P.R.; Giannakopoulos, P.; Canuto, A. Review: Contact Sport-Related Chronic Traumatic Encephalopathy in the Elderly: Clinical Expression and Structural Substrates. *Neuropathol. Appl. Neurobiol.* **2011**, *37*, 570–584. [CrossRef] [PubMed]
47. Franz, C.K.; Joshi, D.; Daley, E.L.; Grant, R.A.; Dalamagkas, K.; Leung, A.; Finan, J.D.; Kiskinis, E. Impact of Traumatic Brain Injury on Amyotrophic Lateral Sclerosis: From Bedside to Bench. *J. Neurophysiol.* **2019**, *122*, 1174–1185. [CrossRef]
48. Heyburn, L.; Sajja, V.S.S.S.; Long, J.B. The Role of TDP-43 in Military-Relevant TBI and Chronic Neurodegeneration. *Front. Neurol.* **2019**, *10*, 680. [CrossRef]
49. Bolton Hall, A.N.; Joseph, B.; Brelsfoard, J.M.; Saatman, K.E. Repeated Closed Head Injury in Mice Results in Sustained Motor and Memory Deficits and Chronic Cellular Changes. *PLoS ONE* **2016**, *11*, e0159442. [CrossRef]
50. Xu, L.; Nguyen, J.V.; Lehar, M.; Menon, A.; Rha, E.; Arena, J.; Ryu, J.; Marsh-Armstrong, N.; Marmarou, C.R.; Koliatsos, V.E. Repetitive Mild Traumatic Brain Injury with Impact Acceleration in the Mouse: Multifocal Axonopathy, Neuroinflammation, and Neurodegeneration in the Visual System. *Exp. Neurol.* **2016**, *275 Pt 3*, 436–449. [CrossRef]
51. Haber, M.; Hutchinson, E.B.; Sadeghi, N.; Cheng, W.H.; Namjoshi, D.; Cripton, P.; Irfanoglu, M.O.; Wellington, C.; Diaz-Arrastia, R.; Pierpaoli, C. Defining an Analytic Framework to Evaluate Quantitative MRI Markers of Traumatic Axonal Injury: Preliminary Results in a Mouse Closed Head Injury Model. *eNeuro* **2017**, *4*. [CrossRef] [PubMed]
52. Vonder Haar, C.; Martens, K.M.; Bashir, A.; McInnes, K.A.; Cheng, W.H.; Cheung, H.; Stukas, S.; Barron, C.; Ladner, T.; Welch, K.A.; et al. Repetitive Closed-Head Impact Model of Engineered Rotational Acceleration (CHIMERA) Injury in Rats Increases Impulsivity, Decreases Dopaminergic Innervation in the Olfactory Tubercle and Generates White Matter Inflammation, Tau Phosphorylation and Degeneration. *Exp. Neurol.* **2019**, *317*, 87–99. [CrossRef] [PubMed]
53. Gazdzinski, L.M.; Mellerup, M.; Wang, T.; Adel, S.A.A.; Lerch, J.P.; Sled, J.G.; Nieman, B.J.; Wheeler, A.L. White Matter Changes Caused by Mild Traumatic Brain Injury in Mice Evaluated Using Neurite Orientation Dispersion and Density Imaging. *J. Neurotrauma* **2020**, *37*, 1818–1828. [CrossRef]
54. Gangolli, M.; Benetatos, J.; Esparza, T.J.; Fountain, E.M.; Seneviratne, S.; Brody, D.L. Repetitive Concussive and Subconcussive Injury in a Human Tau Mouse Model Results in Chronic Cognitive Dysfunction and Disruption of White Matter Tracts, But Not Tau Pathology. *J. Neurotrauma* **2019**, *36*, 735–755. [CrossRef]
55. Zetterberg, H.; Smith, D.H.; Blennow, K. Biomarkers of Mild Traumatic Brain Injury in Cerebrospinal Fluid and Blood. *Nat. Rev. Neurol.* **2013**, *9*, 201–210. [CrossRef]
56. Shahim, P.; Zetterberg, H.; Tegner, Y.; Blennow, K. Serum Neurofilament Light as a Biomarker for Mild Traumatic Brain Injury in Contact Sports. *Neurology* **2017**, *88*, 1788–1794. [CrossRef] [PubMed]
57. Guedes, V.A.; Kenney, K.; Shahim, P.; Qu, B.-X.; Lai, C.; Devoto, C.; Walker, W.C.; Nolen, T.; Diaz-Arrastia, R.; Gill, J.M.; et al. Exosomal Neurofilament Light: A Prognostic Biomarker for Remote Symptoms after Mild Traumatic Brain Injury? *Neurology* **2020**, *94*, e2412–e2423. [CrossRef]
58. Laverse, E.; Guo, T.; Zimmerman, K.; Foiani, M.S.; Velani, B.; Morrow, P.; Adejuwon, A.; Bamford, R.; Underwood, N.; George, J.; et al. Plasma Glial Fibrillary Acidic Protein and Neurofilament Light Chain, but Not Tau, Are Biomarkers of Sports-Related Mild Traumatic Brain Injury. *Brain Commun.* **2020**, *2*, fcaa137. [CrossRef]
59. Cheng, W.H.; Stukas, S.; Martens, K.M.; Namjoshi, D.R.; Button, E.B.; Wilkinson, A.; Bashir, A.; Robert, J.; Cripton, P.A.; Wellington, C.L. Age at Injury and Genotype Modify Acute Inflammatory and Neurofilament-Light Responses to Mild CHIMERA Traumatic Brain Injury in Wild-Type and APP/PS1 Mice. *Exp. Neurol.* **2018**, *301*, 26–38. [CrossRef]
60. Strong, M.J.; Volkening, K.; Hammond, R.; Yang, W.; Strong, W.; Leystra-Lantz, C.; Shoesmith, C. TDP43 Is a Human Low Molecular Weight Neurofilament (HNFL) MRNA-Binding Protein. *Mol. Cell. Neurosci.* **2007**, *35*, 320–327. [CrossRef]
61. Moisse, K.; Mepham, J.; Volkening, K.; Welch, I.; Hill, T.; Strong, M.J. Cytosolic TDP-43 Expression Following Axotomy Is Associated with Caspase 3 Activation in NFL-/- Mice: Support for a Role for TDP-43 in the Physiological Response to Neuronal Injury. *Brain Res.* **2009**, *1296*, 176–186. [CrossRef]
62. Moisse, K.; Volkening, K.; Leystra-Lantz, C.; Welch, I.; Hill, T.; Strong, M.J. Divergent Patterns of Cytosolic TDP-43 and Neuronal Progranulin Expression Following Axotomy: Implications for TDP-43 in the Physiological Response to Neuronal Injury. *Brain Res.* **2009**, *1249*, 202–211. [CrossRef]
63. Armstrong, R.A. White Matter Pathology in Sporadic Frontotemporal Lobar Degeneration with TDP-43 Proteinopathy. *Clin. Neuropathol.* **2017**, *36*, 66–72. [CrossRef] [PubMed]
64. Kumar, S.; Phaneuf, D.; Cordeau, P.; Boutej, H.; Kriz, J.; Julien, J.-P. Induction of Autophagy Mitigates TDP-43 Pathology and Translational Repression of Neurofilament MRNAs in Mouse Models of ALS/FTD. *Mol. Neurodegener.* **2021**, *16*, 1. [CrossRef]
65. Morganti-Kossmann, M.C.; Rancan, M.; Stahel, P.F.; Kossmann, T. Inflammatory Response in Acute Traumatic Brain Injury: A Double-Edged Sword. *Curr. Opin. Crit. Care* **2002**, *8*, 101–105. [CrossRef]
66. Hanrahan, F.; Campbell, M. Neuroinflammation. In *Translational Research in Traumatic Brain Injury*; Laskowitz, D., Grant, G., Eds.; Frontiers in Neuroscience; CRC Press/Taylor and Francis Group: Boca Raton, FL, USA, 2016; ISBN 978-1-4665-8491-4.
67. Block, M.L.; Hong, J.-S. Chronic Microglial Activation and Progressive Dopaminergic Neurotoxicity. *Biochem. Soc. Trans.* **2007**, *35*, 1127–1132. [CrossRef]
68. Gao, H.-M.; Hong, J.-S. Why Neurodegenerative Diseases Are Progressive: Uncontrolled Inflammation Drives Disease Progression. *Trends Immunol.* **2008**, *29*, 357–365. [CrossRef]

69. Namjoshi, D.R.; Cheng, W.H.; McInnes, K.A.; Martens, K.M.; Carr, M.; Wilkinson, A.; Fan, J.; Robert, J.; Hayat, A.; Cripton, P.A.; et al. Merging Pathology with Biomechanics Using CHIMERA (Closed-Head Impact Model of Engineered Rotational Acceleration): A Novel, Surgery-Free Model of Traumatic Brain Injury. *Mol. Neurodegener.* **2014**, *9*, 55. [CrossRef] [PubMed]
70. Chen, H.; Desai, A.; Kim, H.-Y. Repetitive Closed-Head Impact Model of Engineered Rotational Acceleration Induces Long-Term Cognitive Impairments with Persistent Astrogliosis and Microgliosis in Mice. *J. Neurotrauma* **2017**, *34*, 2291–2302. [CrossRef] [PubMed]
71. Kerschensteiner, D.; Guido, W. Organization of the Dorsal Lateral Geniculate Nucleus in the Mouse. *Vis. Neurosci.* **2017**, *34*, E008. [CrossRef]
72. May, P.J. The Mammalian Superior Colliculus: Laminar Structure and Connections. *Prog. Brain Res.* **2006**, *151*, 321–378. [CrossRef]
73. Hofbauer, A.; Dräger, U.C. Depth Segregation of Retinal Ganglion Cells Projecting to Mouse Superior Colliculus. *J. Comp. Neurol.* **1985**, *234*, 465–474. [CrossRef]
74. Ellis, E.M.; Gauvain, G.; Sivyer, B.; Murphy, G.J. Shared and Distinct Retinal Input to the Mouse Superior Colliculus and Dorsal Lateral Geniculate Nucleus. *J. Neurophysiol.* **2016**, *116*, 602–610. [CrossRef]
75. Sophie Su, Y.; Veeravagu, A.; Grant, G. Neuroplasticity after Traumatic Brain Injury. In *Translational Research in Traumatic Brain Injury*; Laskowitz, D., Grant, G., Eds.; Frontiers in Neuroscience; CRC Press/Taylor and Francis Group: Boca Raton, FL, USA, 2016; ISBN 978-1-4665-8491-4.
76. Tarsa, L.; Goda, Y. Synaptophysin Regulates Activity-Dependent Synapse Formation in Cultured Hippocampal Neurons. *Proc. Natl. Acad. Sci. USA* **2002**, *99*, 1012–1016. [CrossRef] [PubMed]
77. Thiel, G. Synapsin I, Synapsin II, and Synaptophysin: Marker Proteins of Synaptic Vesicles. *Brain Pathol. Zurich Switz.* **1993**, *3*, 87–95. [CrossRef] [PubMed]
78. Schirmer, L.; Merkler, D.; König, F.B.; Brück, W.; Stadelmann, C. Neuroaxonal Regeneration Is More Pronounced in Early Multiple Sclerosis than in Traumatic Brain Injury Lesions. *Brain Pathol. Zurich Switz.* **2013**, *23*, 2–12. [CrossRef] [PubMed]
79. Kokotos, A.C.; Harper, C.B.; Marland, J.R.K.; Smillie, K.J.; Cousin, M.A.; Gordon, S.L. Synaptophysin Sustains Presynaptic Performance by Preserving Vesicular Synaptobrevin-II Levels. *J. Neurochem.* **2019**, *151*, 28–37. [CrossRef] [PubMed]
80. Barnes, C.A. Memory Deficits Associated with Senescence: A Neurophysiological and Behavioral Study in the Rat. *J. Comp. Physiol. Psychol.* **1979**, *93*, 74–104. [CrossRef]
81. Bashir, A.; Abebe, Z.A.; McInnes, K.A.; Button, E.B.; Tatarnikov, I.; Cheng, W.H.; Haber, M.; Wilkinson, A.; Barron, C.; Diaz-Arrastia, R.; et al. Increased Severity of the CHIMERA Model Induces Acute Vascular Injury, Sub-Acute Deficits in Memory Recall, and Chronic White Matter Gliosis. *Exp. Neurol.* **2020**, *324*, 113116. [CrossRef] [PubMed]
82. Mouzon, B.; Chaytow, H.; Crynen, G.; Bachmeier, C.; Stewart, J.; Mullan, M.; Stewart, W.; Crawford, F. Repetitive Mild Traumatic Brain Injury in a Mouse Model Produces Learning and Memory Deficits Accompanied by Histological Changes. *J. Neurotrauma* **2012**, *29*, 2761–2773. [CrossRef] [PubMed]
83. McAteer, K.M.; Corrigan, F.; Thornton, E.; Turner, R.J.; Vink, R. Short and Long Term Behavioral and Pathological Changes in a Novel Rodent Model of Repetitive Mild Traumatic Brain Injury. *PLoS ONE* **2016**, *11*, e0160220. [CrossRef] [PubMed]
84. Kane, M.J.; Angoa-Pérez, M.; Briggs, D.I.; Viano, D.C.; Kreipke, C.W.; Kuhn, D.M. A Mouse Model of Human Repetitive Mild Traumatic Brain Injury. *J. Neurosci. Methods* **2012**, *203*, 41–49. [CrossRef] [PubMed]
85. Paxinos, G.; Franklin, K.B.J. *The Mouse Brain in Stereotaxic Coordinates*; Academic Press: San Diego, CA, USA, 2001; ISBN 978-0-12-547636-2.
86. Underhill, H.R.; Rostomily, R.C.; Mikheev, A.M.; Yuan, C.; Yarnykh, V.L. Fast Bound Pool Fraction Imaging of the in Vivo Rat Brain: Association with Myelin Content and Validation in the C6 Glioma Model. *NeuroImage* **2011**, *54*, 2052–2065. [CrossRef] [PubMed]
87. Khodanovich, M.Y.; Sorokina, I.V.; Glazacheva, V.Y.; Akulov, A.E.; Nemirovich-Danchenko, N.M.; Romashchenko, A.V.; Tolstikova, T.G.; Mustafina, L.R.; Yarnykh, V.L. Histological Validation of Fast Macromolecular Proton Fraction Mapping as a Quantitative Myelin Imaging Method in the Cuprizone Demyelination Model. *Sci. Rep.* **2017**, *7*, 46686. [CrossRef] [PubMed]
88. Ruijter, J.M.; Thygesen, H.H.; Schoneveld, O.J.L.M.; Das, A.T.; Berkhout, B.; Lamers, W.H. Factor Correction as a Tool to Eliminate Between-Session Variation in Replicate Experiments: Application to Molecular Biology and Retrovirology. *Retrovirology* **2006**, *3*, 2. [CrossRef] [PubMed]

International Journal of
Molecular Sciences

MDPI

Review

Current Trends in Neurodegeneration: Cross Talks between Oxidative Stress, Cell Death, and Inflammation

Tapan Behl [1], Rashita Makkar [1], Aayush Sehgal [1], Sukhbir Singh [1], Neelam Sharma [1], Gokhan Zengin [2], Simona Bungau [3,4,*], Felicia Liana Andronie-Cioara [5,*], Mihai Alexandru Munteanu [6], Mihaela Cristina Brisc [6], Diana Uivarosan [7] and Ciprian Brisc [6]

1 Chitkara College of Pharmacy, Chitkara University, Punjab 140401, India; tapanbehl31@gmail.com (T.B.); rashitamakker32@gmail.com (R.M.); aayushsehgal00@gmail.com (A.S.); sukhbir.singh@chitkara.edu.in (S.S.); neelam.mdu@gmail.com (N.S.)
2 Department of Biology, Faculty of Science, Selcuk University Campus, Konya 42130, Turkey; biyologzengin@gmail.com
3 Department of Pharmacy, Faculty of Medicine and Pharmacy, University of Oradea, 410028 Oradea, Romania
4 Doctoral School of Biological and Biomedical Sciences, University of Oradea, 410073 Oradea, Romania
5 Department of Psycho-Neuroscience and Recovery, Faculty of Medicine and Pharmacy, University of Oradea, 410073 Oradea, Romania
6 Department of Medical Disciplines, Faculty of Medicine and Pharmacy, University of Oradea, 410073 Oradea, Romania; mihaimunteanual@yahoo.com (M.A.M.); briscristina@yahoo.com (M.C.B.); brisciprian@gmail.com (C.B.)
7 Department of Preclinical Disciplines, Faculty of Medicine and Pharmacy, University of Oradea, 410073 Oradea, Romania; diana.uivarosan@gmail.com
* Correspondence: sbungau@uoradea.ro (S.B.); felicia_cioara@yahoo.com (F.L.A.-C.)

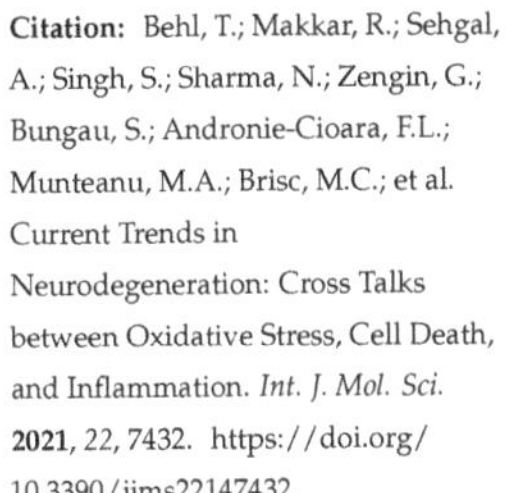

Citation: Behl, T.; Makkar, R.; Sehgal, A.; Singh, S.; Sharma, N.; Zengin, G.; Bungau, S.; Andronie-Cioara, F.L.; Munteanu, M.A.; Brisc, M.C.; et al. Current Trends in Neurodegeneration: Cross Talks between Oxidative Stress, Cell Death, and Inflammation. *Int. J. Mol. Sci.* **2021**, *22*, 7432. https://doi.org/10.3390/ijms22147432

Academic Editor: Giuseppe Lazzarino and Anne Vejux

Received: 19 May 2021
Accepted: 9 July 2021
Published: 11 July 2021

Abstract: The human body is highly complex and comprises a variety of living cells and extracellular material, which forms tissues, organs, and organ systems. Human cells tend to turn over readily to maintain homeostasis in tissues. However, postmitotic nerve cells exceptionally have an ability to regenerate and be sustained for the entire life of an individual, to safeguard the physiological functioning of the central nervous system. For efficient functioning of the CNS, neuronal death is essential, but extreme loss of neurons diminishes the functioning of the nervous system and leads to the onset of neurodegenerative diseases. Neurodegenerative diseases range from acute to chronic severe life-altering conditions like Parkinson's disease and Alzheimer's disease. Millions of individuals worldwide are suffering from neurodegenerative disorders with little or negligible treatment available, thereby leading to a decline in their quality of life. Neuropathological studies have identified a series of factors that explain the etiology of neuronal degradation and its progression in neurodegenerative disease. The onset of neurological diseases depends on a combination of factors that causes a disruption of neurons, such as environmental, biological, physiological, and genetic factors. The current review highlights some of the major pathological factors responsible for neuronal degradation, such as oxidative stress, cell death, and neuroinflammation. All these factors have been described in detail to enhance the understanding of their mechanisms and target them for disease management.

Keywords: neurodegeneration; neuroinflammation; apoptosis; necrosis; cell death; oxidative stress

1. Introduction

The human brain is the most complex organ, controlling all the amazing things we do by regulating several molecular pathways. It comprises billions of cells, called neurons, which control the proper functioning of our body. Neurons stimulate and transmit signals that enable us to talk, move, think, and accomplish everything we do. The brain cells are closely interconnected with each other. Therefore, the slightest miscommunications within cells in a particular area can lead to a disruption in other activities controlled by brain, causing major brain disorders. Neurodegenerative diseases can also be characterized by progressive loss in the functioning of the brain due to an accumulation of toxic proteins that exhibit clinical syndromes [1,2]. Therefore, brain disorders should not be taken lightly, as they can result in widespread problems [3], ultimately leading to neuronal death and shrinkage. The word "neurodegenerative" is formed of two parts—"neuro," which means brain, and "degenerative," which means dying or breaking down. Inadequate communication among brain cells lead to devastating effects, influencing several activities of an individual such as movement, memory, speech intelligence, and many more [4,5]. Neurodegenerative diseases are highly complex, and their etiology is sometimes very hard to predict.

Different areas of the brain encounter different types of neurodegenerative diseases, and they are described in Table 1. Examples of neurodegenerative diseases include Parkinson's disease, Huntington disease, Alzheimer's disease, amyotrophic lateral sclerosis, and many more.

Table 1. Types of neurodegenerative diseases according to the brain region affected.

Brain Region Affected	Types of Neurodegenerative Diseases
Basal ganglia	Parkinson's disease Huntington disease Alzheimer's disease Frontotemporal degeneration
Thalamus	Alzheimer's disease Frontotemporal degeneration Multiple sclerosis
Cerebellum	Multiple sclerosis Multiple systemic atrophy dystonia Alzheimer's disease Spinocerebellar ataxia
Cerebral cortex	Frontotemporal dementia Alzheimer's disease Tremors Parkinson's disease Huntington disease Amyotrophic lateral sclerosis Neuro psychiatric disorders
Brain stem	Frontotemporal lobar degeneration Parkinson's disease Huntington disease Frontotemporal dementia Amyotrophic lateral sclerosis Spinocerebellar ataxia

The symptoms of neurodegenerative diseases are mainly encountered in older groups of people [6]. This group of people is highly vulnerable to memory loss, which results in a poor quality of life and loss of personality [7–9]. With the increasing population worldwide, the incidence of neurological diseases is also increasing. According to published data, it has been determined that new cases of Parkinson's disease and Alzheimer's disease have increased abruptly over the span of the last 30 years. Around the globe, more than 10 million people are suffering from Parkinson's disease and more than 5.4 million people are living with Alzheimer's disease, indicating that neurodegenerative diseases are the leading cause of death worldwide and are highly prevalent in populations of people 60 years of age [10]. The massive increase in neurodegenerative diseases can be contributing to an increase in the prevalence of amyotrophic lateral sclerosis (ALS) disease. In addition, it has been determined that a large population of the elderly age group is estimated to be suffering from Huntington disease.

The prevalence of neurodegenerative diseases among all the genders, races, and geographical areas is increasing with increasing population all over the world [11]. These diseases are highly complex and difficult to cure; therefore, it has become necessary to develop newer medications with effective therapeutic strategies to overcome them. Simulated models comprising everything from unicellular organisms to the most complex functioning have been developed, and have proven to be useful tools in the research and development of new therapeutic medications by exploring the underlined advanced neurological pathways [12]. The collection of biomarkers or therapeutic targets provides greater insight to the pathophysiology of neurological diseases and can also contribute to researching new medications [13,14].

Figure 1 describes some of the common factors that are responsible for the initiation and progression of neurological diseases and provide greater insight into the pathophysiology.

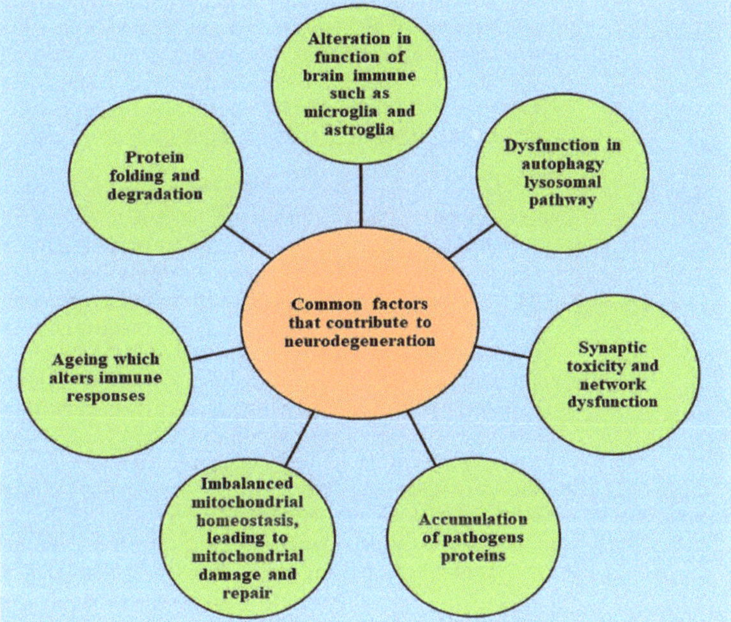

Figure 1. Some of the common factors responsible for the initiation/progression of neurological diseases.

The current review explains the most common pathways that are responsible for the initiation and progression of commonly occurring neurodegenerative disorders.

2. Role of Cell Death in the Onset of Neurodegeneration

Neurological disorders are mainly characterized by increased degradation in the functioning of neurons due to the destruction of synapses and axons, eventually leading to nerve cell death. An understanding of the mechanism that leads to the homeostasis of cellular elements and neurodegradation is highly important for developing novel therapeutic treatments for the diseases [15–19]. The healthy cells in the human body transform to preserve the normal homeostasis of tissues; however, post-mitotic neurons harbor very little capacity to regenerate and their survival is essential to ensure the proper functioning of the nervous system [20]. The death of neurons promotes the development of nervous system; however, if occurring in excess, it leads to declined functioning of nervous system and causes the progression of neurodegenerative diseases, which can be indicated by a range of acute insults, from stroke and traumatic brain injury (TBI) to enduring critical conditions such as Parkinson's disease [21], Alzheimer's disease [22], and amyotrophic lateral sclerosis [23].

Several studies have been conducted to understand the neuropathology behind the chronic conditions of these diseases, and stereotypical patterns of neurodegeneration have been identified in different regions of the central nervous system, which correspond to the disease severity clinically [24]. Several other pathological pathways, such as impairment in axonal transport and synaptic function, oxidative stress, dysfunction of lysosomes and mitochondria [25,26], activation of microglial cells and protein aggregation, also contribute to neuronal damage [27]. Other factors such as genetics, age, and environmental factors influence the disruption of neuronal homeostasis and aggravate the existing neurodegeneration by activating the signaling of different molecules, ultimately causing cell death and declined functioning of the nervous system [28].

With the advancement of technology in recent years, the understanding of pathology and genetic changes invoked in neurodegenerative diseases has significantly improved but is still unsatisfying. Due to complex biology, the connection between the origin and execution of the death of neurons is still lucid [2]. The pathway involved in cell death and the mechanism responsible for its activation is still under question, and unraveling it is important to drive the development of new target-oriented therapeutic medications. There are several pathways that regulate the cell death of neurons, which are explained below.

2.1. Exhibition of Necroptosis, Apoptosis, and Related Processes in Neuronal Death

Apoptosis can be defined as a type of programmed cell death that is mainly associated with the disintegration of deoxy ribonucleic acid (DNA) and the destruction of several nuclear and cytoskeletal proteins, which leads to immune cell-mediated phagocytosis. The protein caspases execute a chain of proteolysis, which ultimately results in cell death [29]. During the development of the central nervous system, excess neurons are removed by neuronal apoptosis, followed by acute insults that range from excited conditions to injuries [30]. The activation of caspase 8 by extrinsic pathway and caspase 9 via intrinsic pathway leads to the activation of death receptors and mitochondrial damage, respectively, and eventually converges to caspase 3, which ultimately initiates downstreaming of neuronal apoptosis. It was established in a published paper that mice that lack caspase 3 expressions present excessive neuronal cells at birth and significantly reduced apoptosis [31]. The damaged mitochondrion in neurons initiates apoptosis with the help of proapoptotic protein Bax, belonging to the bcl-2 family, by activation of the mitochondrial membrane and release of cytochrome C and further caspase 9 activation [32]. This pathway mainly regulates apoptosis of the neurons even when the chief stimuli has not impaired the functioning of mitochondria. This indicates that the permeability of mitochondria broadly contributes to the apoptosis of neurons. It has been discovered that the neurons that lack Bax protein display no activation of caspases and are strongly protected from neurodegeneration even after prolonged insults [33].

Necroptosis is a form of programmed cell death that is independent of caspases and has been recently discovered to explain cellular characteristics of necrosis. Interferons, tumor necrosis factor, toll-like receptors, and signaling of other proteins and infections act as stimuli and induce necroptosis [34]. Apart from apoptosis and necroptosis, several other pathways that cause cell death have also been identified, such as iron-independent necrosis, known as ferroptosis; lysosomal cell death, known as auto lysis; and death by self-eating, known as autophagy. All these mechanisms are interrelated and cause neuronal cell death.

2.2. Cell Death Regulation in Response to Neuronal Stress

The signaling of the c-Jun amino terminal kinase (JNK) pathway induces traumatic responses that lead to the death of neurons. The role of the JNK signaling pathway has been characterized as the functioning of neurons, including responses that mediate neuronal injury [35]. Another protein, namely, dual leucine zipper kinase (DLK), regulates

JNK signaling and upstream neuronal functioning. It is mainly triggered by neurotrophic growth factors, oxidative stress, injuries associated with neurons and axons, and misfolded proteins [36]. DLK proteins lead to JNK phosphorylation and translocate to the nucleus. This phosphorylation further leads to transcriptional stress and culminates in CNS apoptosis. The DLK proteins control the degeneration of neurons in the developmental phase, but if uncontrolled it progresses to conditions of injury in neurons and neurodegeneration [37].

The neuronal cells that lack signaling by DLK proteins fail to generate a normal transcriptional injury response and further attenuate the activation of caspases, and hence, are strongly protected from destruction upon exposure to stress [38]. The activity of protein kinase R-like endoplasmic reticulum kinase (PERK) can be hampered by the signaling of DLK proteins to prevent the phosphorylation and translation of eukaryotic initiation factor-2α in neurons post-acute injury [39]. The signaling of PERK is a component of unfolded protein responses and its activation by DLK proteins significantly demonstrates its participation in the destruction of neurons. Therefore, the signaling of proteins like DLK, JNK, and PERK induces stress responses that further lead to the signaling of pathways, leading to cell death [40].

2.3. Programmed Destruction of Neurons by the Degeneration of Axons

The degeneration of axons can be mainly portrayed by forfeited connectivity in neurons [41]. This is the earliest clinical feature that has been discovered in most neurodegenerative diseases [42]. The degeneration of axons is a dynamic activity that depends upon the signaling of local axons and their transcriptional regulation within the body of neuron cells [43]. Axon degeneration, in conjunction with the apoptosis of neurons, further aggravates the condition. Several pathways such as the withdrawal of trophic factors, DLK/JNK signaling pathways [44], and several chemotherapeutic compounds appear to be involved in axonal degeneration, whereas the DLK pathway is considered to be the governing pathway that leads to axonal degeneration by activating the signaling of caspase proteins and ultimately neuronal cell death [45]. Figure 2 briefly mentions all the factors that are involved in the pathway of cell death via processes of neuronal apoptosis, the formation of a cascade of proteolysis that induces mitochondrial stress and damage, and the signaling of various proteins such as c-Jun amino terminal kinase (JNK) and dual leucine zipper kinase (DLK), which leads to translocation to the nucleus and ultimately causes cell death.

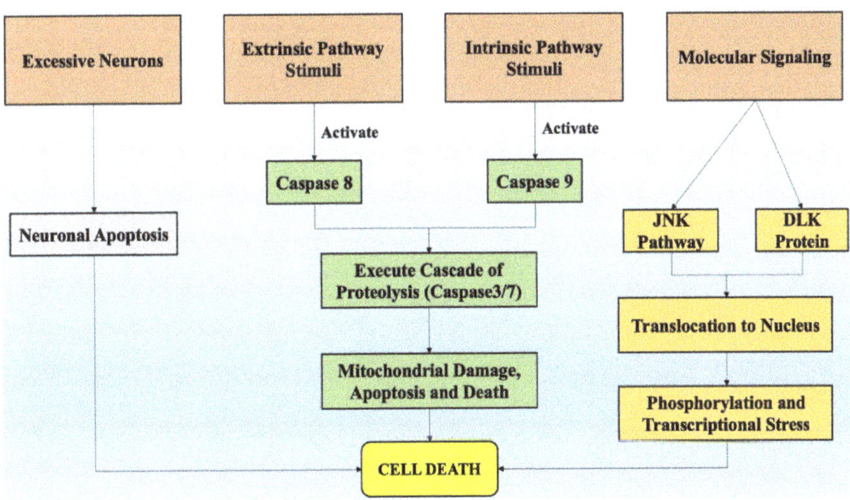

Figure 2. The factors that are involved in the pathways of cell death. Legend: JNK—c-Jun amino-terminal kinase; DLK—dual leucine zipper kinase.

Apart from the mechanisms stated above, there are several other factors that lead to neuronal cell death in diseases, such as protein homeostasis, [20] mitochondrial dysfunction, neuroinflammation, selective vulnerability, and linking of initiators and executioners of neurodegenerative diseases. All these factors run parallel and induce stress intracellularly in neurons, which leads to their death and chronic neurodegenerative disorders. These are also guided by accretion of the proteins that have misfolded or are defective in the endolysosomal system and oxidative stress.

The most common trait involved in the etiology of neurological disorders like Alzheimer's disease and Parkinson's disease includes the aggregation of toxic misfolded proteins, which leads to synaptic dysfunction and the destruction of neurons [46] For instance, the extracellular β-amyloid plaques observed in Alzheimer's disease in conjunction with intracellular neurofibrillary tangles formed by the hyperphosphorylation of the tau protein is the contributing factor in the pathophysiology of Alzheimer's disease [47]. The precise mechanism that triggers the conversion of normal proteins into misfolded proteins is lucid; however, the misfolded proteins recruit healthy proteins into lethal oligomers and accelerate their aggregation. These aggregated proteins sequesters from their colony and prevent the functioning that is necessary for their survival [48].

During normal physiological conditions, proteins, with the help of endoplasmic reticulum chaperones, are folded into their native form while the misfolded proteins are engulfed by autophagosomes, lysosomes, and proteosomes, a process called protein homeostasis. In conditions of stress, the misfolded proteins are generated in bulk and their aggregation leads to toxic functioning due to loss of endogenous functioning, which drives neuron cell death. Protein homeostasis is a critical process that governs the overall health of neurons. Mitochondrial dysfunction and neuroinflammation also lead to cytotoxicity in the brain and play a major role in the pathophysiology of neurodegenerative diseases. These processes mainly include the release of cytotoxic inflammatory cytokines in the brain, which causes phagocytosis of damaged neurons [49].

A selective region in brain and subsets of neurons that are highly exposed to cytotoxic agents lead to disease progression, a phenomenon called selective vulnerability. For instance, the pyramidal neurons in the entorhinal cortex region of the brain are initially targeted in Alzheimer's disease. Hence, different pathophysiology in different regions of the brain mediates neurodegeneration, and thorough knowledge of neuron dysfunction processes and cell death can open opportunities for the development of more targeted therapies and enhance the quality of life of the affected populations.

3. Oxidative Stress and Its Role in Neurodegeneration

The most vital entity for all living organisms is oxygen. Oxygen plays an important role in the physiological functioning of the body, as it is involved in several processes, such as tissue formation, and is a basic element for the growth of every cell. It is highly crucial in inducing gene transcription and signal transduction [50]. However, in excess it may produce detrimental outcomes. The conditions of stress occur mainly due to an imbalance between concentrations of reactive oxygen species (hydroxyl free radical, oxygen) and antioxidants [51]. The imbalance arises mainly due to two factors, which include the excessive synthesis of reactive oxygen species or a disturbance in the antioxidant system of the body [52]. Mitochondria supplies adenosine triphosphate (ATP) to the cells by breaking down glucose molecules, a process called oxidative phosphorylation, and by the synthesis of several other essential biological molecules. The proteins and enzymes required in oxidative phosphorylation are mainly programmed by the DNA of the mitochondria. Apart from the production of ATP via electron transport chain and oxidative phosphorylation, mitochondria are also involved in producing molecules that have a tendency to overcome oxidative stress via apoptotic mechanisms and other functions in cells [53].

The enrichment of mitochondria with enzymes involved in redox reaction and dysfunction of mitochondria is the principal cause of oxidative stress and excessive production of reactive oxygen species in the cellular environment. A phospholipid called cardiolipin is found in the membrane of mitochondria and is specifically involved with proteins in the electron transport chain. Any abnormality in the oxidative phosphorylation process causes the dysfunction of mitochondria cells and the generation of reactive oxygen species (ROS) made up of univalent oxygen molecules. The most important molecule in reactive oxygen species is nitric oxide, which regulates the production and relaxation of the muscle cells of vasculature, leukocyte adhesion, platelet aggregation, angiogenesis, thrombosis, hemodynamics, vascular tone, and many more [54].

The major sources of free radicals include redox metabolism in mitochondria, the metabolism of phospholipids, and several other proteolytic pathways. For normal physiological functioning of the body, the concentration of reactive oxygen species must be low. High concentrations and excessive exposure to ROS leads to the destruction of macromolecules such as DNA, proteins, and lipids, which ultimately cause necrosis and apoptotic cell death [55].

Oxidative stress is also the key element that regulates aging and several other neurological conditions. The chemical integrity of the brain controls the higher functioning of the central nervous system. The human brain is most susceptible to oxidative stress, as it devours a huge proportion of oxygen and is highly rich in lipids [56]. Higher oxygen consumption produces an abundant number of ROS. The membrane of the neurons is made up of polyunsaturated fatty acids, which are also prone to reactive oxygen species [57].

Several neurodegenerative diseases such as Alzheimer's disease, amyotrophic lateral sclerosis, Huntington disease, and Parkinson's disease are a result of alterations in biochemical and biomolecular components mainly due to oxidative stress. Hydrogen peroxide, a highly reactive hydroxyl radical and super oxide anion, are some of the negative oxygen species that are involved in neurodegeneration. Therefore, a dire need arises to understand the principal role of oxidative stress in the etiology of neurodegeneration. Apart from reactive oxygen species, reactive nitrogen species such as nitric oxide also possesses destructive effects on neurons [58]. It is important to understand the pathophysiology involved in cell death via ROS and initiate treatment by targeting the specific pathways involved to combat these diseases.

Oxidative stress is the predominant factor responsible for the pathogenesis of numerous chronic diseases such as diabetes, rheumatoid arthritis, obesity, etc., and almost all neurological disorders. Scavenging of free radicals by antioxidants can prevent aging and oxidative stress-mediated pathological conditions. The factors that cause oxidative stress and the physiological changes that come as a result of increased reactive oxygen species are presented in Figure 3.

An insufficiency in the antioxidant-medicated defense system leads to increased concentrations of oxidative species, which is highly problematic, as neuron cells are made up of polyunsaturated lipids, which are extremely susceptible to oxidation and create deleterious effects on biomolecules [59]. Changes in proteins, lipids, enzymes, DNA, ribonucleic acid (RNA), and other biomolecules under stress conditions can be used as biomarkers due to their highly sensitive nature and thus can be helpful in the determination of oxidative stress [60]. Reactive oxygen species such as OH, ONOO, etc. can be generated by altering the functioning of proteins and lipids.

The heterocyclic bases of DNA and RNA, mainly guanine, is extremely prone to oxidative damage and can be attacked by reactive oxygen species, which leads to the formation of 8-hydroxy-2-deoxyguanosine and 8-hydroxyguanine. These modified bases, when increased to a certain extent in the brain, produce significant damage and are the principle cause of Parkinson's disease [61]. The carbonylation and nitration of proteins is also predominantly witnessed in brains of Alzheimer's patients.

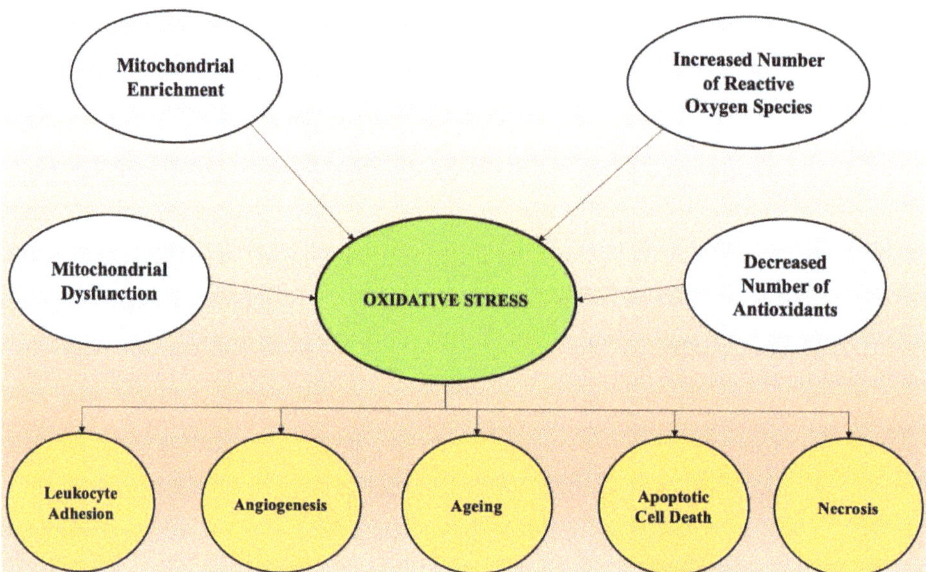

Figure 3. The factors that cause oxidative stress and the physiological changes that come as a result of increased reactive oxygen species.

Lipids form the plasma membrane of the neuron cells and act as a barrier between extracellular and intracellular spaces, and hence play a major role in the functioning of neuron cells [62]. However, these lipids are easily attacked by free radicals and go through lipid peroxidation, which decreases the fluidity of the membrane and causes leakage, which encourages the entry of substances that are usually not allowed to pass intracellularly but now cross these barriers through specific channels, thereby mutilating enzymes, membrane proteins, and receptors [63].

A significant role of pro-oxidants in neurodegenerative disorders has also been established, apart from the endogenous and exogenous sources of oxidative stress. Rich sources of polyphenols such as vegetables and fruits play an important part in generating defense antioxidant mechanisms. Awareness to increase such ingredients in the diet on a routine basis can be highly beneficial. However, the occurrence of neurodegenerative diseases and other chronic health problems is not solely dependent upon antioxidants, and there may be several other factors that induce oxidative stress and have been identified as pro-oxidants [64]. Pro-oxidants are any xenobiotic or endobiotic species that increases the production of reactive oxygen species or reactive nitrogen species by inhibiting the antioxidant system and triggering oxidative stress in tissues and cells. Exogeneous prooxidants include dietary supplements, pathogens, and toxicant drugs, whereas endogenous prooxidants include environmental pollution, metabolites of drugs, anxiety, climate change, and many more.

Heavy metals also play an important role in oxidative stress and have harmful effects on the reproductive system, respiratory system, central nervous system, cardiovascular system, and vital organs of the body [65]. These metal ions accumulate in the environment of the human body and combine with other xenobiotics and exert deteriorating effects on the immune system and hematology. Exposure to heavy metal ions such as lead and mercury and a deficiency in necessary metal ions like zinc and selenium lead to an increase in oxidative stress and produce deleterious effects by generating abnormality in the redox functioning of the cell [66].

The role of mercury in biological processes is minimal but its presence and accumulation in the body can produce harmful effects. Oxidative stress, which is mediated by the presence of mercury, can lead to the damage of biomolecular membranes, which further initiates the synthesis of hydrogen peroxide [67]. The neurotoxic effects of mercury have been discovered, and it has been determined that excessive mercury causes tremors and memory issues in the brain, accompanied by altered hearing and vision. Another heavy metal, lead, is also toxic to humans because of its nature to induce oxidative stress [68]. The damage produced by lead significantly depends upon the dose, route and duration of exposure, health status, and age of the individual. Lead has a high affinity for the -SH group, which is mainly present in amino acids and other metal cofactors that cause a significant reduction in the enzymes responsible for antioxidant activity, thereby elevating oxidative stress and causing failure of vital organs.

Another heavy metal, arsenic, is also prone to binding with the -SH group of glutathione, leading to the production of hydrogen peroxide. Arsenic is also responsible for the inhibition of glucose absorption in cells and the oxidation of fatty acids [69].

Recently, the role of ferroptosis has also emerged as being involved in inducing oxidative stress, which causes neuron cell damage and cell apoptosis [70,71]. The process of ferroptosis is mainly dependent upon the production of soluble and lipid reactive oxygen species by enzymatic reactions that mediate the transitioning of iron metal. Ferroptosis-mediated oxidative stress and mitochondrial dysfunction causes the death of neurons as well as cell apoptosis and targeting them can be an efficient strategy to prevent neurodegenerative diseases [72].

3.1. Mechanism of Oxidative Stress

Reactive oxygen species are mainly produced by a dysfunction in the mitochondria. The range of dysfunction through mitochondria is very wide and comprises a decrease in ATP production, altered regulation of calcium, brain iron accumulation, respiratory chain dysfunction, perturbation in the dynamics of mitochondria, and deregulation of mitochondria clearance [73].

In patients with Parkinson's disease, several genes have been identified that strongly confirm the association of mitochondrial function with oxidative stress in the pathogenesis of the disease. The loss in functioning of the DJ-1 gene causes oxidative stress and exerts a neuroprotective role in response to its antioxidant characteristic in mitochondria [74]. Mutations in the recessive form of the PINK1 gene leads to the onset of Parkinson's disease. PINK1 is a mitochondria kinase that regulates cytosolic calcium, and deficiency in this gene leads to impairment of calcium influx in mitochondria and causes mitochondrial calcium overload, which leads to the opening of the mitochondrial permeability transition pore (MPTP) and permits the translocation of proapoptotic molecules to cytosol from mitochondria [75].

Apart from Parkinson's disease, the dysfunction of mitochondria has also been identified in the pathology of Alzheimer's disease and other common neurodegenerative disorders. The ROS are produced via the oxidation of nicotinamide adenine dinucleotide phosphate (NADPH) and selective neural degeneration. The crucial role of NADPH oxidase in neurodegenerative disease has been confirmed in several animal models. The active form of NADPH oxidase relocates proteins to membranes to enhance healthy neuronal activity and requires the breaching of ion/anion channels to compensate the charge. In microglial cells activated by beta-amyloid proteins, the superoxide production is inhibited [76]. The stimulation of NADPH oxidase by beta-amyloid proteins in combination with the abundant synthesis of nitric oxide can cause the destruction of the surrounding cells. The beta-amyloid proteins also activate NADPH oxidase by causing the entry of calcium into the astrocytes, thereby preventing their entry into the neuronal cells. After NADPH oxidase actuation [77], the mitochondrial membrane is depolarized, which, combined with calcium, leads to the opening of MPTP and changes the membrane structure by phospholipase C stimulation. This signal of oxidative stress is further carried to neighboring neuronal cells that are highly exposed compared to astrocytes. [78]. When the neurons have multiple synapses and long axons, they have high energy requirements for transportation through axons and long-term plasticity.

To fulfil such big energy requirements, a high demand for ATP is rendered, thereby increasing levels of reactive oxygen species in conditions of stress, which imparts a higher degree of neurodegeneration. For example, the neurons in the hippocampus region of the brain generate greater levels of superoxide anions and comprise increased expression of antioxidant and reactive oxygen species that produce genes [79]. The amplified oxidative stress in intrinsic surroundings leads to the dysfunction of mitochondria, which further generates more free radicals, thereby exaggerating the oxidative stress cycle.

Figure 4 represents the process of oxidative stress-mediated neurodegeneration. Based on external and internal factors, the production of reactive oxygen species increases in the body, which causes an overall increase in the oxidative stress. This oxidative stress causes aberrant phosphorylation of proteins, leading to synthesis of misfolded proteins, which alters the normal functioning of neuronal transmission. Due to altered protein structures, they get aggregated and lose the ability to transport, thereby hampering the synaptic activity and causing the initiation and progression of neurodegeneration.

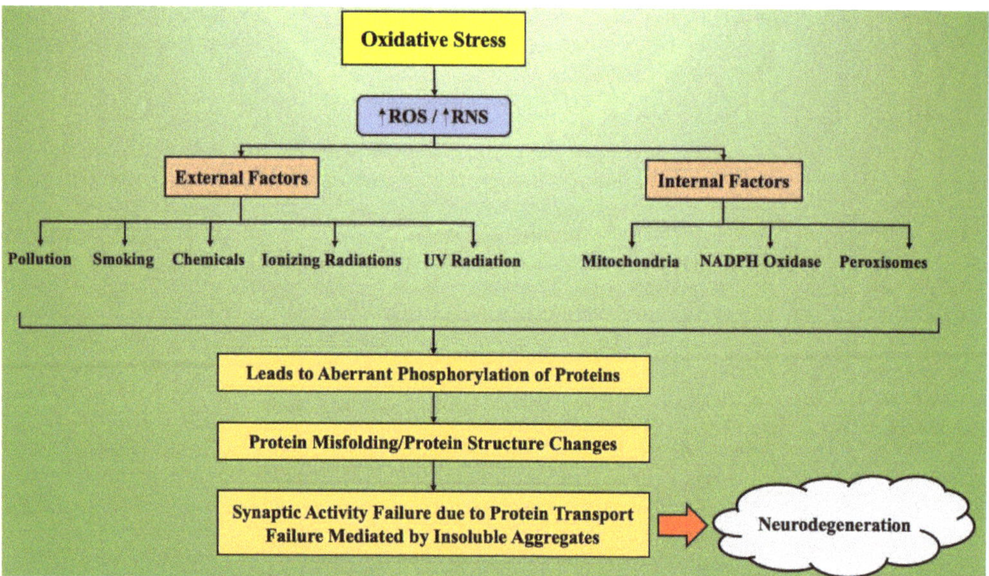

Figure 4. Synthesizing the process of oxidative stress-mediated neurodegeneration. Legend: ROS—reactive oxygen species; RNS—reactive nitrogen species; UV—ultraviolet; NADP—nicotinamide adenine dinucleotide.

3.2. *Targeting of Oxidative Stress in Neurodegenerative Diseases*

For the management of oxidative stress-mediated neurodegeneration, it is essential that antioxidant therapies be administered that target reactive oxygen species and suppress oxidative stress, thereby decreasing the intensity of neurodegeneration. Antioxidants are endogenous or exogenous molecules that antagonize the effects produced by reactive oxygen species and oxidative stress in the cellular system. These compounds neutralize the effect of reactive oxygen species and engulf any free radicals, thereby inhibiting conditions of oxidative stress [80].

By altering the foods in our diet, we can enhance the intake of antioxidants by consuming food products rich in flavonoids, lipoic acid, phenolic compounds, beta carotene, and ascorbic acid (Vitamin C). These naturally occurring antioxidants prevent the protein and lipid oxidation process in the body, thereby suppressing the generation of ROS and upstreaming the therapeutic barrier to oxidative stress. The ROS initiate excitotoxicity and modulate the glutamate receptors by over-activating them.

The drugs that target these receptors can be efficiently used in upstreaming the antioxidant profile and in the management of neurodegenerative diseases by blocking these specific receptors. For instance, Alzheimer's disease can be slowed down by administering memantine, which targets N-methyl-D-aspartate receptor and imparts modest benefits to the patient suffering from a moderate to severe range of the disease. An upcoming therapeutic aspect for the prevention of neuronal cell death and the regulation of oxidative stress includes vaccination of the individual against the potential toxic proteins that are formed in several types of neurodegenerative disorders. A promising example is vaccination against the beta-amyloid protein, which is mainly found in patients with Alzheimer's disease. Vaccination against it prevents the formation of plaque and subsequent neuroinflammation.

4. Role of Neuroinflammation in the Onset of Neurodegeneration

Recent studies have demonstrated a bridge between chronic inflammation and neurodegeneration. Apoptosis or programmed cell death and necrosis lead to neuronal cell death in the brain [81]. An increased burden of neurodegenerative conditions on the health care system and a lack of effective treatments available pose an urgent need to identify new drug targets. The most common feature that has been found in several neurodegenerative diseases such as Parkinson's disease and Alzheimer's disease is chronic neuroinflammation. Glial cells have been identified as the mediators of neurodegeneration and are responsible for the onset and progression of these diseases.

Neuronal health is mainly monitored by the nervous system's immune cells, called microglia [82]. These cells are activated upon any injury to neurons or infections, which further produces proinflammatory factors (M1 phenotype) or anti-inflammatory factors (M2 phenotype). For a healthy functioning of the human brain, it is necessary to have the correct balance between anti-inflammatory mediators, which allow for the repair and healing of tissues, and proinflammatory mediators, which clear the cellular debris and misfolded protein aggregation is maintained [83]. The activation of microglia cells in Alzheimer's and Parkinson's disease is mainly tilted towards the M1 phenotype, which leads to an exaggeration of inflammation and accelerates the progression of the disease [84]. The new therapies for the management of neurodegeneration include induction of the M2 phenotype and deactivation of the M1 phenotype in the brain.

Astrocytes are another type of glial cells that are present in the brain and regulate the maintenance and maturation of neuronal cells. They are highly sensitive and respond to injury very quickly, as do microglia cells [85]. Depending upon their activation, they can be neuroprotective by stimulating repair and reducing inflammation or be neurotoxic by promoting inflammation and contributing to the death of neuronal cells. Activated astrocytes also act as a proinflammatory factor in Parkinson's disease and Alzheimer's disease [86]. Their role has been implicated in the breakdown of the blood–brain barrier, thereby encouraging the infiltration of immune cells into the brain, which increases neuronal death by excessive stimulation and impairment of the uptake of neurotransmitter glutamate. Another type of glial cell, called oligodendrocytes, also poses a significant role in neurodegeneration [87]. Oligodendrocytes form a sheath of myelin around the nerve fibers, which permits the efficient and rapid transmission of electrical impulses across neurons and thus induces signal transmission. Damage in these cells has been attributed to the progression of multiple sclerosis and other neurodegenerative diseases in which the immune system attacks oligodendrocytes and damages the myelin sheath, thereby reducing levels of myelin proteins in brain tissue [88].

4.1. Sources of Neuroinflammation

The mechanisms involved in neuroinflammation by biomolecular or cellular pathways are alike in aging and metabolic diseases such as depression, dementia, diabetes, and hypertension, as well as cerebral insult conditions such as stroke, and contribute silently to neuroinflammation. In elderly groups of patients, the mechanism of inflammation has been strongly associated with the pathogenesis of dementia and functional impairment [89]. Local and systemic inflammation in CNS significantly contributes to the development of small cerebral vessel diseases such as vascular dementia, which is mainly hypothesized as microvascular changes that cause a state of chronic hyper perfusion and lead to the death of oligodendrocytes and consecutively cause the destruction of myelin fibers, which increases inflammation in the affected brain region [90,91].

Other markers of inflammation such as C-reactive protein (CRP) can help with the prediction of clinical and subclinical atherosclerosis and play a major role in the identification of brain hemorrhage and conditions of stroke. Various other metabolic disorders such as diabetes [92], obesity, and hypertension also lead to dysfunction of the adipose tissue and induce low-grade inflammation, thereby worsening neuroinflammation in patients with a history of stroke [93]. Another factor that induces the synthesis of proinflammatory cytokines is aging processes. The inflammation triggered due to aging in the brain manifests chronic activation of parenchymal and perivascular macrophages and increases expression of microglial cells, thereby increasing the number of astrocytes. This chronic activation and signaling of pro-inflammatory cytokines increases the vulnerability of the patient to psychiatric disorders [94].

Females that are obese are mainly observed to have elevated concentrations of interleukin-6, CRP, and adipokines [95]. All these pro-inflammatory cytokines contribute positively to manifesting symptoms of depression and anxiety. Upon surgical removal of fat tissues, an alleviation in anxiety was noted. Further metabolic diseases such as hypertension and obesity accompanied with older age groups pose as a prevalent risk factor in the initiation of cognitive dysfunction, including depression, dementia, and other neurodegenerative disorders [96].

The biological mechanism responsible for depression is still lucid, and patients are administered conventional anti-depressant therapy for the correction of the disease; however, it has been noted that neuroinflammation suppresses the beneficial effects of the therapy in one-third of patients, leading to resistance in treatment [97]. The putative mechanism that links inflammation and neurodegeneration includes elevated pro-inflammatory cytokines, mainly IL-6 or IL-8 [98], oxidative stress, higher glutamate, and uncoupling of endothelial nitric oxide synthase. In severe psychiatric illnesses such as major depressive disorder, evidence of neurovascular dysfunction was determined to be indirectly associated with increased concentrations of inflammatory cytokines in the periphery [99].

Hyperactivated immune response results in an inflammatory cytokine rush and poses problems to the central nervous system [100]. Immune responses are generally triggered upon exposure to certain infectious agents such as viruses, bacteria, or any other associated pathogens [101]. A pathogen can enter the CNS through two different hypothetical pathways, including hematogenous dissemination, through which the pathogen gains direct access to the brain via the blood–brain barrier, and neuronal retrograde dissemination [102,103]. Different pathogens interact in distinct ways and lead to the activation of macrophages and further stimulation of inflammatory cytokines and neuroinflammation.

4.2. *Neuroinflammatory Process*

Through emerging evidence, the role of neuroinflammation in neurological diseases has been well documented, and by controlling the key factors that are responsible for the immune and inflammation processes, can serve as a key to the prevention and delay of most late-onset CNS disorders [104]. Poor quality of life and unhealthy food habits have led to the onset and progression of several diseases such as stroke, hypertension, diabetes, obesity, etc., which causes alteration in lipid hormones, adipokines, and inflammatory cytokines, thereby inducing adverse regulatory responses [105]. These metabolic disorders also regulate chronic activation of the pro-inflammatory cytokines even in the CNS, which makes the population vulnerable to neurological disorders.

A correlation between neurological disorders such as anxiety and depression and proinflammatory cytokines has been developed, which indicates their chief role in the instigation of cognitive dysfunction, thereby favoring neurodegradation. An increase in the level of peripheral inflammatory markers such as interleukin (IL)-6 and IL-8 is associated with mortality from suicide and depression. The complement cascade of the activation of microglial cells in response to pruning synapses is a biological mechanism that commonly occurs in neuroinflammatory processes and the development of a healthy brain. The generation of neurotrophic factors can be compromised by the activation of the immune system and lead to microglial cell-mediated secretion of cytotoxic factors.

Microglial cells are the main agents involved in neuroinflammation, with the second being astroglia cells [106]. Astrocytes are responsive to all forms of CNS insults due to the reactive astrogliosis process. The activation of astrocytes is highly complex, multistage, and a pathogen-specific reaction, and mainly aims for neuroprotection and recovering the injured and damaged neuronal tissue. It is evident that the sustained inflammatory responses in the central nervous system support the major contribution of astrocytes and microglia to neurological disease progression, thereby implying their chief role as effectors of neuroinflammation in neuron dysfunction and cell death [107]. Endothelial cells, neurons, and many other cells act as receptors for cytokines and inflammatory mediators, thereby activating the signaling of inflammatory responses in the brain.

4.3. Neuroinflammation Mediated by Microglial Cells

Microglial cells are widely distributed throughout the region of the brain and spinal cord and are mainly present in the substantia nigra and hippocampus region of the brain. These cells account for 5–20% of the population of all glia cells in the central nervous system and chiefly represent the immune response, as they possess the ability to perform phagocytosis, manifest the secretion of cytotoxic factors, and act like antigen-presenting cells. These cells originate in the primitive yolk sac during hematopoiesis and later rise and migrate to the developing neural tube [108]. These cells comprise the major cellular component of the immune system of the brain. In healthy conditions, microglia cells initiate quick responses to infection and several other stimuli, thereby modulating inflammation and protecting the environment of the brain.

Microglial cells sense different types of stimuli such as bacterial, viral, or fungal infections; antibodies; compliment factors; cytokines; and chemokines that threaten the homeostasis of the CNS and become activated [109]. These cells comprise two main functions: immune defense, by acting as sentinels and detecting the first signs of pathogen invasion and tissue damage; and maintenance by remodeling, repairing, and supporting the tissues of the central nervous system. In response to environmental and toxic factors, microglial cells induce inflammatory and immune responses in the CNS by becoming functionally polarized and expressing specific functioning reactions by activating pro-inflammatory cytokines and expressing the surface of receptors through the release of numerous soluble factors [110]. The stimulus can polarize microglial cells toward the distinctive functional roles, which causes the expression of different proteins and cytokines.

Activated inflammatory phenotype microglial cells upregulate the release of proinflammatory cytokines such as IL-1β, IL-6, IL-12, IL-23, inducible nitric oxide synthase (iNOS), and tumor necrosis factor (TNF- α), whereas activated anti-inflammatory phenotype microglial cells lead to the upregulation of arginase-1 [111], insulin-like growth factor (IGF-1), mannose receptor (CD206), chitinase 3, and triggering receptor expressed on myeloid cells 2 (TREM2). All of these inflammatory factors contribute to the process of microglial cell activation, which further leads to the production of cytokines and other inflammatory mediators, hence contributing to apoptotic cell death and neurodegeneration.

4.4. Neuroinflammation Mediated by Astrocyte Cells

These cells are heterogeneous in type and are highly abundant in the CNS [112]. Based on the stage of development, localization, and subtype, the morphology of the cells can change. For instance, the astrocytes in white matter are fibrous and exhibit long unbranched processes, whereas the astrocytes in grey matter are protoplasmic and possess short branches. These cells act as sensitive markers for the detection of any disease in the neuronal tissue and are responsible for the crucial functioning of the central nervous system [113]. They play a primary role in the transmission of signals in the synapses and control neural circuits. Astrocytes control the CNS environment by maintaining ion homeostasis, blood flow, and pH regulation and controlling oxidative stress. Based on the activities performed by astrocytes, these cells, in combination with microglial cells, act as the main factor in neuroinflammation [114]. In addition, astrocytes rapidly detect damage signals after any injury and undergo changes in their functioning and morphology in response to control and remove brain insults, but if not controlled, the response may have deleterious consequences.

The pathway that leads to the activation of astrocytes is still lucid and several factors that trigger the activation of these cells are involved in the neuropathology of neurodegenerative diseases [104]. It has been demonstrated that in patients with Alzheimer's disease, astrocytes are activated by the presence of amyloid proteins, thereby causing local inflammation. The internal activation of astrocytes causes the transcription of nuclear factor kappa B (NF-κB), which further regulates the secretion and adhesion of chemokines and causes the infiltration of lymphocytes in the cerebrospinal fluid, thereby increasing inflammatory reactions and leading to neurodegeneration [115]. Blocking the activity of NF-κB protein in astrocytes can significantly reduce neuroinflammation, and it is suggested as a potential therapy for Alzheimer's disease and several other associated neurological disorders.

Figure 5 depicts the exposure to certain environmental toxins, chemicals, or genetic factors that can lead to the activation of anti-myelin T-lymphocytes. These T-lymphocytes, upon activation, initiate immune responses and inflammation as defense factors. Prolonged exposure to these toxins can induce increased production of inflammatory mediators and cytokines and generate hyper-immune response, which causes the infiltration of inflammatory cytokines and lymphocytes into the cerebrospinal fluid and initiates neuroinflammation. Microglial cells (plural: microglia) are the chief neuronal cells responsible for immune responses and inflammation in the central nervous system. Exposure to certain toxicants can induce alterations in healthy microglial cells, leading to their hypertrophy and dystrophy based on the intensity of the stress and thereby causing neuroinflammation and ultimately neurodegradation.

4.5. Strategies to Fight off Neuroinflammation in Neurodegenerative Diseases

With progressive understanding of the pathophysiology of neurogenerative diseases, the role of neuroinflammation in the progression of neurodegenerative diseases is highly acknowledged. Most of the pathologies of CNS are characterized by an early activation of microglial cells, which further initiate an early acute reparative phase via the effective removal of threatening compounds and toxic agents. However, this response mostly fails to completely remove all the threats and results in a vicious cycle of unresolved cytotoxic inflammation [116]. This has led to the development of therapies that target arresting such vicious cycles by resolving immune responses and recruiting systemic monocyte-derived macrophages. Below are some of the ways that can aid in fighting neuroinflammation and prevent neurodegenerative diseases [117]. The recruitment of CNS-specific T-cells, myeloid cells, microglial cells, and CNS-infiltrating monocyte-derived macrophages can be used in neuroprotection and to prevent neuroinflammation. Circulating myeloid cells possess neuroprotective effects and support CNS during conditions of neuroinflammation and neurodegradation [118]. It has also been observed that infiltrating blood-derived macrophages exhibit the capacity for phagocytosis and possess more neurotropic and anti-inflammatory effects than microglial cells.

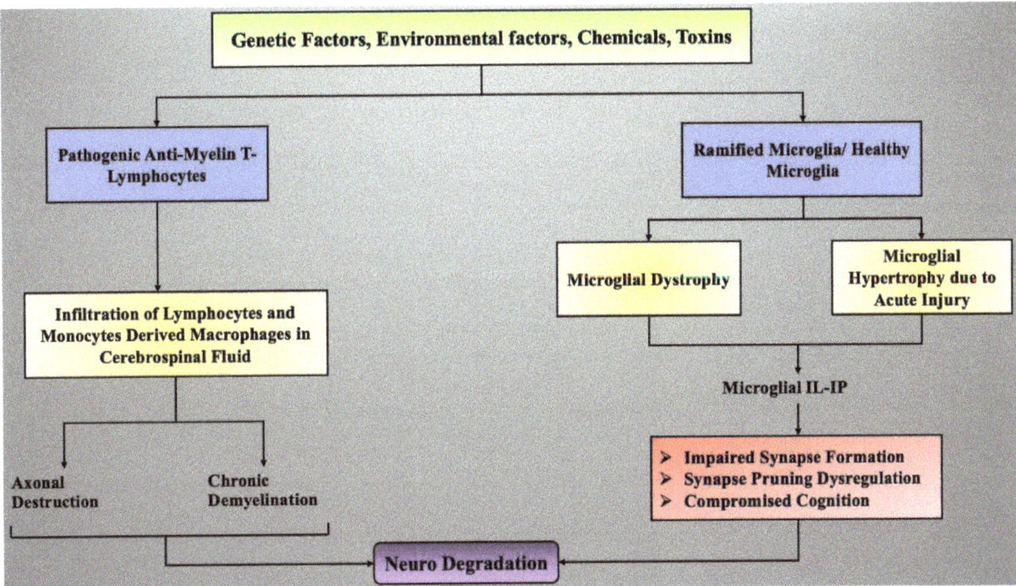

Figure 5. Environmental factors, toxins, chemicals, or genetic factors that can lead to the activation of anti-myelin T-lymphocytes. Legend: IL—interleukin; IP—inducible protein.

5. Conclusions

The number of patients suffering from neurological disorders is increasing at an alarming rate. Several pathologies have been identified that cause the onset of neurological diseases; however, there is still a significant gap in the understanding and targeting of them for the development of medications for prevention and cure. With advancing technologies, the understanding of molecular mechanisms in populations exposed to vulnerable neurodegenerative diseases is also improving, which will lead to deeper insights into the functional interplay between several factors such as oxidative stress, neuronal cell death, and neuroinflammation in the occurrence of neurological diseases, which has been described in the article. Eventually, studying pathological mechanisms can lead to the targeted derived production of therapies that will improve global health and invoke management of neurological disorders, thereby improving the quality of life of individuals.

Author Contributions: All authors contributed equally to this paper. All authors have read and agreed to the published version of the manuscript.

Funding: This review received no external funding.

Institutional Review Board Statement: Not applicable.

Informed Consent Statement: Not applicable.

Acknowledgments: The authors would like to thank Chitkara College of Pharmacy, Chitkara University, Punjab, India, and University of Oradea, Oradea, Romania, for providing the various resources for the completion of the current article.

Conflicts of Interest: The authors declare no conflict of interest.

References

1. Gan, L.; Cookson, M.R.; Petrucelli, L.; La Spada, A.R. Converging pathways in neurodegeneration, from genetics to mechanisms. *Nat. Neurosci.* **2018**, *21*, 1300–1309. [CrossRef] [PubMed]
2. Hardy, J.; Orr, H. The genetics of neurodegenerative diseases. *J. Neurochem.* **2006**, *97*, 1690–1699. [CrossRef] [PubMed]

3. Goedert, M.; Eisenberg, D.S.; Crowther, R.A. Propagation of tau aggregates and neurodegeneration. *Annu. Rev. Neurosci.* **2017**, *40*, 189–210. [CrossRef] [PubMed]
4. Gitler, A.D.; Dhillon, P.; Shorter, J. Neurodegenerative disease: Models, mechanisms, and a new hope. *Dis. Model. Mech.* **2017**, *10*, 499–502. [CrossRef]
5. Jellinger, K.A. Basic mechanisms of neurodegeneration: A critical update. *J. Cell. Mol. Med.* **2010**, *14*, 457–487. [CrossRef] [PubMed]
6. Jack, C.R., Jr.; Wiste, H.J.; Weigand, S.D.; Rocca, W.A.; Knopman, D.S.; Mielke, M.M.; Lowe, V.J.; Senjem, M.L.; Gunter, J.L.; Preboske, G.M.; et al. Age-specific population frequencies of cerebral β-amyloidosis and neurodegeneration among people with normal cognitive function aged 50–89 years: A cross-sectional study. *Lancet Neurol.* **2014**, *13*, 997–1005. [CrossRef]
7. Cuervo, A.M. Autophagy and aging: Keeping that old broom working. *Trends Genet.* **2008**, *24*, 604–612. [CrossRef] [PubMed]
8. Craik, F. Memory Changes in Normal Aging. *Curr. Dir. Psychol. Sci.* **1994**, *3*, 155–158. [CrossRef]
9. Bozoki, A.; Giordani, B.; Heidebrink, J.L.; Berent, S.; Foster, N.L. Mild cognitive impairments predict dementia in nondemented elderly patients with memory loss. *Arch. Neurol.* **2001**, *58*, 411–416. [CrossRef]
10. Deuschl, G.; Beghi, E.; Fazekas, F.; Varga, T.; Christoforidi, K.A.; Sipido, E.; Bassetti, C.L.; Vos, T.; Feigin, V.L. The burden of neurological diseases in Europe: An analysis for the Global Burden of Disease Study 2017. *Lancet Public Health* **2020**, *5*, E551–E567. [CrossRef]
11. Alladi, S.; Hachinski, V. World dementia: One approach does not fit all. *Neurology* **2018**, *91*, 264–270. [CrossRef] [PubMed]
12. Taylor, J.P.; Hardy, J.; Fischbeck, K.H. Toxic proteins in neurodegenerative disease. *Science* **2002**, *296*, 1991–1995. [CrossRef] [PubMed]
13. Trojanowski, J.Q.; Hampel, H. Neurodegenerative disease biomarkers: Guideposts for disease prevention through early diagnosis and intervention. *Prog. Neurobiol.* **2011**, *95*, 491–495. [CrossRef] [PubMed]
14. Makkar, R.; Behl, T.; Bungau, S.; Kumar, A.; Arora, S. Understanding the Role of Inflammasomes in Rheumatoid Arthritis. *Inflammation* **2020**, *43*, 2033–2047. [CrossRef]
15. Behl, T.; Kumar, C.; Makkar, R.; Gupta, A.; Sachdeva, M. Intercalating the role of microRNAs in cancer: As enemy or protector. *Asian Pac. J. Cancer Prev. APJCP* **2020**, *21*, 593–598. [CrossRef]
16. Palace, J. Inflammation versus neurodegeneration: Consequences for treatment. *J. Neurol. Sci.* **2007**, *259*, 46–49. [CrossRef] [PubMed]
17. Adlimoghaddam, A.; Neuendorff, M.; Roy, B.; Albensi, B.C. A review of clinical treatment considerations of donepezil in severe Alzheimer's disease. *CNS Neurosci. Ther.* **2018**, *24*, 876–888. [CrossRef] [PubMed]
18. Gupta, A.; Shah, K.; Oza, M.J.; Behl, T. Reactivation of p53 gene by MDM2 inhibitors: A novel therapy for cancer treatment. *Biomed. Pharmacother.* **2019**, *109*, 484–492. [CrossRef]
19. Makkar, R.; Behl, T.; Bungau, S.; Zengin, G.; Mehta, V.; Kumar, A.; Uddin, M.; Ashraf, G.M.; Abdel-Daim, M.M.; Arora, S.; et al. Nutraceuticals in neurological disorders. *Int. J. Mol. Sci.* **2020**, *21*, 4424. [CrossRef] [PubMed]
20. Menzies, F.M.; Fleming, A.; Caricasole, A.; Bento, C.F.; Andrews, S.P.; Ashkenazi, A.; Füllgrabe, J.; Jackson, A.; Sanchez, M.J.; Karabiyik, C.; et al. Autophagy and neurodegeneration: Pathogenic mechanisms and therapeutic opportunities. *Neuron* **2017**, *93*, 1015–1034. [CrossRef] [PubMed]
21. Burbulla, L.F.; Song, P.; Mazzulli, J.R.; Zampese, E.; Wong, Y.C.; Jeon, S.; Santos, D.P.; Blanz, J.; Obermaier, C.D.; Strojny, C. Dopamine oxidation mediates mitochondrial and lysosomal dysfunction in Parkinson's disease. *Science* **2017**, *357*, 1255–1261. [CrossRef] [PubMed]
22. Kundaje, A.; Meuleman, W.; Ernst, J.; Bilenky, M.; Yen, A.; Heravi-Moussavi, A.; Kheradpour, P.; Zhang, Z.; Wang, J.; Ziller, M.J. Integrative analysis of 111 reference human epigenomes. *Nature* **2015**, *518*, 317–330. [CrossRef] [PubMed]
23. Taylor, J.P.; Brown, R.H.; Cleveland, D.W. Decoding ALS: From genes to mechanism. *Nature* **2016**, *539*, 197–206. [CrossRef]
24. Lui, H.; Zhang, J.; Makinson, S.R.; Cahill, M.K.; Kelley, K.W.; Huang, H.-Y.; Shang, Y.; Oldham, M.C.; Martens, L.H.; Gao, F. Progranulin deficiency promotes circuit-specific synaptic pruning by microglia via complement activation. *Cell* **2016**, *165*, 921–935. [CrossRef] [PubMed]
25. Abisambra, J.F.; Jinwal, U.K.; Blair, L.J.; O'Leary, J.C.; Li, Q.; Brady, S.; Wang, L.; Guidi, C.E.; Zhang, B.; Nordhues, B.A. Tau accumulation activates the unfolded protein response by impairing endoplasmic reticulum-associated degradation. *J. Neurosci.* **2013**, *33*, 9498–9507. [CrossRef] [PubMed]
26. Ashrafi, G.; Schlehe, J.S.; LaVoie, M.J.; Schwarz, T.L. Mitophagy of damaged mitochondria occurs locally in distal neuronal axons and requires PINK1 and Parkin. *J. Cell Biol.* **2014**, *206*, 655–670. [CrossRef] [PubMed]
27. Atkin, J.D.; Farg, M.A.; Walker, A.K.; McLean, C.; Tomas, D.; Horne, M.K. Endoplasmic reticulum stress and induction of the unfolded protein response in human sporadic amyotrophic lateral sclerosis. *Neurobiol. Dis.* **2008**, *30*, 400–407. [CrossRef] [PubMed]
28. Bejanin, A.; Schonhaut, D.R.; La Joie, R.; Kramer, J.H.; Baker, S.L.; Sosa, N.; Ayakta, N.; Cantwell, A.; Janabi, M.; Lauriola, M. Tau pathology and neurodegeneration contribute to cognitive impairment in Alzheimer's disease. *Brain* **2017**, *140*, 3286–3300. [CrossRef]
29. Okouchi, M.; Ekshyyan, O.; Maracine, M.; Aw, T.Y. Neuronal apoptosis in neurodegeneration. *Antioxid. Redox Signal.* **2007**, *9*, 1059–1096. [CrossRef] [PubMed]

30. Benarroch, E.E. Acquired axonal degeneration and regeneration: Recent insights and clinical correlations. *Neurology* **2015**, *84*, 2076–2085. [CrossRef]
31. Hara, T.; Nakamura, K.; Matsui, M.; Yamamoto, A.; Nakahara, Y.; Suzuki-Migishima, R.; Yokoyama, M.; Mishima, K.; Saito, I.; Okano, H. Suppression of basal autophagy in neural cells causes neurodegenerative disease in mice. *Nature* **2006**, *441*, 885–889. [CrossRef] [PubMed]
32. Deckwerth, T.L.; Elliott, J.L.; Knudson, C.M.; Johnson, E.M., Jr.; Snider, W.D.; Korsmeyer, S.J. BAX is required for neuronal death after trophic factor deprivation and during development. *Neuron* **1996**, *17*, 401–411. [CrossRef]
33. Mousavi, S.H.; Tavakkol-Afshari, J.; Brook, A.; Jafari-Anarkooli, I. Role of caspases and Bax protein in saffron-induced apoptosis in MCF-7 cells. *Food Chem. Toxicol.* **2009**, *47*, 1909–1913. [CrossRef] [PubMed]
34. Zhang, S.; Tang, M.-B.; Luo, H.-Y.; Shi, C.-H.; Xu, Y.-M. Necroptosis in neurodegenerative diseases: A potential therapeutic target. *Cell Death Dis.* **2017**, *8*, e2905. [CrossRef] [PubMed]
35. Borsello, T.; Forloni, G. JNK signalling: A possible target to prevent neurodegeneration. *Curr. Pharm. Des.* **2007**, *13*, 1875–1886. [CrossRef] [PubMed]
36. Sengupta Ghosh, A.; Wang, B.; Pozniak, C.D.; Chen, M.; Watts, R.J.; Lewcock, J.W. DLK induces developmental neuronal degeneration via selective regulation of proapoptotic JNK activity. *J. Cell Biol.* **2011**, *194*, 751–764. [CrossRef] [PubMed]
37. Huntwork-Rodriguez, S.; Wang, B.; Watkins, T.; Ghosh, A.S.; Pozniak, C.D.; Bustos, D.; Newton, K.; Kirkpatrick, D.S.; Lewcock, J.W. JNK-mediated phosphorylation of DLK suppresses its ubiquitination to promote neuronal apoptosis. *J. Cell Biol.* **2013**, *202*, 747–763. [CrossRef] [PubMed]
38. Castellani, R.J.; Plascencia-Villa, G.; Perry, G. The amyloid cascade and Alzheimer's disease therapeutics: Theory versus observation. *Lab. Investig.* **2019**, *99*, 958–970. [CrossRef] [PubMed]
39. Larhammar, M.; Huntwork-Rodriguez, S.; Jiang, Z.; Solanoy, H.; Ghosh, A.S.; Wang, B.; Kaminker, J.S.; Huang, K.; Eastham-Anderson, J.; Siu, M. Dual leucine zipper kinase-dependent PERK activation contributes to neuronal degeneration following insult. *eLife* **2017**, *6*, e20725. [CrossRef] [PubMed]
40. Robitaille, H.; Proulx, R.; Robitaille, K.; Blouin, R.; Germain, L. The mitogen-activated protein kinase kinase kinase dual leucine zipper-bearing kinase (DLK) acts as a key regulator of keratinocyte terminal differentiation. *J. Biol. Chem.* **2005**, *280*, 12732–12741. [CrossRef] [PubMed]
41. Burke, R.E.; O'Malley, K. Axon degeneration in Parkinson's disease. *Exp. Neurol.* **2013**, *246*, 72–83. [CrossRef]
42. Ciryam, P.; Kundra, R.; Freer, R.; Morimoto, R.I.; Dobson, C.M.; Vendruscolo, M. A transcriptional signature of Alzheimer's disease is associated with a metastable subproteome at risk for aggregation. *Proc. Natl. Acad. Sci. USA* **2016**, *113*, 4753–4758. [CrossRef] [PubMed]
43. Cashman, C.R.; Höke, A. Mechanisms of distal axonal degeneration in peripheral neuropathies. *Neurosci. Lett.* **2015**, *596*, 33–50. [CrossRef] [PubMed]
44. Codolo, G.; Plotegher, N.; Pozzobon, T.; Brucale, M.; Tessari, I.; Bubacco, L.; de Bernard, M. Triggering of inflammasome by aggregated α–synuclein, an inflammatory response in synucleinopathies. *PLoS ONE* **2013**, *8*, e55375. [CrossRef] [PubMed]
45. Shin, J.E. The Roles of PHR and DLK in Axon Development and Post-Injury Responses. Ph.D. Thesis, Washington University in St. Louis, St. Louis, MO, USA, 2012.
46. Bellucci, A.; Navarria, L.; Zaltieri, M.; Falarti, E.; Bodei, S.; Sigala, S.; Battistin, L.; Spillantini, M.; Missale, C.; Spano, P. Induction of the unfolded protein response by α-synuclein in experimental models of Parkinson's disease. *J. Neurochem.* **2011**, *116*, 588–605. [CrossRef] [PubMed]
47. Davies, P.; Maloney, A. Selective loss of central cholinergic neurons in Alzheimer's disease. *Lancet* **1976**, *308*, 1403. [CrossRef]
48. Dawson, T.M.; Dawson, V.L. Mitochondrial mechanisms of neuronal cell death: Potential therapeutics. *Annu. Rev. Pharmacol. Toxicol.* **2017**, *57*, 437–454. [CrossRef]
49. Golpich, M.; Amini, E.; Mohamed, Z.; Azman Ali, R.; Mohamed Ibrahim, N.; Ahmadiani, A. Mitochondrial dysfunction and biogenesis in neurodegenerative diseases: Pathogenesis and treatment. *CNS Neurosci. Ther.* **2017**, *23*, 5–22. [CrossRef] [PubMed]
50. Owen, L.; Sunram-Lea, S.I. Metabolic agents that enhance ATP can improve cognitive functioning: A review of the evidence for glucose, oxygen, pyruvate, creatine, and L-carnitine. *Nutrients* **2011**, *3*, 735–755. [CrossRef] [PubMed]
51. Gill, S.S.; Tuteja, N. Reactive oxygen species and antioxidant machinery in abiotic stress tolerance in crop plants. *Plant Physiol. Biochem.* **2010**, *48*, 909–930. [CrossRef]
52. Egea, G.; Jimenez-Altayo, F.; Campuzano, V. Reactive Oxygen Species and Oxidative Stress in the Pathogenesis and Progression of Genetic Diseases of the Connective Tissue. *Antioxidants* **2020**, *9*, 1013. [CrossRef] [PubMed]
53. Devi, L.; Raghavendran, V.; Prabhu, B.M.; Avadhani, N.G.; Anandatheerthavarada, H.K. Mitochondrial import and accumulation of α-synuclein impair complex I in human dopaminergic neuronal cultures and Parkinson disease brain. *J. Biol. Chem.* **2008**, *283*, 9089–9100. [CrossRef] [PubMed]
54. Lin, M.T.; Beal, M.F. Mitochondrial dysfunction and oxidative stress in neurodegenerative diseases. *Nature* **2006**, *443*, 787–795. [CrossRef]
55. Halliwell, B. Oxidative stress and neurodegeneration: Where are we now? *J. Neurochem.* **2006**, *97*, 1634–1658. [CrossRef] [PubMed]
56. Dröge, W. Free radicals in the physiological control of cell function. *Physiol. Rev.* **2002**, *82*, 47–95. [CrossRef] [PubMed]
57. Praticò, D. Evidence of oxidative stress in Alzheimer's disease brain and antioxidant therapy: Lights and shadows. *Ann. N. Y. Acad. Sci.* **2008**, *1147*, 70–78. [CrossRef]

58. Dalfó, E.; Portero-Otín, M.; Ayala, V.; Martínez, A.; Pamplona, R.; Ferrer, I. Evidence of oxidative stress in the neocortex in incidental Lewy body disease. *J. Neuropathol. Exp. Neurol.* **2005**, *64*, 816–830. [CrossRef] [PubMed]
59. Beal, M.F. Oxidatively modified proteins in aging and disease. *Free Radic. Biol. Med.* **2002**, *32*, 797–803. [CrossRef]
60. Seet, R.C.; Lee, C.-Y.J.; Lim, E.C.; Tan, J.J.; Quek, A.M.; Chong, W.-L.; Looi, W.-F.; Huang, S.-H.; Wang, H.; Chan, Y.-H. Oxidative damage in Parkinson disease: Measurement using accurate biomarkers. *Free Radic. Biol. Med.* **2010**, *48*, 560–566. [CrossRef] [PubMed]
61. Schapira, A.H. Mitochondria in the aetiology and pathogenesis of Parkinson's disease. *Lancet Neurol.* **2008**, *7*, 97–109. [CrossRef]
62. Lev, N.; Ickowicz, D.; Melamed, E.; Offen, D. Oxidative insults induce DJ-1 upregulation and redistribution: Implications for neuroprotection. *Neurotoxicology* **2008**, *29*, 397–405. [CrossRef] [PubMed]
63. Gandhi, S.; Wood-Kaczmar, A.; Yao, Z.; Plun-Favreau, H.; Deas, E.; Klupsch, K.; Downward, J.; Latchman, D.S.; Tabrizi, S.J.; Wood, N.W. PINK1-associated Parkinson's disease is caused by neuronal vulnerability to calcium-induced cell death. *Mol. Cell* **2009**, *33*, 627–638. [CrossRef]
64. Aruoma, O.I. Assessment of potential prooxidant and antioxidant actions. *J. Am. Oil Chem. Soc.* **1996**, *73*, 1617–1625. [CrossRef]
65. Ercal, N.; Gurer-Orhan, H.; Aykin-Burns, N. Toxic metals and oxidative stress part I: Mechanisms involved in metal-induced oxidative damage. *Curr. Top. Med. Chem.* **2001**, *1*, 529–539. [CrossRef]
66. Cuypers, A.; Vangronsveld, J.; Clijsters, H. The chemical behaviour of heavy metals plays a prominent role in the induction of oxidative stress. *Free Radic. Res.* **1999**, *31*, 34–38. [CrossRef]
67. Farina, M.; Avila, D.S.; Da Rocha, J.B.T.; Aschner, M. Metals, oxidative stress and neurodegeneration: A focus on iron, manganese and mercury. *Neurochem. Int.* **2013**, *62*, 575–594. [CrossRef]
68. Aschner, M.; Aschner, J.L. Mercury neurotoxicity: Mechanisms of blood-brain barrier transport. *Neurosci. Biobehav. Rev.* **1990**, *14*, 169–176. [CrossRef]
69. Prakash, C.; Soni, M.; Kumar, V. Mitochondrial oxidative stress and dysfunction in arsenic neurotoxicity: A review. *J. Appl. Toxicol.* **2016**, *36*, 179–188. [CrossRef] [PubMed]
70. Wu, C.; Zhao, W.; Yu, J.; Li, S.; Lin, L.; Chen, X. Induction of ferroptosis and mitochondrial dysfunction by oxidative stress in PC12 cells. *Sci. Rep.* **2018**, *8*, 1–11. [CrossRef]
71. Barnham, K.J.; Masters, C.L.; Bush, A.I. Neurodegenerative diseases and oxidative stress. *Nat. Rev. Drug Discov.* **2004**, *3*, 205–214. [CrossRef]
72. Gandhi, S.; Abramov, A.Y. Mechanism of oxidative stress in neurodegeneration. *Oxidative Med. Cell. Longev.* **2012**, *2012*. [CrossRef] [PubMed]
73. Emerit, J.; Edeas, M.; Bricaire, F. Neurodegenerative diseases and oxidative stress. *Biomed. Pharmacother.* **2004**, *58*, 39–46. [CrossRef] [PubMed]
74. Federico, A.; Cardaioli, E.; Da Pozzo, P.; Formichi, P.; Gallus, G.N.; Radi, E. Mitochondria, oxidative stress and neurodegeneration. *J. Neurol. Sci.* **2012**, *322*, 254–262. [CrossRef] [PubMed]
75. Piccoli, C.; Sardanelli, A.; Scrima, R.; Ripoli, M.; Quarato, G.; D'Aprile, A.; Bellomo, F.; Scacco, S.; De Michele, G.; Filla, A. Mitochondrial respiratory dysfunction in familiar parkinsonism associated with PINK1 mutation. *Neurochem. Res.* **2008**, *33*, 2565–2574. [CrossRef]
76. Kim, G.H.; Kim, J.E.; Rhie, S.J.; Yoon, S. The role of oxidative stress in neurodegenerative diseases. *Exp. Neurobiol.* **2015**, *24*, 325–340. [CrossRef]
77. Glass, C.K.; Saijo, K.; Winner, B.; Marchetto, M.C.; Gage, F.H. Mechanisms underlying inflammation in neurodegeneration. *Cell* **2010**, *140*, 918–934. [CrossRef] [PubMed]
78. Ito, Y.; Ofengeim, D.; Najafov, A.; Das, S.; Saberi, S.; Li, Y.; Hitomi, J.; Zhu, H.; Chen, H.; Mayo, L. RIPK1 mediates axonal degeneration by promoting inflammation and necroptosis in ALS. *Science* **2016**, *353*, 603–608. [CrossRef]
79. Esteves, A.R.; Arduíno, D.M.; Swerdlow, R.H.; Oliveira, C.R.; Cardoso, S.M. Oxidative stress involvement in α-synuclein oligomerization in Parkinson's disease cybrids. *Antioxid. Redox Signal.* **2009**, *11*, 439–448. [CrossRef] [PubMed]
80. Drouin-Ouellet, J.; Cicchetti, F. Inflammation and neurodegeneration: The story 'retolled'. *Trends Pharmacol. Sci.* **2012**, *33*, 542–551. [CrossRef]
81. McGeer, P.L.; McGeer, E.G. Inflammation and neurodegeneration in Parkinson's disease. *Parkinsonism Relat. Disord.* **2004**, *10*, S3–S7. [CrossRef]
82. Gosselin, D.; Skola, D.; Coufal, N.G.; Holtman, I.R.; Schlachetzki, J.C.; Sajti, E.; Jaeger, B.N.; O'Connor, C.; Fitzpatrick, C.; Pasillas, M.P.; et al. An environment-dependent transcriptional network specifies human microglia identity. *Science* **2017**, *356*, eaal3222. [CrossRef]
83. Esiri, M.M. The interplay between inflammation and neurodegeneration in CNS disease. *J. Neuroimmunol.* **2007**, *184*, 4–16. [CrossRef] [PubMed]
84. Dringen, R. Oxidative and antioxidative potential of brain microglial cells. *Antioxid. Redox Signal.* **2005**, *7*, 1223–1233. [CrossRef]
85. Colombo, E.; Farina, C. Astrocytes: Key regulators of neuroinflammation. *Trends Immunol.* **2016**, *37*, 608–620. [CrossRef]
86. Rafieian-Kopaei, M.; Baradaran, A.; Rafieian, M. Oxidative stress and the paradoxical effects of antioxidants. *J. Res. Med. Sci.* **2013**, *18*, 628.
87. McManus, R.M.; Heneka, M.T. Role of neuroinflammation in neurodegeneration: New insights. *Alzheimer's Res. Ther.* **2017**, *9*, 1–7. [CrossRef] [PubMed]

88. Gelders, G.; Baekelandt, V.; Van der Perren, A. Linking neuroinflammation and neurodegeneration in Parkinson's disease. *J. Immunol. Res.* **2018**, *2018*, 4784268. [CrossRef]
89. Ellwardt, E.; Zipp, F. Molecular mechanisms linking neuroinflammation and neurodegeneration in MS. *Exp. Neurol.* **2014**, *262*, 8–17. [CrossRef]
90. Di Filippo, M.; Chiasserini, D.; Tozzi, A.; Picconi, B.; Calabresi, P. Mitochondria and the link between neuroinflammation and neurodegeneration. *J. Alzheimer's Dis.* **2010**, *20*, S369–S379. [CrossRef]
91. Schwartz, M.; Baruch, K. The resolution of neuroinflammation in neurodegeneration: Leukocyte recruitment via the choroid plexus. *EMBO J.* **2014**, *33*, 7–22. [CrossRef]
92. Kempuraj, D.; Thangavel, R.; Selvakumar, G.P.; Zaheer, S.; Ahmed, M.E.; Raikwar, S.P.; Zahoor, H.; Saeed, D.; Natteru, P.A.; Iyer, S. Brain and peripheral atypical inflammatory mediators potentiate neuroinflammation and neurodegeneration. *Front. Cell. Neurosci.* **2017**, *11*, 216. [CrossRef]
93. van Dijk, G.; van Heijningen, S.; Reijne, A.C.; Nyakas, C.; van der Zee, E.A.; Eisel, U.L. Integrative neurobiology of metabolic diseases, neuroinflammation, and neurodegeneration. *Front. Neurosci.* **2015**, *9*, 173. [CrossRef] [PubMed]
94. Ward, R.J.; Dexter, D.T.; Crichton, R.R. Ageing, neuroinflammation and neurodegeneration. *Front. Biosci. (Schol. Ed.)* **2015**, *7*, 189–204. [CrossRef] [PubMed]
95. Niranjan, R. Recent advances in the mechanisms of neuroinflammation and their roles in neurodegeneration. *Neurochem. Int.* **2018**, *120*, 13–20. [CrossRef] [PubMed]
96. Peixoto, C.A.; de Oliveira, W.H.; da Racho Araújo, S.M.; Nunes, A.K.S. AMPK activation: Role in the signaling pathways of neuroinflammation and neurodegeneration. *Exp. Neurol.* **2017**, *298*, 31–41. [CrossRef] [PubMed]
97. Venigalla, M.; Sonego, S.; Gyengesi, E.; Sharman, M.J.; Münch, G. Novel promising therapeutics against chronic neuroinflammation and neurodegeneration in Alzheimer's disease. *Neurochem. Int.* **2016**, *95*, 63–74. [CrossRef] [PubMed]
98. Vila, M.; Przedborski, S. Targeting programmed cell death in neurodegenerative diseases. *Nat. Rev. Neurosci.* **2003**, *4*, 365–375. [CrossRef]
99. Cacquevel, M.; Lebeurrier, N.; Chéenne, S.; Vivien, D. Cytokines in neuroinflammation and Alzheimer's disease. *Curr. Drug Targets* **2004**, *5*, 529–534. [CrossRef] [PubMed]
100. Zusso, M.; Stokes, L.; Moro, S.; Giusti, P. Neuroinflammation and Its Resolution: From Molecular Mechanisms to Therapeutic Perspectives. *Front. Pharmacol.* **2020**, *11*, 480. [CrossRef]
101. Schomberg, D.; Ahmed, M.; Miranpuri, G.; Olson, J.; Resnick, D.K. Neuropathic pain: Role of inflammation, immune response, and ion channel activity in central injury mechanisms. *Ann. Neurosci.* **2012**, *19*, 125–132. [CrossRef]
102. Björkqvist, M.; Wild, E.J.; Tabrizi, S.J. Harnessing immune alterations in neurodegenerative diseases. *Neuron* **2009**, *64*, 21–24. [CrossRef]
103. Martin, L.J. Mitochondrial and cell death mechanisms in neurodegenerative diseases. *Pharmaceuticals* **2010**, *3*, 839–915. [CrossRef]
104. Chitnis, T.; Weiner, H.L. CNS inflammation and neurodegeneration. *J. Clin. Investig.* **2017**, *127*, 3577–3587. [CrossRef]
105. Witte, M.E.; Geurts, J.J.; de Vries, H.E.; van der Valk, P.; van Horssen, J. Mitochondrial dysfunction: A potential link between neuroinflammation and neurodegeneration? *Mitochondrion* **2010**, *10*, 411–418. [CrossRef]
106. Kempuraj, D.; Thangavel, R.; Natteru, P.; Selvakumar, G.; Saeed, D.; Zahoor, H.; Zaheer, S.; Iyer, S.; Zaheer, A. Neuroinflammation induces neurodegeneration. *J. Neurol. Neurosurg. Spine* **2016**, *1*, 1003. [PubMed]
107. Chen, W.W.; Zhang, X.; Huang, W.J. Role of neuroinflammation in neurodegenerative diseases. *Mol. Med. Rep.* **2016**, *13*, 3391–3396. [CrossRef]
108. von Mässenhausen, A.; Tonnus, W.; Linkermann, A. Cell death pathways drive necroinflammation during acute kidney injury. *Nephron* **2018**, *140*, 144–147. [CrossRef]
109. Duprez, L.; Wirawan, E.; Berghe, T.V.; Vandenabeele, P. Major cell death pathways at a glance. *Microbes Infect.* **2009**, *11*, 1050–1062. [CrossRef] [PubMed]
110. Najjar, S.; Pearlman, D.M.; Alper, K.; Najjar, A.; Devinsky, O. Neuroinflammation and psychiatric illness. *J. Neuroinflamm.* **2013**, *10*, 1–24. [CrossRef] [PubMed]
111. Mansilla, S.; Llovera, L.; Portugal, J. Chemotherapeutic targeting of cell death pathways. *Anticancer Agents Med. Chem.* **2012**, *12*, 226–238. [CrossRef] [PubMed]
112. Jenkins, V.K.; Timmons, A.K.; McCall, K. Diversity of cell death pathways: Insight from the fly ovary. *Trends Cell Biol.* **2013**, *23*, 567–574. [CrossRef] [PubMed]
113. Arya, R.; White, K. Cell death in development: Signaling pathways and core mechanisms. In *Seminars in Cell & Developmental Biology*; Academic Press: Cambridge, MA, USA, 2015; pp. 12–19.
114. Bredesen, D.E.; Rao, R.V.; Mehlen, P. Cell death in the nervous system. *Nature* **2006**, *443*, 796–802. [CrossRef] [PubMed]
115. Hirsch, E.C.; Vyas, S.; Hunot, S. Neuroinflammation in Parkinson's disease. *Parkinsonism Relat. Disord.* **2012**, *18*, S210–S212. [CrossRef]
116. Lyman, M.; Lloyd, D.G.; Ji, X.; Vizcaychipi, M.P.; Ma, D. Neuroinflammation: The role and consequences. *Neurosci. Res.* **2014**, *79*, 1–12. [CrossRef] [PubMed]
117. Rosenberg, G.A. Matrix metalloproteinases in neuroinflammation. *Glia* **2002**, *39*, 279–291. [CrossRef]
118. Heneka, M.T.; Carson, M.J.; El Khoury, J.; Landreth, G.E.; Brosseron, F.; Feinstein, D.L.; Jacobs, A.H.; Wyss-Coray, T.; Vitorica, J.; Ransohoff, R.M. Neuroinflammation in Alzheimer's disease. *Lancet Neurol.* **2015**, *14*, 388–405. [CrossRef]

International Journal of
Molecular Sciences

Review

Neuroinflammation as a Common Denominator of Complex Diseases (Cancer, Diabetes Type 2, and Neuropsychiatric Disorders)

Serena Asslih, Odeya Damri and Galila Agam *

Department of Clinical Biochemistry and Pharmacology and Psychiatry Research Unit, Faculty of Health Sciences, Ben-Gurion University of the Negev and Mental Health Center, Beer-Sheva 8461144, Israel; serenaa@post.bgu.ac.il (S.A.); odeyad@post.bgu.ac.il (O.D.)
* Correspondence: galila@bgu.ac.il; Tel.: +972-52-5706388

Abstract: The term neuroinflammation refers to inflammation of the nervous tissue, in general, and in the central nervous system (CNS), in particular. It is a driver of neurotoxicity, it is detrimental, and implies that glial cell activation happens prior to neuronal degeneration and, possibly, even causes it. The inflammation-like glial responses may be initiated in response to a variety of cues such as infection, traumatic brain injury, toxic metabolites, or autoimmunity. The inflammatory response of activated microglia engages the immune system and initiates tissue repair. Through translational research the role played by neuroinflammation has been acknowledged in different disease entities. Intriguingly, these entities include both those directly related to the CNS (commonly designated neuropsychiatric disorders) and those not directly related to the CNS (e.g., cancer and diabetes type 2). Interestingly, all the above-mentioned entities belong to the same group of "complex disorders". This review aims to summarize cumulated data supporting the hypothesis that neuroinflammation is a common denominator of a wide variety of complex diseases. We will concentrate on cancer, type 2 diabetes (T2DM), and neuropsychiatric disorders (focusing on mood disorders).

Keywords: neuroinflammation; central nervous system (CNS); cancer; type 2 diabetes (T2DM); neuropsychiatric disorders; mood disorders; glia; cytokines; oxidative stress; blood-brain barrier (BBB)

Citation: Asslih, S.; Damri, O.; Agam, G. Neuroinflammation as a Common Denominator of Complex Diseases (Cancer, Diabetes Type 2, and Neuropsychiatric Disorders). *Int. J. Mol. Sci.* **2021**, *22*, 6138. https://doi.org/10.3390/ijms22116138

Academic Editor: Anne Vejux

Received: 12 April 2021
Accepted: 31 May 2021
Published: 7 June 2021

Publisher's Note: MDPI stays neutral with regard to jurisdictional claims in published maps and institutional affiliations.

1. Introduction

Inflammation is a vital host response to the loss of cellular and tissue homeostasis. Inflammation plays a major role in the pathogenesis of metabolic and behavioral abnormalities during cancer, diabetes type II, and neuropsychiatric disorders.

Neuroinflammation is a term used when there is an inflammation of the nervous tissue, in general, and of the central nervous system (CNS), in particular. Neuroinflammation has been suggested as one of the main drivers of neurotoxic symptoms. It is detrimental, and implies that glial cell activation happens prior to neuronal degeneration and, possibly, causes it.

In neuroinflammation, inflammation-like glial responses that do not reproduce classic characteristics of inflammation in the periphery may be initiated in response to a variety of cues, including infection, traumatic brain injury, toxic metabolites, or autoimmunity [1]. Activation of microglia results in their morphological and phenotypical changes and in the release of inflammatory mediators such as cytokines and chemokines [1]. Molfino et al. [2] further suggest that neuroinflammation is both triggered and perpetuated by the activation of microglia, that the release of inflammatory mediators occurs within hypothalamic areas and that the inflammatory response of activated microglia serves to further engage the immune system and initiate tissue repair.

Complex diseases are diseases caused by a combination of genetic, environmental, and lifestyle factors, most of which have not yet been identified. Thus, these diseases do not obey the Mendelian pattern of inheritance, and genetic factors represent only part of the risk associated with complex disease phenotypes. A vast majority of diseases fall into this category, including Alzheimer's disease, scleroderma, asthma, Parkinson's disease, multiple sclerosis, osteoporosis, connective tissue diseases, kidney diseases, autoimmune diseases, and more. Intriguingly, a common denominator of many of these disorders is neuroinflammation.

2. Neuroinflammation in Cancer

Inflammation predisposes the development of cancer and promotes all stages of tumorigenesis. Namely, "pro-tumorigenic inflammation promotes cancer by blocking anti-tumor immunity, shaping the tumor microenvironment (TME), and by exerting direct tumor-promoting signals and functions onto epithelial and cancer cells" [3].

2.1. Inflammatory Cytokines in Cancer

Cancer is repeatedly accompanied by depression and cognitive difficulties, the etiology of which remains largely unknown. It is conceivable that the mere realization of the disease might cause changes in mood or that mood problems arise from the toxic side effects of chemotherapy and radiation. On the other hand, tumors themselves might have biological effects on the function of the CNS. In some rodent models, tumor and non-tumor cells in the TME (e.g., leukocytes, fibroblasts, endothelial cells) secrete inflammatory mediators such as interleukin (IL)-1, tumor necrosis factor (TNF)-α, IL-6, IL-8, IFN-α, IL-10, IL-12, TGF-β, and CXCR4 [4,5]. The latter promote tumor development, and, in some cases, transduced, among other tissues, into the brain, leading to neuroinflammation, which, in turn, influences behavior [6]. For example, as shown in rats [7], tumors by themselves are enough to induce impairments in working memory and CNS inflammation, cause a chronic increase in blood cytokine levels and in the expression of brain cytokines, enhance negative or positive feedback on glucocorticoid production, and as a result, generate depressive-like behavior.

Using a large cohort of newly diagnosed breast cancer patients, Patel and colleagues investigated the presence of cancer-related symptoms prior to onset of treatment and the association between symptoms, including neurocognitive performance, and circulating pro-inflammatory cytokine levels [8]. Tumor-induced memory impairment was found to be accompanied by increased expression of hippocampal TNFα mRNA in the brain (though not upregulation in peripheral plasma TNFα). They report that higher plasma levels of a marker of TNFα production was associated with poorer verbal memory but not with impaired executive functioning or processing speed performance. They also found higher IL-1 receptor antagonist (IL-1ra) but not IL-6 levels. Their results suggest that elevated pro-inflammatory cytokines elicited by the underlying disease may be sufficient to induce impaired memory performance.

Several studies [9–11] showed that cancer patients suffer from a high prevalence of depression, anxiety, and cognitive disorders. The fact that these disorders are common among other populations afflicted with chronic inflammatory disease stimulated discussion of potential shared biological mechanisms of neuroinflammation and depressed mood to precede major changes resulting, later, in diagnosis of cancer [12]. Potentially, two mechanisms may be involved—cytokine-related or glucocorticoid responses-related. The relative contribution of the two has yet to be understood. The increase observed in behavioral despair in the absence of important, measurable disease behaviors, indicates a selective effect of tumors on affective behaviors [13]. Together, increases in the production of hippocampal cytokines and GR gene expression suggest that the hippocampus may be a neural substrate in which the endocrine and the inflammation-related cytokines merge to induce depressive behavior in chronic disease, in general, and in cancer, in particular [7]. According to Molfino et al. [14] inflammation and cytokines affect the CNS, and the interaction between inflammatory mediators and the CNS may occur at the periphery, and may play a role in the activation of host inflammatory response which may lead to cancer development. At the periphery, tumor growth might be sensed by the vagal nerve, perhaps by sensing the release of the pro-inflammatory cytokines. Several studies [11,15,16] emphasized that the majority of the postulated mechanisms of mental comorbidities within the cancer context (and other chronic inflammatory diseases) logically focuses on neuroinflammatory pathways. Peripheral tumors and their microenvironment provide the source of various cytokines and, potentially, use neural and/or humoral signaling pathways similar to peripheral infection to gain access to the brain. In addition, Dantzer et al., and Quan and Banks [16,17] focused on models of acute illness. The canonical acute bacterial infection model, a single, sub-toxic i.p. injection of a lipopolysaccharide (LPS) component of the cell wall of the gram-negative bacteria *E. coli*, causes pro-inflammatory cytokine production in the peritoneal cavity. Then, through both neural and humoral signaling pathways, cytokine production rapidly ensues in the brain (hippocampus, hypothalamus, forebrain) and stimulates the production or activation of other inflammatory effectors (IDO, iNOS, Nf-kB, COX). Three other studies [18–20] mentioned that the role of these inflammatory markers in cytokine-induced behavioral changes are consistent with clinical research in depressed patients. In summary, they claim that LPS treatment elicits acute sickness behaviors (e.g., lethargy, social withdrawal, fever, anorexia) akin to "somatic" or "vegetative" symptoms of depression and subsequent affective-like behaviors including impaired learning and memory. Experimental manipulation of cytokines in these models (e.g., pharmacologic blockade, cytokine gene knockout) suggests that brain-production of cytokines is necessary and sufficient for subsequent behavioral changes. A recent study in rats bearing bone cancer [21] reported elevated levels of IL-1β, IL-6, and TNF-α and of their respective receptors in the rats' periaqueductal gray brain region. Prolonged microglia activation leads to the release of the pro-inflammatory cytokines IL-1β, IL-6 and TNF-α, which initiates a pro-inflammatory cascade and subsequently contributes to neuronal damage and losses [22]. Interestingly, blockade of the receptors alleviated the cancer-induced hyperalgesia. This recent report corroborates the previous above-mentioned studies, but further research is still required to improve basic scientific understanding of how activation of pro-inflammatory cytokine networks by cancer cells may increase cancer-related symptoms, to guide clinical interventions.

2.2. Cancer, Cytokines and Stress

Psychosocial stress is highly prevalent in cancer patients and can increase neuroinflammation. Therefore, stress is considered a likely contributor to neurotoxic symptoms. According to Pyter, Brydon, and Woon [13,23,24], in newly diagnosed cancer patients, acute psychological stress is often elevated. Plausible causes are that patients undergo staging and other medical testing, make treatment decisions, cope with current or anticipated physical symptoms, and grapple with existential concerns, all considered to have a major role in neuroinflammation, a predictor of neurotoxic symptoms, both during and after treatment, which may not only affect performance on neurocognitive tests but may themselves activate pro-inflammatory pathways. Indeed, psychological stress may interact with inflammatory pathways to synergistically increase cognitive changes and other behavioral symptoms. Psychological stress may also have direct effects on the CNS, including decreased neurogenesis and hippocampal volume. The latter could be exacerbated by additional biological insults of cancer and its treatment.

The growing tumor is sensed by the brain via neural, humoral, and inflammatory input. These signals activate the behavioral and metabolic response to stress by activating microglial cells. In turn, microglial activation triggers and perpetuates neuroinflammation by the release of inflammatory mediators within hypothalamic areas. Experimental data suggest that neuroinflammation may contribute to tumor growth and aggressiveness by modulating the peripheral immune response through autonomic output [2].

In breast cancer survivors, Kesler et al. reported an association between lower left hippocampal volume (measured by MRI), higher levels of circulating TNF-α, and lower levels of IL-6 [25]. Similarly, Jenkins et al. reported higher soluble TNF receptor (sTNFR)- 2 and IL-6 levels associated with decreased gray matter volume in specific regions in eight breast cancer patients who underwent chemotherapy during the study [26].

Based on the above, it is tempting to suggest that some of the variance in human mood disorders is comparable to that in cancer, attributable to the effects of tumors by themselves on emotional states. These potential interactions between stress and physiological reactions to cancer warrant further research.

2.3. Cancer, Mitochondria, and Inflammation

Mitochondria are responsible for cellular energy production by changing adenosine diphosphate (ADP) into adenosine triphosphate (ATP) through aerobic respiration. This conversion occurs in a series of redox reactions in the electron-transport chain-enzyme compounds, labeled complex I–IV. Changes in any of these complexes lead to changes in energy output. Hence, dysfunction in mitochondria and in their DNA (mtDNA) creates reduced cellular energy, which has been linked with neurologic symptoms [27]. Mitochondria are especially vulnerable to oxidative stress, a disturbed balance between reactive oxygen species (ROS), and antioxidants. In turn, mitochondrial dysfunction can lead to ROS overproduction, inducing a downward spiral in cellular functioning [28]. Cancer treatment, inflammation, and stress can all affect mitochondrial function by increasing ROS levels, thereby destroying mitochondria [29]. For example, radiation therapy was shown to increase oxidative stress markers in breast cancer patients with severe acute skin reactions to radiation. In another example, chemo-radiation seemed to normalize tumor-related changes in mtDNA expression in the liver of a rodent model of head and neck cancer, while causing severe changes in mtDNA expression in the brain [30]. This suggests that cancer treatment specifically affects cellular energy production in the CNS. In addition, Lacourt and Heijnen [31] argued that mitochondrial dysfunction is the mechanism leading to neurotoxic symptoms. Both inflammation and stress hormones, as well as cancer treatment, can promote mitochondrial dysfunction, resulting in reduced cellular energy. It should be noted that during cancer many disruptions occur in the physiological functioning of brain areas controlling energy homeostasis [31]. In particular, increased hypothalamic expression and release of mediators of neural inflammation play a major role in this process [2].

In summary, mitochondrial dysfunction may be a final common outcome of cancer, cancer therapy, inflammation, ROS, and stress that leads to neurotoxic symptoms. Mitochondria-protecting drugs preventing cancer therapy-related toxicities may point to promising avenues for treatment of neuro-toxicities in cancer patients. Establishment of these drugs in clinical settings, in parallel to the early implementation of stress-reduction interventions shortly after diagnosis, should be considered to prevent the long-term neurotoxic symptoms that plague so many cancer survivors.

2.4. Chemotherapy and Inflammation

Cancer therapy may trigger an inflammatory response through several pathways, including direct immune changes in the TME, tumor cell death, and damage to healthy tissue. Associations between symptoms and inflammatory markers such as IL-6, TNF-α, and C-reactive protein (CRP) have been observed for every treatment modality, both in cross-sectional and in longitudinal designs [32]. In patients who recently completed chemotherapy (compared with treatment-naïve patients), Smith et al. [33] observed reduced blood mononuclear cell DNA methylation associated with higher plasma concentrations of sTNFR2 and IL-6. In turn, sTNFR2 was associated with fatigue, suggesting a transient effect of chemotherapy on inflammation and subsequent fatigue. In paclitaxel-treated mice, Loman et al. [34] found increased fatigue and decreased cognitive performance in parallel with reduced microglial immunoreactivity, increased circulating chemokine (CXCL1) expression, as well as a transient increase in brain gene expression of pro-inflammatory cytokines (IL-1β, TNF-α, IL-6) and CXCL1 in a brain region-dependent manner. The study implied that the brain-gut-microbiota axis was involved in the neuroinflammation induced by the chemotherapeutic agent.

2.5. The HPA Axis in Cancer in Relation with Inflammation

Cortisol is released by the hypothalamic-pituitary-adrenal (HPA) axis in response to psychosocial stress. Preliminary evidence indicates that tumors can affect endocrine function [12]. Studies by Pyter et al. [13,35] showed that corticosterone (the primary glucocorticoid in most animals) levels are generally higher in tumor-bearing rodents than in tumor-free controls, and that the hormone's responsivity to stress is reduced in these animals. Moreover, they reported that growth rates of tumors can be predicted by the production of cytokines, from the constitutive HPA function and according to depressive behavior. The same group made similar observations in patients, although they might have been caused by a combination of psychosocial stress and the tumor. Bower reported a blunted cortisol response in breast cancer survivors along with enhanced inflammatory response to psychological stress [36].

Moderate change in the HPA axis has also been found in a variety of CNS-related disorders caused by tumors [37]. In parallel, tumor-bearing mice developed depressive-like behavior along with higher plasma levels of corticosterone, the stress-related hormone [38].

In summary, although the mechanism of the interaction between neuroinflammation, systemic inflammation, and tumor growth has not yet been unraveled in full, the relationship between neuroinflammation and cancer-related neuro-toxicities is well-established. Association between inflammatory markers and neuro-toxicities seems to exist even when inflammation is low. It cannot be ruled out that inflammation is only an important mediator in individuals with a genetic vulnerability for exaggerated inflammatory responses to internal (tumor, stress) and external (cancer therapy) stressors.

3. Neuroinflammation in Type 2 Diabetes Mellitus (T2DM)

This section presents data and interpretation of the effect of T2DM on brain structure and function in relation to neuroinflammation. Brain tissue of T2DM patients exhibiting cognitive impairment contains deposits of amylin [islet amyloid polypeptide (IAPP)], a peptide hormone synthesized and co-secreted with insulin by pancreatic β cells [39]. Amylin deposition occurs following chronic over-secretion of amylin (hyper-amylinemia), common in humans with obesity or pre-diabetes insulin resistance. It is toxic, causes pancreatic islet inflammation and is thought to contribute to the development of T2DM [40,41]. Hyper-amylinemia and its consequent oligomerization mediate an inflammatory response, inducing neurological defects [39]. Using a rat model of overexpression of human amylin in the pancreas (the HIP rat) Luchsinger and Nelson et al. [42,43] observed psychomotor speed disturbances followed by full-blown T2DM (blood glucose > 10 nM) accompanied by a significant drop in cognition and memory. Janciauskiene and Ahren and Westwell-Roper et al. [44,45] reported elevation of the pro-inflammatory cytokines TNF-α and IL-6 and down-regulation of the anti-inflammatory cytokine IL-10 in brains from HIP rats. These data support the notion that "human" hyper-amylinemia promotes accumulation of brain oligomerized amylin which, in turn, might trigger an inflammatory response leading to neurological deficits. Using the same HIP rat, Srodulski et al. [39] further describe significantly reduced exploratory drive and impairment in the rotarod test, implicating that infusion of amylin decreases ambulation and locomotion ability in rats. The decline in long-term memory suggests a direct impact of hyper-amylinemia on hippocampal neurons. The authors also found activated microglia, particularly gathering around the small blood vessels in areas positive for amylin infiltration, implicative of neuroinflammation. The latter corroborate the study by Bahniwal et al [46]. They investigated the effects of elevated glucose concentrations (up to 30.5 mM) on functions of cultured human astrocytes in the presence of inflammatory stimuli. Using primary human astrocytes and U-118 MG astrocytoma cells, they found that high glucose increased mRNA expression of IL-6 and secretion of both IL-6 and IL-8 by astrocytes. High glucose also increased the susceptibility of undifferentiated human SH-SY5Y neuronal cells to injury by hydrogen peroxide.

Hyperglycemia in T2DM could contribute to worsening cognitive impairment [47]. This was seen in a rat model of vascular dementia caused by an impaired supply of blood to the brain, and by chronic neuronal inflammation caused by the activation of brain microglia and astrocytes, which might further contribute to neuronal loss [48–50].

To conclude, T2DM might involve neuroinflammation. Hence, neuroinflammation inhibitors might be novel drugs for this disease. In the case of T2DM induced by brain injury, the list of such targets may also include antioxidants and neurotrophic factors. Since the studies described above imply that brain amylin accumulation might be a pathological substrate for diabetic patients with cognitive decline, reducing blood amylin levels might be another direction for drug design in T2DM. Additional studies are required to clarify the link between brain amylin pathology and impaired cognition, and to search for drugs that will protect the brain in T2DM patients.

4. Neuroinflammation in Mood Disorders

There is growing comprehension of the role of the immune system, in general, and neuroinflammation, in particular, in the pathophysiology of mood disorders. Intriguingly, a comprehensive review and meta-analysis reported that despite a high level of heterogeneity, both monotherapy and add-on anti-inflammatory treatment result in a beneficial effect on depressive symptoms. Nonsteroidal anti-inflammatory drugs (NSAIDs) had a better antidepressant effect [51]. According to Yang et al. [52] neuroinflammation and mitochondrial dysfunction are among the characteristics of psychiatric disorders. Both can lead to increased oxidative stress by excessive release of harmful ROS and reactive nitrogen species (RNS), which further promote neuronal damage and subsequent inflammation.

The following sections review findings of high cytokine levels as well as possible causes for neuroinflammation in psychiatric disorders.

4.1. High Proinflammatory Cytokine Levels

Cumulative data detailed below support the hypothesis that at least in a sub-population of patients afflicted with unipolar depression, the pathophysiology and neurobiological mechanisms underlying resistance to conventional antidepressants stem from high cytokine levels. It also suggests that specific cytokines and their activators and regulators play a role in depression. Indeed, major depression disorder (MDD) and depressed mood are linked with pro-inflammatory cytokines released during periods of disturbance [18]. As nicely elaborated by Jeon and Kim [53], the cytokine hypothesis of depression suggests that cytokine production is initially activated by stress and sympathetic nerve system activation. In turn, cytokines play an important role by acting via neurotransmitter depletion pathways, neuroendocrine pathways, and neural plasticity pathways. There are multiple interactions between these pathways, suggesting existence of a complex model for pathogenesis of depression.

In mouse models of depressive-like behavior, several groups reported hallmark features of neuroinflammation, i.e., microglia and astrocyte reactive morphology, microglial proliferation, increased levels of proinflammatory cytokines, and upregulation of translocator protein (TSPO, a clinical biomarker widely accepted as a surrogate of neuroinflammation that involves an activated state of brain microglia) [54–56]. In humans, positron emission tomography (PET) studies [57,58] enhanced TSPO uptake in several brain regions indicated neuroinflammation in depressed patients.

IL-1β was found to be present in abnormally high levels in plasma, CSF, and post-mortem brain tissue of individuals with mood disorders and its levels correlated positively with the severity of depression [59,60]. Accordingly, mRNA levels of proinflammatory cytokines and other related innate immune system proteins were found to be elevated in peripheral blood cells in mood disorder patients [61–66], and their plasma and CSF levels were found to be higher during acute depressive episodes, suggesting the pathogenic function of cytokines [62,67]. A meta-analysis based on 29 studies of serum cytokines [68] indicated increased sIL-2R, IL-6, and TNF-α levels, supporting the notion that elevated cytokine proinflammatory levels contribute to the pathophysiology of depression.

A recent review [69] raised the question whether a specific inflammatory profile underlies suicide risk. The authors summarized that although most studies showed a link between abnormally higher IL-1β, IL-6, TNF-α, transforming growth factor (TGF)-β1, vascular endothelial growth factor (VEGF), kynurenic acid (KYN), and lower IL-2, IL-4, and interferon (IFN)-γ levels in specific brain regions and suicidal behavior, the contribution of MDD as a mediator of the link between these cytokine abnormalities and suicidal behavior could not be excluded. Thus, obviously, additional studies to clarify if, and which immune pathways underlie suicidal behavior are needed.

As for neuroinflammation, in general, and cytokine levels, in particular, in bipolar disorder (BD), a meta-analysis of 30 studies [70] found significantly elevated plasma concentrations of IL-4, IL-6, IL-10, soluble IL-2 receptor (sIL-2R), sIL-6R, TNF α, sTNFR1, and IL-1 receptor antagonist; IL-1β and IL-6 tended to show higher values in patients. While concentrations of IL-2, IL-4, sIL-6R, and INF-γ were unrelated to medication status, phasic difference was observed for TNF-α, sTNFR1, sIL-2R, IL-6, and IL-1RA, but not for IL-4 and IL-10. The data point at a "cytokine storm" in BD. Nevertheless, the authors of a recent systematic review of 51 articles that measured inflammatory markers in postmortem BD brain samples attested that an absolute statement cannot be concluded whether neuroinflammation is present in BD since a large number of studies did not evaluate the presence of infiltrating peripheral immune cells in the CNS parenchyma, cytokine levels and microglia activation in the same postmortem brain sample [71]. The authors claim that "Future analyses should rectify these potential sources of heterogeneity and reach a consensus regarding the inflammatory markers in postmortem BD brain".

4.2. Glia Pathology

Astroglia and oligodendroglia are essential in neural metabolic homeostasis to maintain behavior and higher cognitive functions. Astroglia and oligodendroglia produce anti-inflammatory cytokines that regulate harmful inflammation [14,72,73]. Animal and human studies report astroglial pathology in psychiatric disorders like MDD and their models [74]. For example, post-mortem studies of MDD subjects implicated reduced oligodendroglial density in the prefrontal cortex and amygdala. Accordingly, it has been suggested that glial loss may contribute to neuroinflammation in psychiatric disorders by several mechanisms [75]. Indeed, using glia-specific genetically modified mice revealed that glial cells such as oligodendrocytes, astrocytes, and microglia affect neuronal function and are involved in the underlying pathobiology of psychiatric disorders [76]. In a model of chronic stress which, as mentioned above, is assumed to be relevant in studying the role of neuroinflammation associated with depression, [77] opted to study whether physiological conditions such as stress enhance susceptibility to inflammation in the substantia nigra, where dopaminergic neuron death occurs in Parkinson's disease. In a rat model of induced stress and inflammation they found higher TNF-α, IL-1β, IL-6, and iNOS levels in the substantia nigra. Likewise, microglial activation was significantly increased in the infralimbic, cingulate, and medial orbital cortices, nucleus accumbens, caudate putamen, amygdala, and hippocampus of the mice brain following unpredictable chronic mild stress—a reliable model to study depression-induced neuroinflammation [78].

Multiple studies, including some mentioned above, use animal models of depression induced by LPS, which also induces neuroinflammation. In the nonhuman-primate brain, LPS-induced systemic inflammation produces a robust increase in the level of TSPO (detected by PET), reflecting the state of neuroinflammation changes [67]. For example, doxycycline prevented and reversed LPS-induced changes in immobility time on the forced swimming test (FST), and in brain IL1β [79]. Likewise, minocycline also attenuated LPS-induced behavioral changes and markers of neuroinflammation in mice [80].

4.3. Increased Oxidative Stress

At the time of microglial activation, pro-inflammatory cytokines and NO production might increase oxidative stress. Namely, pro-inflammatory cytokines and high NO levels may promote ROS formation which, in turn, accelerates lipid peroxidation, damaging membrane phospholipids and their membrane-bound monoamine neurotransmitter receptors and depleting endogenous antioxidants. The consequence of increased production of pro-inflammatory cytokines via stimulation of NF-κB and enhanced microglial activation caused by the increase in ROS products might be increased prevalence of psychiatric disorders [81]. Indeed, studying MMD patients' fibroblasts, Scapagnini et al. [82] reported an increase in oxidative stress independent of glutathione levels. Moreover, diseases such as MDD, BD, and schizophrenia might go through increased oxidative stress due to mitochondrial dysfunction. Consistent with the high prevalence of psychiatric disturbances in primary mitochondrial disorders, there are reports [82–84] of abnormalities in mitochondrial DNA in these disorders. Alternatively, as data in Ott et al. [82,83] imply, there might be mechanistic links among neuroinflammation, mitochondrial dysfunction, and oxidative stress, meriting further investigation of these intersecting pathogenic pathways in human psychiatric disorders.

4.4. BBB Dysfunction

MDD-related clinical and experimental studies indicate indirectly that increased oxidation might contribute to endothelial dysfunction. Moreover, oxidation-mediated endothelial dysfunction might contribute to the pathophysiology of BBB dysfunction in psychiatric disorders [85,86]. In a prolonged learned helplessness depression model in mice, the non-recovered group had, within 4 weeks, higher hippocampal levels of TNFα, IL-17A, and IL-23, increased permeability of the BBB and lower levels of the BBB tight junction protein claudin-5 and the tight junction receptors occludin and zonula occludens protein 1 (ZO1), as compared with mice that recovered and with control mice, [87].

As for the BBB in BD, in a recent study [88], bipolar patients and control subjects matched for sex, age, and metabolic status underwent contrast-enhanced dynamic MRI scanning to quantitate their BBB leakage. Nearly 30% of the patients exhibited significantly higher percentages of brain volume with BBB leakage. This subgroup had more severe depression and anxiety and a more chronic course of illness.

4.5. The Microbiota-Gut-Brain Axis

The microbiota-gut-brain axis (microbiome) is a dynamic matrix of tissues and organs including the brain, glands, gut, immune cells, and gastrointestinal microbiota that communicate in a complex multidirectional manner to maintain homeostasis. It is, thus, regarded as a modulator of various central processes affecting changes in neuroinflammation, as well as neurotransmission and behavior, including stress adaption and immune response. Therefore, gut microbiome dysbiosis might be detrimental, contributing to the development of a number of CNS disorders such as aberrant anxiety and fear responses, despair and anhedonia via mechanisms not yet unraveled. This triggered largely preclinical animal studies investigating the influence of the microbiome, searching for mechanisms by which the microbiome may affect mental health. Some of the studies demonstrate encouraging results in the treatment of depression (for review see [89]), while studies in clinical cohorts have, mostly, been diagnostic in nature, and warrant further ones with pre- and pro-biotic interventions (for review see [90]).

4.6. Microbiota and the BBB

A variety of neuropsychiatric disorders [anxiety, depression, autism spectrum disorders (ASDs), Parkinson's disease, Alzheimer's disease, and schizophrenia] have been related to microbial-induced BBB dysfunction [91,92], although the mechanism by which the microbiota affects BBB is unknown. Mediation by gut-derived neurotransmitters and bacterial metabolites is conceivable. Rodent models pointed at a link between microbiota dysbiosis and increased permeability of the BBB, further accompanied by behavioral changes [93], while a pathogen-free gut microbiota restored BBB functionality [91].

To summarize this chapter, the data reviewed support the notion that neuroinflammation is a dominant factor which plays a role in the pathophysiology of psychiatric disorders. The field awaits more multidisciplinary efforts to improve basic scientific understanding of whether antidepressant effects might be achieved using therapies for mood disorders that influence neuroinflammation.

5. Conclusions

Although cancer, T2DM, and mood disorders, all belonging to the entity of complex disorders, represent diseases of different symptomatology residing in different tissues, perplexingly, they all exhibit neuroinflammation as a common denominator. This awareness may be translated into practice. It may be suggested that drugs known to alleviate neuroinflammation, such as aspirin or lithium, may be repurposed as add-on treatment in these disorders.

Author Contributions: S.A. wrote the original draft; O.D. carried out the initial review and editing; G.A. conceptualized the idea of the review, supervised S.A. and O.D. and finalized the manuscript. All authors have read and agreed to the published version of the manuscript.

Funding: The preparation of this review received no external funding.

Institutional Review Board Statement: Not applicable.

Informed Consent Statement: Not applicable.

Data Availability Statement: Not applicable.

Conflicts of Interest: The authors declare no conflict of interest.

References

1. Arroyo, D.S.; Gaviglio, E.A.; Peralta Ramos, J.M.; Bussi, C.; Rodriguez-Galan, M.C.; Iribarren, P. Autophagy in inflammation, infection, neurodegeneration and cancer. *Int. Immunopharmacol.* **2014**, *18*, 55–65. [CrossRef]
2. Molfino, A.; Gioia, G.; Rossi Fanelli, F.; Laviano, A. Contribution of Neuroinflammation to the Pathogenesis of Cancer Cachexia. *Mediat. Inflamm.* **2015**, *2015*, 801685. [CrossRef]
3. Greten, F.R.; Grivennikov, S.I. Inflammation and Cancer: Triggers, Mechanisms, and Consequences. *Immunity* **2019**, *51*, 27–41. [CrossRef]
4. Esquivel-Velazquez, M.; Ostoa-Saloma, P.; Palacios-Arreola, M.I.; Nava-Castro, K.E.; Castro, J.I.; Morales-Montor, J. The role of cytokines in breast cancer development and progression. *J. Interferon. Cytokine Res.* **2015**, *35*, 1–16. [CrossRef] [PubMed]
5. Mumm, J.B.; Oft, M. Cytokine-based transformation of immune surveillance into tumor-promoting inflammation. *Oncogene* **2008**, *27*, 5913–5919. [CrossRef] [PubMed]
6. Schrepf, A.; Lutgendorf, S.K.; Pyter, L.M. Pre-treatment effects of peripheral tumors on brain and behavior: Neuroinflammatory mechanisms in humans and rodents. *Brain Behav. Immun.* **2015**, *49*, 1–17. [CrossRef] [PubMed]
7. Jeon, S.W.; Kim, Y.K. The role of neuroinflammation and neurovascular dysfunction in major depressive disorder. *J. Inflamm. Res.* **2018**, *11*, 179–192. [CrossRef]
8. Patel, S.K.; Wong, A.L.; Wong, F.L.; Breen, E.C.; Hurria, A.; Smith, M.; Kinjo, C.; Paz, I.B.; Kruper, L.; Somlo, G.; et al. Inflammatory Biomarkers, Comorbidity, and Neurocognition in Women with Newly Diagnosed Breast Cancer. *J. Natl. Cancer Inst.* **2015**, *107*. [CrossRef]
9. Evans, J.R.; Fletcher, A.E.; Wormald, R.P. Depression and anxiety in visually impaired older people. *Ophthalmology* **2007**, *114*, 283–288. [CrossRef]
10. Haroon, E.; Raison, C.L.; Miller, A.H. Psychoneuroimmunology meets neuropsychopharmacology: Translational implications of the impact of inflammation on behavior. *Neuropsychopharmacology* **2012**, *37*, 137–162. [CrossRef] [PubMed]
11. Lee, B.N.; Dantzer, R.; Langley, K.E.; Bennett, G.J.; Dougherty, P.M.; Dunn, A.J.; Meyers, C.A.; Miller, A.H.; Payne, R.; Reuben, J.M.; et al. A cytokine-based neuroimmunologic mechanism of cancer-related symptoms. *Neuroimmunomodulation* **2004**, *11*, 279–292. [CrossRef] [PubMed]
12. Ahmad, M.H.; Rizvi, M.A.; Fatima, M.; Mondal, A.C. Pathophysiological implications of neuroinflammation mediated HPA axis dysregulation in the prognosis of cancer and depression. *Mol. Cell. Endocrinol.* **2021**, *520*, 111093. [CrossRef] [PubMed]
13. Pyter, L.M.; Pineros, V.; Galang, J.A.; McClintock, M.K.; Prendergast, B.J. Peripheral tumors induce depressive-like behaviors and cytokine production and alter hypothalamic-pituitary-adrenal axis regulation. *Proc. Natl. Acad. Sci. USA* **2009**, *106*, 9069–9074. [CrossRef]
14. Molfino, A.; Rossi-Fanelli, F.; Laviano, A. The interaction between pro-inflammatory cytokines and the nervous system. *Nat. Rev. Cancer* **2009**, *9*, 224. [CrossRef] [PubMed]

15. Cleeland, C.S.; Bennett, G.J.; Dantzer, R.; Dougherty, P.M.; Dunn, A.J.; Meyers, C.A.; Miller, A.H.; Payne, R.; Reuben, J.M.; Wang, X.S.; et al. Are the symptoms of cancer and cancer treatment due to a shared biologic mechanism? A cytokine-immunologic model of cancer symptoms. *Cancer* **2003**, *97*, 2919–2925. [CrossRef]
16. Dantzer, R.; O'Connor, J.C.; Freund, G.G.; Johnson, R.W.; Kelley, K.W. From inflammation to sickness and depression: When the immune system subjugates the brain. *Nat. Rev. Neurosci.* **2008**, *9*, 46–56. [CrossRef]
17. Quan, N.; Banks, W.A. Brain-immune communication pathways. *Brain Behav. Immun.* **2007**, *21*, 727–735. [CrossRef]
18. Dantzer, R.; Aubert, A.; Bluthe, R.M.; Gheusi, G.; Cremona, S.; Laye, S.; Konsman, J.P.; Parnet, P.; Kelley, K.W. Mechanisms of the behavioural effects of cytokines. *Adv. Exp. Med. Biol.* **1999**, *461*, 83–105. [CrossRef]
19. Pugh, C.R.; Johnson, J.D.; Martin, D.; Rudy, J.W.; Maier, S.F.; Watkins, L.R. Human immunodeficiency virus-1 coat protein gp120 impairs contextual fear conditioning: A potential role in AIDS related learning and memory impairments. *Brain Res.* **2000**, *861*, 8–15. [CrossRef]
20. Pugh, C.R.; Nguyen, K.T.; Gonyea, J.L.; Fleshner, M.; Wakins, L.R.; Maier, S.F.; Rudy, J.W. Role of interleukin-1 beta in impairment of contextual fear conditioning caused by social isolation. *Behav Brain Res.* **1999**, *106*, 109–118. [CrossRef]
21. Zhang, J.; Wang, L.; Wang, H.; Su, Z.; Pang, X. Neuroinflammation and central PI3K/Akt/mTOR signal pathway contribute to bone cancer pain. *Mol. Pain* **2019**, *15*, 1744806919830240. [CrossRef] [PubMed]
22. Perry, V.H.; Nicoll, J.A.; Holmes, C. Microglia in neurodegenerative disease. *Nat. Rev. Neurol.* **2010**, *6*, 193–201. [CrossRef] [PubMed]
23. Brydon, L.; Walker, C.; Wawrzyniak, A.; Whitehead, D.; Okamura, H.; Yajima, J.; Tsuda, A.; Steptoe, A. Synergistic effects of psychological and immune stressors on inflammatory cytokine and sickness responses in humans. *Brain Behav. Immun.* **2009**, *23*, 217–224. [CrossRef]
24. Woon, F.L.; Sood, S.; Hedges, D.W. Hippocampal volume deficits associated with exposure to psychological trauma and posttraumatic stress disorder in adults: A meta-analysis. *Prog. Neuropsychopharmacol. Biol. Psychiatry* **2010**, *34*, 1181–1188. [CrossRef] [PubMed]
25. Kesler, S.; Janelsins, M.; Koovakkattu, D.; Palesh, O.; Mustian, K.; Morrow, G.; Dhabhar, F.S. Reduced hippocampal volume and verbal memory performance associated with interleukin-6 and tumor necrosis factor-alpha levels in chemotherapy-treated breast cancer survivors. *Brain Behav. Immun.* **2013**, *30*, S109–S116. [CrossRef] [PubMed]
26. Jenkins, V.; Shilling, V.; Deutsch, G.; Bloomfield, D.; Morris, R.; Allan, S.; Bishop, H.; Hodson, N.; Mitra, S.; Sadler, G.; et al. A 3-year prospective study of the effects of adjuvant treatments on cognition in women with early stage breast cancer. *Br. J. Cancer* **2006**, *94*, 828–834. [CrossRef]
27. Parikh, S. The neurologic manifestations of mitochondrial disease. *Dev. Disabil. Res. Rev.* **2010**, *16*, 120–128. [CrossRef]
28. Zorov, D.B.; Juhaszova, M.; Sollott, S.J. Mitochondrial reactive oxygen species (ROS) and ROS-induced ROS release. *Physiol. Rev.* **2014**, *94*, 909–950. [CrossRef]
29. Picard, M.; Gentil, B.J.; McManus, M.J.; White, K.; St Louis, K.; Gartside, S.E.; Wallace, D.C.; Turnbull, D.M. Acute exercise remodels mitochondrial membrane interactions in mouse skeletal muscle. *J. Appl. Physiol.* **2013**, *115*, 1562–1571. [CrossRef]
30. Jeanneteau, F.; Arango-Lievano, M. Linking Mitochondria to Synapses: New Insights for Stress-Related Neuropsychiatric Disorders. *Neural Plast.* **2016**, *2016*, 3985063. [CrossRef]
31. Lacourt, T.E.; Heijnen, C.J. Mechanisms of Neurotoxic Symptoms as a Result of Breast Cancer and Its Treatment: Considerations on the Contribution of Stress, Inflammation, and Cellular Bioenergetics. *Curr. Breast Cancer Rep.* **2017**, *9*, 70–81. [CrossRef]
32. Schmidt, M.E.; Meynkohn, A.; Habermann, N.; Wiskemann, J.; Oelmann, J.; Hof, H.; Wessels, S.; Klassen, O.; Debus, J.; Potthoff, K.; et al. Resistance Exercise and Inflammation in Breast Cancer Patients Undergoing Adjuvant Radiation Therapy: Mediation Analysis From a Randomized, Controlled Intervention Trial. *Int. J. Radiat. Oncol. Biol. Phys.* **2016**, *94*, 329–337. [CrossRef]
33. Smith, A.K.; Conneely, K.N.; Pace, T.W.; Mister, D.; Felger, J.C.; Kilaru, V.; Akel, M.J.; Vertino, P.M.; Miller, A.H.; Torres, M.A. Epigenetic changes associated with inflammation in breast cancer patients treated with chemotherapy. *Brain Behav. Immun.* **2014**, *38*, 227–236. [CrossRef] [PubMed]
34. Loman, B.R.; Jordan, K.R.; Haynes, B.; Bailey, M.T.; Pyter, L.M. Chemotherapy-induced neuroinflammation is associated with disrupted colonic and bacterial homeostasis in female mice. *Sci. Rep.* **2019**, *9*, 16490. [CrossRef] [PubMed]
35. Pyter, L.M.; Cochrane, S.F.; Ouwenga, R.L.; Patel, P.N.; Pineros, V.; Prendergast, B.J. Mammary tumors induce select cognitive impairments. *Brain Behav. Immun.* **2010**, *24*, 903–907. [CrossRef] [PubMed]
36. Bower, J.E.; Ganz, P.A.; Aziz, N.; Olmstead, R.; Irwin, M.R.; Cole, S.W. Inflammatory responses to psychological stress in fatigued breast cancer survivors: Relationship to glucocorticoids. *Brain Behav. Immun.* **2007**, *21*, 251–258. [CrossRef] [PubMed]
37. Soygur, H.; Palaoglu, O.; Akarsu, E.S.; Cankurtaran, E.S.; Ozalp, E.; Turhan, L.; Ayhan, I.H. Interleukin-6 levels and HPA axis activation in breast cancer patients with major depressive disorder. *Prog. Neuropsychopharmacol. Biol. Psychiatry* **2007**, *31*, 1242–1247. [CrossRef]
38. Kubera, M.; Obuchowicz, E.; Goehler, L.; Brzeszcz, J.; Maes, M. In animal models, psychosocial stress-induced (neuro)inflammation, apoptosis and reduced neurogenesis are associated to the onset of depression. *Prog. Neuropsychopharmacol. Biol. Psychiatry* **2011**, *35*, 744–759. [CrossRef]
39. Srodulski, S.; Sharma, S.; Bachstetter, A.B.; Brelsfoard, J.M.; Pascual, C.; Xie, X.S.; Saatman, K.E.; Van Eldik, L.J.; Despa, F. Neuroinflammation and neurologic deficits in diabetes linked to brain accumulation of amylin. *Mol. Neurodegener.* **2014**, *9*, 30. [CrossRef]

40. Robertson, R.P. Chronic oxidative stress as a central mechanism for glucose toxicity in pancreatic islet beta cells in diabetes. *J. Biol. Chem.* **2004**, *279*, 42351–42354. [CrossRef]
41. Lenzen, S. Oxidative stress: The vulnerable beta-cell. *Biochem. Soc. Trans.* **2008**, *36*, 343–347. [CrossRef]
42. Luchsinger, J.A. Type 2 diabetes and cognitive impairment: Linking mechanisms. *J. Alzheimers Dis.* **2012**, *30* (Suppl. 2), S185–S198. [CrossRef]
43. Nelson, P.T.; Smith, C.D.; Abner, E.A.; Schmitt, F.A.; Scheff, S.W.; Davis, G.J.; Keller, J.N.; Jicha, G.A.; Davis, D.; Wang-Xia, W.; et al. Human cerebral neuropathology of Type 2 diabetes mellitus. *Biochim. Biophys. Acta* **2009**, *1792*, 454–469. [CrossRef] [PubMed]
44. Janciauskiene, S.; Ahren, B. Fibrillar islet amyloid polypeptide differentially affects oxidative mechanisms and lipoprotein uptake in correlation with cytotoxicity in two insulin-producing cell lines. *Biochem. Biophys. Res. Commun.* **2000**, *267*, 619–625. [CrossRef]
45. Westwell-Roper, C.; Dai, D.L.; Soukhatcheva, G.; Potter, K.J.; van Rooijen, N.; Ehses, J.A.; Verchere, C.B. IL-1 blockade attenuates islet amyloid polypeptide-induced proinflammatory cytokine release and pancreatic islet graft dysfunction. *J. Immunol.* **2011**, *187*, 2755–2765. [CrossRef] [PubMed]
46. Bahniwal, M.; Little, J.P.; Klegeris, A. High Glucose Enhances Neurotoxicity and Inflammatory Cytokine Secretion by Stimulated Human Astrocytes. *Curr. Alzheimer Res.* **2017**, *14*, 731–741. [CrossRef]
47. Soderbom, G.; Zeng, B.Y. The NLRP3 inflammasome as a bridge between neuro-inflammation in metabolic and neurodegenerative diseases. *Int. Rev. Neurobiol.* **2020**, *154*, 345–391. [CrossRef] [PubMed]
48. Kimm, H.; Lee, P.H.; Shin, Y.J.; Park, K.S.; Jo, J.; Lee, Y.; Kang, H.C.; Jee, S.H. Mid-life and late-life vascular risk factors and dementia in Korean men and women. *Arch. Gerontol. Geriatr.* **2011**, *52*, e117–e122. [CrossRef]
49. Xu, W.; Caracciolo, B.; Wang, H.X.; Winblad, B.; Backman, L.; Qiu, C.; Fratiglioni, L. Accelerated progression from mild cognitive impairment to dementia in people with diabetes. *Diabetes* **2010**, *59*, 2928–2935. [CrossRef] [PubMed]
50. Xu, W.; Qiu, C.; Gatz, M.; Pedersen, N.L.; Johansson, B.; Fratiglioni, L. Mid- and late-life diabetes in relation to the risk of dementia: A population-based twin study. *Diabetes* **2009**, *58*, 71–77. [CrossRef]
51. Kohler, O.; Benros, M.E.; Nordentoft, M.; Farkouh, M.E.; Iyengar, R.L.; Mors, O.; Krogh, J. Effect of anti-inflammatory treatment on depression, depressive symptoms, and adverse effects: A systematic review and meta-analysis of randomized clinical trials. *JAMA Psychiatry* **2014**, *71*, 1381–1391. [CrossRef]
52. Yang, Y.; Ouyang, Y.; Yang, L.; Beal, M.F.; McQuibban, A.; Vogel, H.; Lu, B. Pink1 regulates mitochondrial dynamics through interaction with the fission/fusion machinery. *Proc. Natl. Acad. Sci. USA* **2008**, *105*, 7070–7075. [CrossRef] [PubMed]
53. Jeon, S.W.; Kim, Y.K. Neuroinflammation and cytokine abnormality in major depression: Cause or consequence in that illness? *World J. Psychiatry* **2016**, *6*, 283–293. [CrossRef]
54. Aschner, M.; Allen, J.W.; Kimelberg, H.K.; LoPachin, R.M.; Streit, W.J. Glial cells in neurotoxicity development. *Annu. Rev. Pharmacol. Toxicol.* **1999**, *39*, 151–173. [CrossRef] [PubMed]
55. McAfoose, J.; Baune, B.T. Evidence for a cytokine model of cognitive function. *Neurosci. Biobehav. Rev.* **2009**, *33*, 355–366. [CrossRef] [PubMed]
56. You, Z.; Luo, C.; Zhang, W.; Chen, Y.; He, J.; Zhao, Q.; Zuo, R.; Wu, Y. Pro- and anti-inflammatory cytokines expression in rat's brain and spleen exposed to chronic mild stress: Involvement in depression. *Behav. Brain Res.* **2011**, *225*, 135–141. [CrossRef] [PubMed]
57. Richards, E.M.; Zanotti-Fregonara, P.; Fujita, M.; Newman, L.; Farmer, C.; Ballard, E.D.; Machado-Vieira, R.; Yuan, P.; Niciu, M.J.; Lyoo, C.H.; et al. PET radioligand binding to translocator protein (TSPO) is increased in unmedicated depressed subjects. *EJNMMI Res.* **2018**, *8*, 57. [CrossRef]
58. Setiawan, E.; Wilson, A.A.; Mizrahi, R.; Rusjan, P.M.; Miler, L.; Rajkowska, G.; Suridjan, I.; Kennedy, J.L.; Rekkas, P.V.; Houle, S.; et al. Role of translocator protein density, a marker of neuroinflammation, in the brain during major depressive episodes. *JAMA Psychiatry* **2015**, *72*, 268–275. [CrossRef]
59. Jones, K.A.; Thomsen, C. The role of the innate immune system in psychiatric disorders. *Mol. Cell. Neurosci.* **2013**, *53*, 52–62. [CrossRef]
60. Soderlund, J.; Olsson, S.K.; Samuelsson, M.; Walther-Jallow, L.; Johansson, C.; Erhardt, S.; Landen, M.; Engberg, G. Elevation of cerebrospinal fluid interleukin-1ss in bipolar disorder. *J. Psychiatry Neurosci.* **2011**, *36*, 114–118. [CrossRef]
61. Carvalho, L.A.; Bergink, V.; Sumaski, L.; Wijkhuijs, J.; Hoogendijk, W.J.; Birkenhager, T.K.; Drexhage, H.A. Inflammatory activation is associated with a reduced glucocorticoid receptor alpha/beta expression ratio in monocytes of inpatients with melancholic major depressive disorder. *Transl. Psychiatry* **2014**, *4*, e344. [CrossRef]
62. Cattaneo, A.; Gennarelli, M.; Uher, R.; Breen, G.; Farmer, A.; Aitchison, K.J.; Craig, I.W.; Anacker, C.; Zunsztain, P.A.; McGuffin, P.; et al. Candidate genes expression profile associated with antidepressants response in the GENDEP study: Differentiating between baseline 'predictors' and longitudinal 'targets'. *Neuropsychopharmacology* **2013**, *38*, 377–385. [CrossRef] [PubMed]
63. Janssen, B.; Vugts, D.J.; Funke, U.; Spaans, A.; Schuit, R.C.; Kooijman, E.; Rongen, M.; Perk, L.R.; Lammertsma, A.A.; Windhorst, A.D. Synthesis and initial preclinical evaluation of the P2X7 receptor antagonist [(1)(1)C]A-740003 as a novel tracer of neuroinflammation. *J. Labelled Comp. Radiopharm.* **2014**, *57*, 509–516. [CrossRef] [PubMed]
64. Padmos, R.C.; Hillegers, M.H.; Knijff, E.M.; Vonk, R.; Bouvy, A.; Staal, F.J.; de Ridder, D.; Kupka, R.W.; Nolen, W.A.; Drexhage, H.A. A discriminating messenger RNA signature for bipolar disorder formed by an aberrant expression of inflammatory genes in monocytes. *Arch. Gen. Psychiatry* **2008**, *65*, 395–407. [CrossRef]

65. Powell, T.R.; McGuffin, P.; D'Souza, U.M.; Cohen-Woods, S.; Hosang, G.M.; Martin, C.; Matthews, K.; Day, R.K.; Farmer, A.E.; Tansey, K.E.; et al. Putative transcriptomic biomarkers in the inflammatory cytokine pathway differentiate major depressive disorder patients from control subjects and bipolar disorder patients. *PLoS ONE* **2014**, *9*, e91076. [CrossRef] [PubMed]
66. Savitz, J.; Frank, M.B.; Victor, T.; Bebak, M.; Marino, J.H.; Bellgowan, P.S.; McKinney, B.A.; Bodurka, J.; Kent Teague, T.; Drevets, W.C. Inflammation and neurological disease-related genes are differentially expressed in depressed patients with mood disorders and correlate with morphometric and functional imaging abnormalities. *Brain Behav. Immun.* **2013**, *31*, 161–171. [CrossRef] [PubMed]
67. Hannestad, J.; Gallezot, J.D.; Schafbauer, T.; Lim, K.; Kloczynski, T.; Morris, E.D.; Carson, R.E.; Ding, Y.S.; Cosgrove, K.P. Endotoxin-induced systemic inflammation activates microglia: [(1)(1)C]PBR28 positron emission tomography in nonhuman primates. *Neuroimage* **2012**, *63*, 232–239. [CrossRef]
68. Liu, Y.; Ho, R.C.; Mak, A. Interleukin (IL)-6, tumour necrosis factor alpha (TNF-alpha) and soluble interleukin-2 receptors (sIL-2R) are elevated in patients with major depressive disorder: A meta-analysis and meta-regression. *J. Affect. Disord.* **2012**, *139*, 230–239. [CrossRef]
69. Serafini, G.; Parisi, V.M.; Aguglia, A.; Amerio, A.; Sampogna, G.; Fiorillo, A.; Pompili, M.; Amore, M. A Specific Inflammatory Profile Underlying Suicide Risk? Systematic Review of the Main Literature Findings. *Int. J. Environ. Res. Public Health* **2020**, *17*, 2393. [CrossRef]
70. Modabbernia, A.; Taslimi, S.; Brietzke, E.; Ashrafi, M. Cytokine alterations in bipolar disorder: A meta-analysis of 30 studies. *Biol. Psychiatry* **2013**, *74*, 15–25. [CrossRef] [PubMed]
71. Giridharan, V.V.; Sayana, P.; Pinjari, O.F.; Ahmad, N.; da Rosa, M.I.; Quevedo, J.; Barichello, T. Postmortem evidence of brain inflammatory markers in bipolar disorder: A systematic review. *Mol. Psychiatry* **2020**, *25*, 94–113. [CrossRef] [PubMed]
72. Bazan, N.G. The docosanoid neuroprotectin D1 induces homeostatic regulation of neuroinflammation and cell survival. *Prostaglandins Leukot. Essent. Fatty Acids* **2013**, *88*, 127–129. [CrossRef] [PubMed]
73. Pereira, A., Jr.; Furlan, F.A. Astrocytes and human cognition: Modeling information integration and modulation of neuronal activity. *Prog. Neurobiol.* **2010**, *92*, 405–420. [CrossRef] [PubMed]
74. Rajkowska, G.; Stockmeier, C.A. Astrocyte pathology in major depressive disorder: Insights from human postmortem brain tissue. *Curr. Drug Targets* **2013**, *14*, 1225–1236. [CrossRef]
75. Cotter, D.R.; Pariante, C.M.; Everall, I.P. Glial cell abnormalities in major psychiatric disorders: The evidence and implications. *Brain Res. Bull.* **2001**, *55*, 585–595. [CrossRef]
76. Yamamuro, K.; Kimoto, S.; Rosen, K.M.; Kishimoto, T.; Makinodan, M. Potential primary roles of glial cells in the mechanisms of psychiatric disorders. *Front. Cell Neurosci.* **2015**, *9*, 154. [CrossRef]
77. de Pablos, R.M.; Herrera, A.J.; Espinosa-Oliva, A.M.; Sarmiento, M.; Munoz, M.F.; Machado, A.; Venero, J.L. Chronic stress enhances microglia activation and exacerbates death of nigral dopaminergic neurons under conditions of inflammation. *J. Neuroinflamm.* **2014**, *11*, 34. [CrossRef]
78. Farooq, R.K.; Isingrini, E.; Tanti, A.; Le Guisquet, A.M.; Arlicot, N.; Minier, F.; Leman, S.; Chalon, S.; Belzung, C.; Camus, V. Is unpredictable chronic mild stress (UCMS) a reliable model to study depression-induced neuroinflammation? *Behav. Brain. Res.* **2012**, *231*, 130–137. [CrossRef]
79. Mello, B.S.; Monte, A.S.; McIntyre, R.S.; Soczynska, J.K.; Custodio, C.S.; Cordeiro, R.C.; Chaves, J.H.; Vasconcelos, S.M.; Nobre, H.V., Jr.; Florenco de Sousa, F.C.; et al. Effects of doxycycline on depressive-like behavior in mice after lipopolysaccharide (LPS) administration. *J. Psychiatr. Res.* **2013**, *47*, 1521–1529. [CrossRef]
80. Henry, C.J.; Huang, Y.; Wynne, A.; Hanke, M.; Himler, J.; Bailey, M.T.; Sheridan, J.F.; Godbout, J.P. Minocycline attenuates lipopolysaccharide (LPS)-induced neuroinflammation, sickness behavior, and anhedonia. *J. Neuroinflamm.* **2008**, *5*, 15. [CrossRef]
81. Barger, S.W.; Horster, D.; Furukawa, K.; Goodman, Y.; Krieglstein, J.; Mattson, M.P. Tumor necrosis factors alpha and beta protect neurons against amyloid beta-peptide toxicity: Evidence for involvement of a kappa B-binding factor and attenuation of peroxide and Ca2+ accumulation. *Proc. Natl. Acad. Sci. USA* **1995**, *92*, 9328–9332. [CrossRef]
82. Scapagnini, G.; Davinelli, S.; Drago, F.; De Lorenzo, A.; Oriani, G. Antioxidants as antidepressants: Fact or fiction? *CNS Drugs* **2012**, *26*, 477–490. [CrossRef]
83. Ott, M.; Gogvadze, V.; Orrenius, S.; Zhivotovsky, B. Mitochondria, oxidative stress and cell death. *Apoptosis* **2007**, *12*, 913–922. [CrossRef] [PubMed]
84. Salim, S.; Chugh, G.; Asghar, M. Inflammation in anxiety. *Adv. Protein Chem. Struct. Biol.* **2012**, *88*, 1–25. [CrossRef]
85. Isingrini, E.; Belzung, C.; Freslon, J.L.; Machet, M.C.; Camus, V. Fluoxetine effect on aortic nitric oxide-dependent vasorelaxation in the unpredictable chronic mild stress model of depression in mice. *Psychosom. Med.* **2012**, *74*, 63–72. [CrossRef] [PubMed]
86. Najjar, S.; Pahlajani, S.; De Sanctis, V.; Stern, J.N.H.; Najjar, A.; Chong, D. Neurovascular Unit Dysfunction and Blood-Brain Barrier Hyperpermeability Contribute to Schizophrenia Neurobiology: A Theoretical Integration of Clinical and Experimental Evidence. *Front. Psychiatry* **2017**, *8*, 83. [CrossRef] [PubMed]
87. Cheng, Y.; Desse, S.; Martinez, A.; Worthen, R.J.; Jope, R.S.; Beurel, E. TNFalpha disrupts blood brain barrier integrity to maintain prolonged depressive-like behavior in mice. *Brain Behav. Immun.* **2018**, *69*, 556–567. [CrossRef]
88. Kamintsky, L.; Cairns, K.A.; Veksler, R.; Bowen, C.; Beyea, S.D.; Friedman, A.; Calkin, C. Blood-brain barrier imaging as a potential biomarker for bipolar disorder progression. *Neuroimage Clin.* **2020**, *26*, 102049. [CrossRef]

89. Capuco, A.; Urits, I.; Hasoon, J.; Chun, R.; Gerald, B.; Wang, J.K.; Kassem, H.; Ngo, A.L.; Abd-Elsayed, A.; Simopoulos, T.; et al. Current Perspectives on Gut Microbiome Dysbiosis and Depression. *Adv. Ther.* **2020**, *37*, 1328–1346. [CrossRef]
90. Rea, K.; Dinan, T.G.; Cryan, J.F. The microbiome: A key regulator of stress and neuroinflammation. *Neurobiol. Stress* **2016**, *4*, 23–33. [CrossRef]
91. Braniste, V.; Al-Asmakh, M.; Kowal, C.; Anuar, F.; Abbaspour, A.; Toth, M.; Korecka, A.; Bakocevic, N.; Ng, L.G.; Kundu, P.; et al. The gut microbiota influences blood-brain barrier permeability in mice. *Sci. Transl. Med.* **2014**, *6*, 263ra158. [CrossRef] [PubMed]
92. Fiorentino, M.; Sapone, A.; Senger, S.; Camhi, S.S.; Kadzielski, S.M.; Buie, T.M.; Kelly, D.L.; Cascella, N.; Fasano, A. Blood-brain barrier and intestinal epithelial barrier alterations in autism spectrum disorders. *Mol. Autism* **2016**, *7*, 49. [CrossRef] [PubMed]
93. Spadoni, I.; Fornasa, G.; Rescigno, M. Organ-specific protection mediated by cooperation between vascular and epithelial barriers. *Nat. Rev. Immunol.* **2017**, *17*, 761–773. [CrossRef] [PubMed]

International Journal of
Molecular Sciences

Review

Neuroinflammation in Post-Ischemic Neurodegeneration of the Brain: Friend, Foe, or Both?

Ryszard Pluta [1,*], Sławomir Januszewski [1] and Stanisław J. Czuczwar [2]

1 Laboratory of Ischemic and Neurodegenerative Brain Research, Mossakowski Medical Research Institute, Polish Academy of Sciences, PL 02-106 Warsaw, Poland; sjanuszewski@imdik.pan.pl
2 Department of Pathophysiology, Medical University of Lublin, PL 20-090 Lublin, Poland; stanislawczuczwar@umlub.pl
* Correspondence: pluta@imdik.pan.pl

Abstract: One of the leading causes of neurological mortality, disability, and dementia worldwide is cerebral ischemia. Among the many pathological phenomena, the immune system plays an important role in the development of post-ischemic degeneration of the brain, leading to the development of neuroinflammatory changes in the brain. After cerebral ischemia, the developing neuroinflammation causes additional damage to the brain cells, but on the other hand it also plays a beneficial role in repair activities. Inflammatory mediators are sources of signals that stimulate cells in the brain and promote penetration, e.g., T lymphocytes, monocytes, platelets, macrophages, leukocytes, and neutrophils from systemic circulation to the brain ischemic area, and this phenomenon contributes to further irreversible ischemic brain damage. In this review, we focus on the issues related to the neuroinflammation that occurs in the brain tissue after ischemia, with particular emphasis on ischemic stroke and its potential treatment strategies.

Keywords: brain ischemia; stroke; neuroinflammation; microglia; astrocytes; T lymphocytes; monocytes; platelets; macrophages; leukocytes; neutrophils

Citation: Pluta, R.; Januszewski, S.; Czuczwar, S.J. Neuroinflammation in Post-Ischemic Neurodegeneration of the Brain: Friend, Foe, or Both? *Int. J. Mol. Sci.* **2021**, *22*, 4405. https://doi.org/10.3390/ijms22094405

Academic Editor: Anne Vejux

Received: 6 April 2021
Accepted: 19 April 2021
Published: 23 April 2021

1. Introduction

Brain ischemia and its consequences in humans are third most frequent cause of disability in 80% of survivors [1], the second most common cause of dementia, and second leading cause of death in the world [2–6]. Every year brain ischemia affects 17 million people worldwide, of whom 6 million die, and the other 5 million are permanently disabled [5,7,8]. The incidence of ischemic stroke in men is around 63 per 100,000 and in women around 59 per 100,000, which suggests that men are more affected by the disease than women [8]. The risk of ischemic stroke is age-related, with about 75% of all cases occurring in patients over 64 years of age, and about 25% of cases occurring in young people, suggesting that the pathology does not only affect the elderly [8]. Worldwide, the number of post-ischemic cases is currently estimated at around 33 million [5,7]. According to forecasts, the number of cases will increase to about 77 million in 2030 [5,7]. If the trend in ischemic stroke incidence continues, there will be about 12 million deaths by 2030, 70 million people will survive a stroke, and more than 200 million disability-adjusted life years will be recorded worldwide annually [5,9]. In 2010, the annual cost of treating stroke patients in Europe was around EUR 64 billion [5]. In the UK, stroke results in therapeutic and social costs of GBP 9 billion per year, with care costs accounting for about 5% of the national health system expenditure [10].

Human and animal studies have revealed that brain ischemia/ischemic stroke are risk factors for Alzheimer's disease [11–13] and vice versa [14,15]. In the first year post-stroke, 4 out of 10 cases have some degree of cognitive impairment [16]. The diagnosis of dementia immediately post-stroke is difficult due to additional deficits in both global and individual cognition, e.g., attention and processing speed, language, memory, and frontal executive functions may be impaired [9]. History of ischemic stroke has been shown to be an important risk factor for the development of dementia [4,9,11,15,17,18]. It has been shown that cerebral ischemia accelerates the onset of dementia by 10 years [19]; in 10% of cases dementia will develop soon after the first stroke, and in about 41% cases it develops after a repeated stroke [9,20]. Within 25 years of post-stroke survival, the estimated development of dementia is approximately 48% [5].

The phenomena following experimental cerebral ischemia and human ischemic stroke are under constant investigation and are revealing interesting new data. Studies conducted over the last 5 years have significantly expanded our understanding of the genetic basis of brain neurodegeneration following ischemia. It is now well known that development of ischemic brain neurodegeneration is caused by a set of genetic changes that lead to neuronal loss in an amyloid and tau protein dependent manner [21–25], with progressive neuroinflammation [26,27] resulting in uncontrolled irreversible brain atrophy [17,28–30] with the development of full-blown dementia [31–33]. Disruption of blood supply to the brain causes neuronal death and, consequently, brain atrophy with progressive dementia. Disruption of many pathways, including oxidative stress, excitotoxicity, neuroinflammation, blood–brain barrier permeability, and others, at least partially explains post-ischemic neurodegeneration of the brain. Post-ischemic damage to neurons causes a significant release of glutamate, leading to over-activation of N-methyl-D-aspartate (NMDA) receptors and a massive Ca^{2+} inflow to neurons, resulting in their death [34]. As a result of ischemia, neurons and astrocytes produce reactive oxygen species (ROS), and the same mechanism reduces glutathione, an essential antioxidant that prevents DNA damage from ROS [35]. Irreversibly damaged brain cells and their remains, without the presence of microorganisms, trigger neuroinflammation after cerebral ischemia [8,36]. Post-ischemic oxidative stress and inflammatory processes, inter alia, cause additional damage to the blood–brain barrier and enable activated blood immune cells, such as T-lymphocytes, platelets, and neutrophils, to reach the ischemic site of the brain [8,37,38]. After the accumulation of activated immune cells from the blood in the ischemic areas of the brain, microglial cells are activated as a result of an increase in extracellular ATP after its release from the membranes of necrotic cells [39]. The activated microglia secretes pro-inflammatory factors such as cytokines and develops phagocytic properties [40]. Microglia activation has beneficial effects as it promotes the generation of growth factors such as brain-derived neurotrophic factors and removes necrotic tissue and ischemic debris, but the release of pro-inflammatory cytokines (such as tumor necrosis factor α (TNF-α)), nitric oxide, and ROS is harmful to the brain tissue after ischemia [8]. Increasing expression of cytokines promotes the expression of adhesion molecules on endothelial cells, which results in additional recruitment of, for example, leukocytes and platelets from the blood to the brain [37,38]. As neuronal death and brain tissue damage increase, there is a further increment in active microglia, infiltrating platelets, and leukocytes, resulting in more pro-inflammatory cytokines as a consequence of feedback [37]. This post-ischemic phenomenon increases both neuronal death and the infarct volume and causes poorer neurological outcomes. Neuroinflammatory changes in the brain are present in all stages of an ischemia episode, from cerebral blood flow arrest to late recirculation processes in ischemic brain tissue [26,27]. Neuroinflammation promotes further brain damage, causing the death of surviving neurons from the primary ischemia, but it also has a beneficial function to aid recovery and develop glial scaring. In this review, we look at the beneficial and harmful roles of neuroinflammation in post-ischemic brain neurodegeneration and possible future therapeutic strategies to reduce pathological responses following ischemia.

Moreover, inflammatory mechanisms are largely portrayed as deleterious to post-ischemic pathology, while in fact many immune processes such as phagocytosis help to reduce the consequences of ischemia. In this review, we strive to delineate the delicate balance between the beneficial and harmful aspects of inflammatory/immune activation in post-ischemic brain neurodegeneration, as a more detailed understanding of these processes is crucial for the development of effective therapies.

2. Neuroinflammation in the Post-Ischemic Brain

Numerous studies have shown an inflammatory response in brain tissue to local or complete ischemia in animals and humans [8,26,27,41–45]. The severity and extent of the neuroinflammation depends on the site, area, course, and type of the ischemic brain injury. Inflammation following ischemic brain injury in rats surviving 2 years after global cerebral ischemia showed different severity of microglia and astrocyte responses in different brain structures. In these animals, the study revealed significant astrocyte activation in the CA1 and CA3 areas of the hippocampus and the dentate gyrus, in the motor and sensory cortex, and in the striatum and thalamus, while microglial activation was only seen in the CA1 and CA3 areas of the hippocampus and in the motor cortex. In areas of the brain sensitive to ischemia, microglia and astrocytes showed increased activation at the same time, while in areas resistant to ischemia, only astrocytes were activated. Thus, there is strong evidence of less intense inflammation in ischemia-resistant areas of the brain. Neuroinflammatory processes are supported by microglia and astrocyte activity for up to 2 years in post-ischemic brain neurodegeneration. The study therefore revealed a chronic effect of brain ischemia on the neuroinflammatory response in the rat brain up to 2 years after the injury [27]. In another study, immunostaining confirmed the presence of T lymphocytes in the ischemic hippocampus and striatum in long-surviving animals after an ischemic episode [26]. The above observations indicate a persistent dysfunction of the blood–brain barrier, which in the long run may still allow T lymphocytes to pass from the blood to the post-ischemic brain. Such processes are supported by microglia activity up to 2 years after ischemia [27]. In addition, these animals showed increased expression of neurogenesis markers and the migration of neuroblasts in the subventricular zone [26]. Thus, the balance of degenerative processes and inflammation surveillance with neurogenesis may be decisive for long-term survival after cerebral ischemia [26]. Brain ischemia induces neuronal necrosis and apoptosis, which triggers an inflammatory response controlled by the release of ROS, cytokines, and chemokines. This process develops not only in the brain but also in the microcirculation and involves several types of cells, such as innate microglia immune cells, adaptive immune cells, and lymphocytes, enhancing neuronal death [8,41]. As a result of neuroinflammation in the brain, the secretion of many cytokines increases both in damaged brain tissue and in peripheral blood. These cytokines are involved in the progression of post-ischemic brain neurodegeneration and influence disease severity and neurological outcomes [8,41].

3. Pro- and Anti-Inflammatory Cytokines and Inflammatory Cells in the Post-Ischemic Brain

3.1. Cytokines

The occurrence of generalized inflammatory changes following brain ischemia is a relatively well-known phenomenon [46–49]. Cytokines such as IL-6 and TNF-α appear to be key mediators of this phenomenon [46–49]. Increased release of pro-inflammatory cytokines has been observed in the blood of people with ischemic stroke and has been correlated with a larger area of cerebral ischemia and worse outcomes [46,47,49]. Increased concentration of IL-6 in blood and cerebrospinal fluid is associated with an increase in neurological symptoms, a greater volume of infarction, and a worse prognosis [49]. In addition, elevated levels of TNF-α in the cerebrospinal fluid and blood in people with stroke have been associated with deteriorating neurological symptoms, an increase in infarct size, and worse clinical outcomes [48]. In contrast to the pro-inflammatory cytokines, IL-10 and IL-4 are anti-inflammatory cytokines. The exact relationship between pro- and anti-inflammatory cytokines and their relevance to clinical outcomes in ischemic stroke patients remain unexplained. However, this balance is disturbed in the early stages of an ischemic stroke [8]. This supports the study of elevated pro-inflammatory IL-6 in the blood at 12 h after cerebral ischemia in stroke patients compared to controls, and this increase has been correlated with severe neurological deficits and worse outcomes [50]. In conclusion, increased IL-6 and reduced IL-10 concentrations are present in the early stroke period and are associated with a degree of neurological deficit and stroke outcome [50]. This observation confirms the interaction between pro- and anti-inflammatory cytokines in the first phases of ischemic stroke, and the advantage of the balance in favor of inflammation results in more severe neurological deficits. Interestingly, at the moment, we cannot precisely define the phenomena modulating this interaction.

3.2. Cells

The involvement of cells other than neurons in the development of post-ischemic brain neurodegeneration has been evaluated in many experimental and clinical studies [1,26,27,37]. When neuroinflammation following brain ischemia begins, it starts with the release of pro-inflammatory factors that involve various cells. First, we observe the involvement of neuroglial cells in the brain. Next, leukocytes, monocytes, and other cells with immune functions enter the ischemic brain tissue. This mechanism may additionally exacerbate post-ischemic brain damage by increasing blood–brain barrier permeability, edema, and progressive neuronal death. The variety of cells involved in this process can have a beneficial or detrimental effect, depending on the post-ischemia period: early or delayed. Ultimately, we need to know which cells are involved in post-ischemic changes and when they are involved to establish ways to control them.

3.2.1. Microglia

Microglia, the resident innate immune cells of the brain accounting for up to 20% of the neuroglial population, undergo morphological and phenotypic changes after brain ischemia [1]. Activated microglia function like macrophages during systemic inflammation and have the ability to remove foreign organisms and cellular debris. At rest, the microglia are referred to as small cells with wide protruding branches. However, after ischemia, microglial cells are activated, they change shape and function, but the exact mechanisms of this phenomenon are still unknown. They are activated after brain ischemia, as a result of which changes in their phenotypes can be observed [8,26,27,42–44,51]. Transient focal brain ischemia in the rat leads to microglia activation in the cerebral cortex of the ischemic hemisphere, and the severity and extent of the injury is reflected in the intensification of microglia activation [52]. The microglia around the post-ischemic parenchyma migrate towards the ischemic lesion and remain in close relationship with the neurons in a process called "capping", that is, after neuron death, the capping helps in early recognition and rapid phagocytic removal of dead neurons [53,54]. They become active a few minutes after the onset of brain ischemia, increasing in number in the following days, reaching a peak on the tenth day after transient local brain ischemia [55]. After ischemic injury, the microglia activating the phenotype become amoeboid and have a functional macrophage character. After this transformation, microglia look like macrophages not only in appearance but also in their behavior, they can release cytokines and secrete extracellular matrix metalloproteinases (MMPs) that are able to damage the blood–brain barrier and thus increase its permeability. This process facilitates the early transfer of leukocytes from the circulation to the ischemic brain, contributing to an increased level of pro-inflammatory factors that aggravates post-ischemic injury. Once activated, the microglial cells can take on two different phenotypes: classic pro-inflammatory (M1) and alternative anti-inflammatory (M2). The M1 phenotype releases the cytokines TNF-α, IL-6, and IL-1β and substances with oxidizing properties, such as nitric oxide [56]. The M2 phenotype has beneficial effects, it causes ischemic brain healing post-ischemia and the release of anti-inflammatory factors, such as IL-4 and IL-10, and secretes many factors with neurotrophic properties capable of preventing the development of neuroinflammation [56]. A recent study showed that microglia depletion by the dual colony-stimulating factor-1 inhibitor, PLX3397, exacerbates brain infarction and neurological deficits [57]. Following a transient focal brain ischemia, microglial depletion enhances leukocyte infiltration, expression of inflammatory factors, and neuronal loss in mice [57]. This pathological phenomenon is dependent not only on lymphocytes and monocytes, but also on astrocyte-mediated inflammatory factors. Hence, the presence of microglial cells prevents astrocytes from secreting inflammatory factors during and after ischemia [57]. Moreover, by supporting the above, the microglial cells produce different neurotrophic factors that animate neurogenesis and plasticity [58]. Thus, following brain ischemia, different subsets of microglial cells have different roles.

3.2.2. Astrocytes

Similar to microglia, astrocytes are housekeeping cells essential to the continuous functioning of the central nervous system. Astrocytes are involved in the physiological and pathological functioning of the brain. They regulate the water–ion balance; secrete neurotrophic factors; and remove unnecessary neurotransmitters, transport products, and waste of cellular metabolism. Astrocytes participate in the structure and function of the blood–brain barrier [59]. Under normal conditions, astrocytes take up excess glutamate from the extracellular space and convert it into glutamine for neurons to reuse, but during brain damage following ischemia, the degree of astrocyte damage affects their glutamate-uptake capacity [1,34,59]. How ischemia affects glutamate uptake by astrocytes is not fully elucidated, but expression of the excitatory amino acid transporter 2 (EAAT2) has been suggested to be impaired post-ischemia [60,61].

Cytokines from neurons and neuroglial cells cause post-ischemic astrocyte hyperplasia. As a result of ischemia, astrocytes release vimentin, nestin, IL-1β, monocyte chemotactic protein-1, and glial fibrillary acidic protein [27,62], which contribute to the development of reactive gliosis and the formation of glial scars after ischemia [27,63]. As a result of the Na^+/K^+ pump dysfunction, astrocytes swell after brain ischemia [62,64,65], which causes an increase in intracranial pressure and a consequent reduction in cerebral blood flow. Activated astrocytes release matrix metalloproteinase-2 (MMP-2) capable of damaging the extracellular matrix [66] and also contribute to the presence of ephrin-A5 in the ischemic brain area, which hinders axonal sprouting [67]. After embolic focal ischemia of the brain in rats, an exaggerated astroglial response is observed in the ischemic injury core from 4 h to 1 day, it peaks on day 4, persists for 28 days, and forms a glial scar [68]. Three days after reversible total brain ischemia in hippocampus astrocytes, significant upregulation of iNOS, glial fibrillary acidic protein (GFAP), and NADPH diaphorease expression was observed [69]. Post-mortem examination of the brain tissue after ischemia of patients who died within 7 days post-stroke showed increased expression of IL-15 in astrocytes [1]. Alternatively, IL-15 knockdown in astrocytes reduced ischemic brain damage in mice after transient local ischemia [70]. Transgenic mice expressing IL-15 with the controlled GFAP promoter exhibited increased cerebral infarction and increased neurological deficits following cerebral ischemia [1]. In addition, GFAP/Vimentin double-knockout mice showed reduced cortical blood flow in the brain and greater lesions following local ischemia [71]. Astrocytes release fibroblast growth factor-2, brain-derived neurotrophic factor, and nerve growth factor, which have neuroprotective properties [71,72]. In addition to their neurotrophic support, structurally, astrocytes by their terminal feet have a strong relationship with the endothelial cells of the brain's capillaries and the pericytes that make up the blood–brain barrier. During brain ischemia, MMP-9 breaks the connection between the terminal feet of astrocytes and endothelial cells by degrading the basal lamina [73]. Consequently, the open blood–brain barrier acts as the main gateway for invasion of the brain by peripheral inflammatory cells.

3.2.3. Neutrophils

Leukocytosis has been found to be a marker of inflammation in response to ischemic stroke. Leukocytosis is associated with a high degree of disability, impairment, and increased mortality [74]. Neutrophils are the first blood-derived immune cells to invade ischemic brain tissue, followed by monocytes. After brain ischemia, neutrophils undergo conformational changes due to the presence of numerous adhesive molecules, which facilitates their migration across the vessel wall into the brain tissue. In addition to blood-derived microglia and macrophages, neutrophils are among the most important leukocytes that infiltrate the post-ischemic brain. A high number of post-ischemic neutrophils in the brain come from the peripheral circulation. Later, neutrophils are attracted to the ischemic region by chemokines and then cause secondary damage to the ischemic tissue by releasing pro-inflammatory mediators, proteases, ROS, and MMPs [75]. These toxic mediators weaken the endothelial cell membrane and the basal lamina leading to permeability of the blood–brain barrier and the development of post-ischemic brain edema. Their onset is fairly early, reaching the brain within half an hour to several hours after ischemia, peaking over the next 3 days, and gradually decreasing over 15 days [8,76]. After five hours of recirculation, neutrophils enter the ischemically-damaged area of the brain [77,78]. Following neutrophil invasion, monocytes then adhere to the vessel wall and migrate towards the ischemic area with maximum involvement within 3–7 days after ischemia [76]. It has been noted that infiltrating neutrophils remain for more than a month in the ischemic areas of the brain and their presence is masked after 3 days by overactivation of microglia/macrophages in the inflammatory area [79]. These cells activate molecules capable of contact with the endothelial cells as early as 15 minutes after brain ischemia, and within 6–8 h they surround the brain's blood vessels and penetrate the brain [80–82]. Neutrophils are believed to block microcirculation in the brain either mechanically or by secreting vasoconstrictors, releasing pro-inflammatory factors, ROS, and enzymes with hydrolytic activities [83–85]. In addition, neutrophils produce MMP-9, which is a protease that damages the blood–brain barrier, enhancing brain edema and causing hemorrhagic transformation of acute ischemic stroke [86]. The size of ischemic infarction and level of neurological deficits positively correlate with an increase in the number and activity of neutrophils, which in turn leads to an increased risk of death [76,87]. In contrast to neutrophils, after brain ischemia, the number of lymphocytes decreases, and thus the neutrophil/lymphocyte ratio increases. This ratio is closely related to the size of the infarct and mortality [87]. Leukocytes, which include neutrophils and T lymphocytes, intensify ischemic brain damage in many ways. First, the neutrophils adhere to the endothelium, which blocks the flow of erythrocytes through the microcirculation, which leads to the no-reflow phenomenon in the brain. Second, on the endothelial surface, activated neutrophils produce proteases, MMPs, and ROS, which significantly damage blood vessels and brain tissue. The consequence of the above phenomena is vasoconstriction and platelet aggregation inside the brain vessels [37,88]. Finally, infiltrating leukocytes further aggravate neuronal damage by activating pro-inflammatory mediators in and around the penumbra and in the core of the infarct [89].

3.2.4. Lymphocytes

T lymphocytes become involved in the later stages of post-ischemic neurodegeneration of the brain. Lymphocytes surround the periphery of the ischemic lesion, and their number increases after 3 days, reaches a maximum after one week, and then decreases after another week [90]. The effect of various T lymphocytes on inflammation and thrombosis, with the consequence of increased brain damage and worsening of neurological deficits, has been demonstrated in studies in mice deficient in T lymphocytes [91–93]. In addition, immunodepletion of $CD4^+$ lymphocyte cells in mice increased neuronal loss and was associated with more severe neurological deficits 7 days after focal brain ischemia [94]. The study showed that mice without $\gamma\delta$ T lymphocyte cells had reduced infarct volume post-ischemia and the same phenomenon was reported in mice after administration of antibodies against the receptors of these cells [95]. Another study found an increase in $CD4^+CD28$ null lymphocyte cells in stroke survivors or those who died from ischemic stroke [96]. This study revealed an association between the high probability of death after a stroke or new ischemic episodes and the $CD4^+CD28$ null lymphocyte cell count, and proposed that the number of these lymphocytes could serve as a warning biomarker against recurrence of an ischemic event or death. Another study found that peripheral frequency of $CD4^+$ lymphocyte cells and $CD4^+CD28$ null lymphocyte cells was significantly higher in patients with acute ischemic stroke than in the control group [97]. The results of this investigation indicate that in acute ischemic stroke, a higher percentage of peripheral $CD4^+CD28$ null lymphocyte cells may be associated with more massive brain damage. Analysis of this study also showed that the percentage of CD^+CD28 null lymphocyte cells may be useful for distinguishing between subtypes of stroke. In addition, genetic study has found an increased expression of activating "pro-inflammatory" killer cell immunoglobulin-like receptor (KIR) genes in people with ischemic stroke, which likely explains the massive development of inflammation in the acute phase of stroke [98]. Accumulating evidence shows that both innate and adaptive immune cells penetrate the brain after ischemia. Up to one month after an experimental ischemic stroke, T lymphocyte cell invasion into the ischemic area has been observed and has been shown to persist for years in post-stroke patients [99]. Up to one month after focal brain ischemia, a significant increase in the number of different subtypes of T lymphocytes was observed in the peri-infarct zone [99]. T lymphocytes entering the brain after ischemia had a close interaction with activated astrocytes and a progressively developed pro-inflammatory phenotype as evidenced by markers of increased lymphocyte activation, pro-inflammatory cytokines TNF-α, INF-γ, IL-10, IL-17, and perforin, with appropriate T-bet and RORc transcription factors [99]. Treg immunodepletion using a specific CD-25 antibody aggravated tissue injury and impaired neurological deficits on day 7 after local brain ischemia in mice [94].

3.2.5. Macrophages

Circulating macrophages are involved in delayed development of the neuroinflammatory mechanism in an ischemic brain. As early as 2 h after ischemia, activated macrophages can be detected in the brain [100]. Between 22 and 46 h after ischemia, both blood-born and brain-resident macrophages are dispersed throughout the ischemic injury in the brain and remain detectable for up to 1 week in mice after a 30 min ischemic injury [100]. In another study, their presence in brain tissue was recorded 4 days after the onset of ischemia, peaking after 7 days and then diminishing [55]. Pathological post-stroke mechanisms aggravate cell damage due to primary cellular events that initiate a vicious pathological cycle of inflammatory mediators that further enhances neuronal death. In summary, all the inflammatory cells described above play an important role both in initiating and enhancing the pathological response following brain ischemia, but also in maintaining homeostasis of brain cells, especially neurons that have survived a primary ischemic event.

4. Interaction of Inflammatory Cells in the Post-Ischemic Brain

The release of IL-6 and TNF-α by area(s) of brain ischemia enhances the invasion of neutrophils into the brain, which intensifies the blood–brain barrier permeability, and an increased number of neutrophils in the blood is associated with the extent of the infarction [101]. Moreover, the role of microglia after brain ischemia depends on the state of cell polarity. The dominance of the M1 phenotype correlates with extensive post-ischemic damage, increased anaerobic glycolysis, and activation of hypoxia-inducible factor 1α. Polarization of the microglial cells to the M1 phenotype and the resulting increased generation of IL-23 favor the recruitment and stimulation of $\gamma\delta$ T lymphocyte cells, which play an unfavorable role within acute ischemic stroke [8]. The results of studies on experimental models of brain ischemia confirm that $\gamma\delta$ T lymphocyte cells are pathogenic for the brain due to IL-17 secretion and stimulation of inflammatory changes [102]. Moreover, the release of anti-inflammatory cytokines by microglia such as IL-10 and TGF-β

promotes the recruitment of regulatory lymphocytes that perform immunomodulatory and immunosuppressive functions in the ischemic brain [26]. Evidence supports a beneficial function of $CD4^{+}$ lymphocyte cells in experimental brain ischemia [103]. Inflammatory cell interactions are very complex and involve intercellular crosstalk by mechanisms that are not fully understood. Evidence suggests that inflammation plays a bivalent role in the development of post-ischemic brain neurodegeneration, promoting neuronal loss and healing of post-ischemic lesions. As a consequence, neuroinflammation may be a promising target for the future treatment of stroke.

5. Pre-Clinical Neuroinflammation Treatment in the Post-Ischemic Brain

Anti-inflammatory actions are aimed at reducing neuroinflammatory reactions by inhibiting the factors and phenomena that increase inflammation, and at the same time stimulating anti-inflammatory factors and phenomena naturally occurring in the body (Table 1). One of the many cytokines most involved in the development of post-ischemic brain neurodegeneration is IL-1 [104]. IL-1 deficient mice showed a significant reduction in infarct volume of approximately 70% after focal brain ischemia compared to that of wild-type mice [104]. It has been shown that inhibition of microglia activation by 2% isoflurane in transient focal brain ischemia in rats reduced the infarct size, attenuated apoptosis, and significantly decreased microglia activation in ischemic penumbra [105]. Within 7 days after focal brain ischemia, edaravone, a free radical scavenger that mimics glutathione peroxidase, reduces microglia activation and early accumulation of oxidative products in rats [106]. Similarly, multiple exposures to hyperbaric oxygen reduced infarct volume by decreasing microglia activation (Table 1) [107]. Inactivation of NF-kB in astrocytes promoted survival of the neurons after ischemic damage in mice [108]. Treg immunodepletion using a specific CD-25 antibody aggravated tissue injury and impaired neurological deficits on day 7 after local brain ischemia in mice (Table 1) [94]. A beneficial effect of administration of a recombinant human interleukin-1 receptor antagonist on the reduction of neuronal death in rats after hypoxia-ischemia and focal cerebral ischemia has also been shown (Table 1) [109,110]. Administration of anti-TNF-α antibody (polyclonal rabbit anti-mouse TNF-α neutralizing antibody) improved neurological outcomes in rats after reversible local brain ischemia [111]. An approximately 40% reduction in infarct volume was observed in IL-10 overexpressing transgenic mice after local brain ischemia with a parallel decrease in caspase 3 levels compared to that of wild-type mice [112]. The administration of insulin-like growth factor-1 reduced infarct volume and improved sensitivity and mobility in mice [113]. In microglia after ischemia, Toll-like receptors (TLR) induce the expression of genes for cytokines with pro-inflammatory properties [114–116]. Knockout mice for TLR4, but not for TLR3 or TLR9, showed a significant reduction in infarct volume after ischemia compared to that of wild-type mice [114]. It has been found that deficiency of T and B lymphocyte cells in animal models of ischemic stroke resulted in a smaller lesion size and reduced neuroinflammation [117]. It has been shown that inhibition of T lymphocyte migration to the brain reduces infarct volume and post-ischemic inflammation [118]. Reduction in infarct size and improvement in neurological outcomes following local brain ischemia were noted after administration of heme oxygenase-1 and in heme oxygenase-1 overexpressing transgenic mice due to anti-inflammatory and anti-apoptotic properties [119,120]. Additionally, heme oxygenase-1 deficient mice have been shown to exhibit increased infarct size following local permanent brain ischemia compared to that of wild-type controls (Table 1) [121].

Table 1. Preclinical studies in the prevention/treatment of neuroinflammation in post-ischemic brain neurodegeneration.

Type of Ischemia	Animal	Target	Benefits	References
Focal	Mouse	IL-1 lack	Reduction of infarct size	[104]
Focal	Rat	Microglia	Reduction of infarct size and apoptosis, microglia activation	[105]
Focal	Rat	Glutathione peroxidase	Reduction of microglia activation, oxidative products	[106]
Focal	Rat	Microglia	Reduction of infarct size	[107]
Hypoxia-ischemia	Rat	IL-1 antibody	Reduction of neurological deficits	[109]
Focal	Rat	IL-1 antibody	Reduction of neuronal death	[110]
Focal	Rat	TNF-α antibody	Improvement in neurological outcome	[111]
Permanent focal	Mouse	IL-10	Reduction of infarct size, decrease in caspase 3	[112]
Permanent focal	Mouse	IGF-1	Reduction of infarct size, improvement in sensitivity and mobility	[113]
Focal	Mouse	TLR4 lack	Reduction of infarct size	[114–116]
Focal	Mouse	T cells lack	Reduction of lesion size and inflammation	[117]
Permanent focal	Mouse	Lymphocytes	Reduction of lymphocytes, neuronal damage, infarct size, and inflammation	[118]
Permanent focal	Mouse	HO-1	Reduction of infarct size and neurological deficits	[119]
Focal	Mouse	HO-1	Reduction if infarct size and neurological deficits	[120]
Permanent focal	Mouse	HO-1 lack	Increase in infarct size	[121]
Focal	Mouse	NF-kB	Support of neuron survival	[108]
Focal	Mouse	Treg antibody	Increase of infarct size and neurological deficits	[94]
Focal	Rat	Neutrophils	Reduction of blood–brain barrier breakdown	[80]

IL-1: Interleukin 1, TNF-α: Tumor necrosis factor-α, IL-10: Interleukin 10, IGF-1: Insulin-like growth factor 1, TLR4: Toll-like receptor 4, T cells: T lymphocytes, HO-1: Heme oxygenase 1.

6. Neuroinflammation: Good or Bad?

Accumulating evidence suggests that neuroinflammation plays a key role in the pathogenesis of ischemic stroke and has become an interesting target for therapeutic intervention. Numerous reports indicate that neuroinflammatory cells play multiphase roles (beneficial and harmful) in which inhibiting the same pathway at the wrong time may exaggerate pathogenesis. Thus, better characterizing the pathophysiology of ischemic stroke together with timed treatment may provide the ultimate protective strategy of benefit.

A large amount of important data regarding the connection of inflammatory cells and their mediators that are released into the ischemic neurodegenerative brain has been accumulated in recent years. However, it is unclear whether the malfunctioning of inflammatory cells initiates the pathophysiological events of post-ischemic neurodegeneration, or whether the dysfunction of the inflammatory cells is a consequence of other adverse changes that occur in the early stages of the disease. Therefore, more data are needed on the origin of the inflammatory cell dysfunction. Recent research suggests that factors released from healthy inflammatory cells may also play a protective role when neurodegenerative conditions occur [122]. In conclusion, it is very interesting, but at the same time extremely difficult to understand how inflammatory cells exert various actions in different tissues under physiological and pathological conditions. Their functional complexity clearly requires an interdisciplinary approach to develop novel therapeutic interventions that will benefit from the multi-faceted nature of inflammatory cells, including their ability to facilitate crosstalk between the systemic environment and the brain.

Processes related to post-ischemic neurodegeneration of the brain are a strong support for the contribution of neuroinflammation to the progress of ischemic neuropathology (Figure 1). However, which aspects of this contribution seem positive or negative are still under debate. In this review, we suggest that there is a delicate balance in the response of neuroinflammation to post-ischemic brain neurodegeneration that may have both beneficial and harmful effects

(Figure 1) [122]. We suggest that certain aspects of neuroinflammation in the post-ischemic brain are necessary and beneficial, and may limit or prevent post-ischemic neuropathology. By exploiting the benefits of post-ischemic neuroinflammation, we can work to find a causal therapy for post-ischemic neurodegeneration. Scientists have focused too long on the negative consequences of astrogliosis and microgliosis, judging the neuroglial cells as excited. However, recent evidence has demonstrated the enormous heterogeneity of neuroglial cells and the variability of their functioning. Although our primary focus is on the involvement of the innate immune system in post-ischemic brain neurodegeneration, evidence suggests that the adaptive immune system also plays a role in this pathology, which is an area that requires further research. Although it is clear that neuroinflammation contributes to the pathogenesis of post-ischemic injury, it is too broad a word to describe the mixture of various elements of the innate immune processes that become active as the disease progresses [122]. Microgliosis involves phagocytosis of amyloid plaques, neurofibrillary tangles, dysfunctional synapses, and the release of trophic factors for cell plasticity and growth (Figure 1) [123]. In contrast, microgliosis increases the amounts of chemokines and cytokines, which become toxic and harmful to neurons (Figure 1). Astrogliosis is also beneficial because it causes the propagation of calcium currents to increase signal conduction and improve repair and protection. In contrast, excessive astrogliosis can increase the amount of various neurotoxic substances. Therefore, the neuroinflammatory and immune responses following ischemic brain injury are both good and bad. Treatment options based on these data suggest early use of antibiotics to prevent infection (in stroke patients) and inhibition of the early inflammatory/immune response, although it may be likely that after brain ischemia develops, it is too late for this treatment, and treatment should focus on enhancing the protective immune response to accelerate the repair of neurons that survived the primary ischemic episode. It is clear that neuroinflammation is strongly involved with post-ischemic neurodegeneration of the brain (Figure 1), but a more thorough understanding of the ischemic-specific immune/inflammatory response will be critical in developing causal therapy for post-ischemic brain injury.

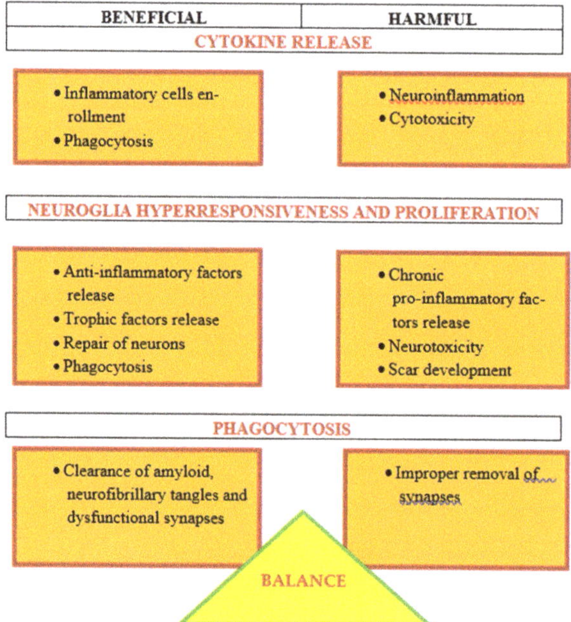

Figure 1. There is a delicate balance between the beneficial and harmful effects of neuroinflammation in post-ischemic brain neurodegeneration. Some neuroinflammation phenomena are protective, such as phagocytosis, but others, such as pro-inflammatory mediators, are detrimental.

7. Conclusions

Neuroinflammation in post-ischemic brain neurodegeneration is extremely complex, with multiphase pro-inflammatory responses without causal treatment. In the case of ischemic neuronal cell damage, the locally and systemically released chemokines, cytokines, and ROS play an important and mutual role of triggering neuroinflammation in the brain. In inflammation of the nervous system, the blood–brain barrier is partially impaired and allows for increased, but not completely uncontrolled, entry of immune cells into the brain. Changes in the perivascular space and chemokine environment, coupled with impairment of the glial limitans, ultimately allow immune cells to infiltrate into the parenchyma of the brain, resulting in impaired brain function and hence exacerbation of clinical disease symptoms. An in-depth understanding of the specific pathological mechanisms used by different types of immune cells to cross the blood–brain barrier would improve therapeutic targeting by inhibiting potentially brain-damaging subsets of immune cells, while leaving the brain's positive immune surveillance largely intact. Post-ischemic brain neurodegeneration with neuroinflammation begins in earnest when monocytes and leukocytes reach the brain, activating resident cells such as astrocytes and microglia and endothelial cells, and releasing another pool of pro-inflammatory mediators. Moreover, the evidence shows that cells involved in neuroinflammation have dual helper and deleterious functions, in fact, the same pathway, inactivated at different times, can increase or decrease ischemic damage to brain tissue. In view of the above, any future therapies should take into account the timing of their application post-ischemia. Based on the latest data, many genes can influence the course, extent of the damaged area, and prognosis in post-ischemic brain neurodegeneration. Neuroinflammation is a complex phenomenon governed by many factors that play a key role not only in the pathogenesis of post-ischemic injury, but also in determining its evolution; therefore, post-ischemic neuroinflammation may be a promising target in developing new therapeutic strategies for neurodegenerative diseases.

In summary, neuroinflammatory cells may express various activities leading eventually to either beneficial of harmful outcomes (Figure 1). No doubt, neuroinflammation is a key player in the development of ischemic neuropathology, as discussed in Section 6. A good example of the ambivalent role of neuroinflammation may be the broadly expressed pleiotropic protein osteopontin, which plays a role in neurodegenerative conditions, including Alzheimer's disease [124]. On one hand, osteopontin is associated with detrimental effects on neurons due to recruiting inflammatory cells to the lesioned area. On the other hand, the protein may promote neuronal repair/regeneration via the inflammatory response [124]. These clearly opposed activities may be partially due to different functional domains of osteopontin that are exposed following MMP or thrombin cleavages [124].

Author Contributions: Conceptualization, R.P. and S.J.C.; methodology, S.J.; software, S.J.; validation, R.P. and S.J.C.; formal analysis, R.P.; investigation, S.J.; resources, S.J.; data curation, R.P.; writing—original draft preparation, R.P.; writing—review and editing, R.P. and S.J.C.; visualization, R.P.; supervision, R.P.; project administration, R.P.; funding acquisition, R.P. and S.J.C. All authors have read and agreed to the published version of the manuscript.

Funding: The authors acknowledge the financial support from the following institutions: The Mossakowski Medical Research Institute, Polish Academy of Sciences, Warsaw, Poland (T3-RP) and the Medical University of Lublin, Lublin, Poland (DS 475/20-SJC).

Institutional Review Board Statement: Not applicable.

Informed Consent Statement: Not applicable.

Data Availability Statement: Not applicable.

Conflicts of Interest: The authors declare no conflict of interest.

References

1. Jayaraj, R.L.; Azimullah, S.; Beiram, R.; Jalal, F.Y.; Rosenberg, G.A. Neuroinflammation: Friend and foe for ischemic stroke. *J. Neuroinflamm.* **2019**, *16*, 142. [CrossRef] [PubMed]
2. Mok, V.C.T.; Lam, B.Y.K.; Wang, Z.; Liu, W.; Au, L.; Leung, E.Y.L.; Chen, S.; Yang, J.; Chu, W.C.W.; Lau, A.Y.L.; et al. Delayed-onset dementia after stroke or transient ischemic attack. *Alzheimer's Dement.* **2016**, *12*, 1167–1176. [CrossRef] [PubMed]
3. Portegies, M.L.; Wolters, F.J.; Hofman, A.; Ikram, M.K.; Koudstaal, P.J.; Ikram, M.A. Prestroke vascular pathology and the risk of recurrent stroke and poststroke dementia. *Stroke* **2016**, *47*, 2119–2122. [CrossRef]
4. Kim, J.H.; Lee, Y. Dementia and death after stroke in older adults during a 10-year follow-up: Results from a competing risk model. *J. Nutr. Health Aging* **2018**, *22*, 297–301. [CrossRef] [PubMed]
5. Pluta, R.; Ułamek-Kozioł, M.; Januszewski, S.; Czuczwar, S. Amyloid pathology in the brain after ischemia. *Folia Neuropathol.* **2019**, *57*, 220–226. [CrossRef] [PubMed]
6. Goulay, R.; Romo, L.M.; Hol, E.M.; Dijkhuizen, R.M. From stroke to dementia: A comprehensive review exposing tight interactions between stroke and amyloid-β formation. *Transl. Stroke Res.* **2020**, *11*, 601–614. [CrossRef]
7. Bejot, Y.; Daubail, B.; Giroud, M. Epidemiology of stroke and transient ischemic attacks: Current knowledge and perspectives. *Rev. Neurol.* **2016**, *172*, 59–68. [CrossRef]
8. Maida, C.D.; Norrito, R.L.; Daidone, M.; Tuttolomondo, A.; Pinto, A. Neuroinflammatory mechanisms in ischemic stroke: Focus on cardioembolic stroke, background, and therapeutic approaches. *Int. J. Mol. Sci.* **2020**, *21*, 6454. [CrossRef]
9. Lo, J.W.; Crawford, J.D.; Desmond, D.W.; Godefroy, O.; Jokinen, H.; Mahinrad, S.; Bae, H.J.; Lim, J.S.; Kohler, S.; Douven, E.; et al. Stroke and Cognition (STROKOG) Collaboration. Profile of and risk factors for post-stroke cognitive impairment in diverse ethno-regional groups. *Neurology* **2019**, *93*, e2257–e2271. [CrossRef]
10. Murphy, S.J.X.; Werring, D.J. Stroke: Causes and clinical features. *Medicine* **2020**, *48*, 9. [CrossRef]
11. Pluta, R. *Ischemia-Reperfusion Pathways in Alzheimer's Disease*; Nova, Science Publishers, Inc.: New York, NY, USA, 2007.
12. Chi, N.F.; Chien, L.N.; Ku, H.L.; Hu, C.J.; Chiou, H.Y. Alzheimer disease and risk of stroke: A population-based cohort study. *Neurology* **2013**, *80*, 705–711. [CrossRef]
13. Tolppanen, A.M.; Lavikainen, P.; Solomon, A.; Kivipelto, M.; Soininen, H.; Hartikainen, S. Incidence of stroke in people with Alzheimer disease: A national register-based approach. *Neurology* **2013**, *80*, 353–358. [CrossRef]
14. Gamaldo, A.; Moghekar, A.; Kilada, S.; Resnick, S.M.; Zonderman, A.B.; O'Brien, R. Effect of a clinical stroke on the risk of dementia in a prospective cohort. *Neurology* **2006**, *67*, 1363–1369. [CrossRef]
15. Pluta, R. *Brain Ischemia: Alzheimer's Disease Mechanisms*; Nova, Science Publishers, Inc.: New York, NY, USA, 2019; p. 311.
16. Sexton, E.; McLoughlin, A.; Williams, D.J.; Merriman, N.A.; Donnelly, N.; Rohde, D.; Hickey, A.; Wren, M.A.; Bennett, K. Systematic review and meta-analysis of the prevalence of cognitive impairment no dementia in the first year post-stroke. *Eur. Stroke J.* **2019**, *4*, 160–171. [CrossRef] [PubMed]
17. Pluta, R.; Ułamek, M.; Jabłoński, M. Alzheimer's mechanisms in ischemic brain degeneration. *Anat. Rec.* **2009**, *292*, 1863–1881. [CrossRef] [PubMed]
18. Kuzma, E.; Lourida, I.; Moore, S.F.; Levine, D.A.; Ukoumunne, O.C.; Llewellyn, D.J. Stroke and dementia risk: A systematic review and meta-analysis. *Alzheimer's Dement.* **2018**, *14*, 1416–1426. [CrossRef]
19. De Ronchi, D.; Palmer, K.; Pioggiosi, P.; Atti, A.R.; Berardi, D.; Ferrari, B.; Dalmonte, E.; Fratiglioni, L. The combined effect of age, education, and stroke on dementia and cognitive impairment no dementia in the elderly. *Dement. Geriatr. Cogn. Disord.* **2007**, *24*, 266–273. [CrossRef]
20. Pendlebury, S.T.; Rothwell, P.M. Prevalence, incidence, and factors associated with pre-stroke and post-stroke dementia: A systematic review and metaanalysis. *Lancet Neurol.* **2009**, *8*, 1006–1018. [CrossRef]
21. Kocki, J.; Ułamek-Kozioł, M.; Bogucka-Kocka, A.; Januszewski, S.; Jabłoński, M.; Gil-Kulik, P.; Brzozowska, J.; Petniak, A.; Furmaga-Jabłońska, W.; Bogucki, J.; et al. Dysregulation of amyloid precursor protein, β-secretase, presenilin 1 and 2 genes in the rat selectively vulnerable CA1 subfield of hippocampus following transient global brain ischemia. *J. Alzheimers Dis.* **2015**, *47*, 1047–1056. [CrossRef]
22. Pluta, R.; Kocki, J.; Ułamek-Kozioł, M.; Petniak, A.; Gil-Kulik, P.; Januszewski, S.; Bogucki, J.; Jabłoński, M.; Brzozowska, J.; Furmaga-Jabłońska, W.; et al. Discrepancy in expression of β-secretase and amyloid-β protein precursor in Alzheimer-related genes in the rat medial temporal lobe cortex following transient global brain ischemia. *J. Alzheimers Dis.* **2016**, *51*, 1023–1031. [CrossRef]
23. Pluta, R.; Kocki, J.; Ułamek-Kozioł, M.; Bogucka-Kocka, A.; Gil-Kulik, P.; Januszewski, S.; Jabłoński, M.; Petniak, A.; Brzozowska, J.; Bogucki, J.; et al. Alzheimer-associated presenilin 2 gene is dysregulated in rat medial temporal lobe cortex after complete brain ischemia due to cardiac arrest. *Pharmacol. Rep.* **2016**, *68*, 155–161. [CrossRef]
24. Pluta, R.; Bogucka-Kocka, A.; Ułamek-Kozioł, M.; Bogucki, J.; Czuczwar, S.J. Ischemic tau protein gene induction as an additional key factor driving development of Alzheimer's phenotype changes in CA1 area of hippocampus in an ischemic model of Alzheimer's disease. *Pharmacol. Rep.* **2018**, *70*, 881–884. [CrossRef]
25. Pluta, R.; Ułamek-Kozioł, M.; Kocki, J.; Bogucki, J.; Januszewski, S.; Bogucka-Kocka, A.; Czuczwar, S.J. Expression of the tau protein and amyloid protein precursor processing genes in the CA3 area of the hippocampus in the ischemic model of Alzheimer's disease in the rat. *Mol. Neurobiol.* **2020**, *57*, 1281–1290. [CrossRef] [PubMed]

26. Sekeljic, V.; Bataveljic, D.; Stamenkovic, S.; Ułamek, M.; Jabłoński, M.; Radenovic, L.; Pluta, R.; Andjus, P.R. Cellular markers of neuroinflammation and neurogenesis after ischemic brain injury in the long-term survival rat model. *Brain Struct. Funct.* **2012**, *217*, 411–420. [CrossRef]
27. Radenovic, L.; Nenadic, M.; Ułamek-Kozioł, M.; Januszewski, S.; Czuczwar, S.J.; Andjus, P.R.; Pluta, R. Heterogeneity in brain distribution of activated microglia and astrocytes in a rat ischemic model of Alzheimer's disease after 2 years of survival. *Aging* **2020**, *12*, 12251–12267. [CrossRef]
28. Hossmann, K.A.; Schmidt-Kastner, R.; Ophoff, B.G. Recovery of integrative central nervous function after one hour global cerebro-circulatory arrest in normothermic cat. *J. Neurol. Sci.* **1987**, *77*, 305–320. [CrossRef]
29. Pluta, R. The role of apolipoprotein E in the deposition of β-amyloid peptide during ischemia–reperfusion brain injury. A model of early Alzheimer's disease. *Ann. N. Y. Acad. Sci.* **2000**, *903*, 324–334. [CrossRef]
30. Jabłoński, M.; Maciejewski, R.; Januszewski, S.; Ułamek, M.; Pluta, R. One year follow up in ischemic brain injury and the role of Alzheimer factors. *Physiol. Res.* **2011**, *60* (Suppl. S1), 113–119. [CrossRef]
31. Kiryk, A.; Pluta, R.; Figiel, I.; Mikosz, M.; Ułamek, M.; Niewiadomska, G.; Jabłoński, M.; Kaczmarek, L. Transient brain ischemia due to cardiac arrest causes irreversible long-lasting cognitive injury. *Behav. Brain Res.* **2011**, *219*, 1–7. [CrossRef]
32. Li, J.; Wang, Y.J.; Zhang, M.; Fang, C.Q.; Zhou, H.D. Cerebral ischemia aggravates cognitive impairment in a rat model of Alzheimer's disease. *Life Sci.* **2011**, *89*, 86–92. [CrossRef]
33. Cohan, C.H.; Neumann, J.T.; Dave, K.R.; Alekseyenko, A.; Binkert, M.; Stransky, K.; Lin, H.W.; Barnes, C.A.; Wright, C.B.; Perez-Pinzon, M.A. Effect of cardiac arrest on cognitive impairment and hippocampal plasticity in middle-aged rats. *PLoS ONE* **2015**, *10*, e0124918. [CrossRef]
34. Pluta, R.; Salinska, E.; Puka, M.; Stafiej, A.; Łazarewicz, J.W. Early changes in extracellular amino acids and calcium concentrations in rabbit hippocampus following complete 15-min cerebral ischemia. *Resuscitation* **1988**, *16*, 193–210. [CrossRef]
35. Won, S.J.; Kim, J.-E.; Cittolin-Santos, G.F.; Swanson, R.A. Assessment at the single-cell level identifies neuronal glutathione depletion as both a cause and effect of ischemia-reperfusion oxidative stress. *J. Neurosci.* **2015**, *35*, 7143–7152. [CrossRef]
36. Mishra, M.; Hedna, V.S. Neuroinflammation after acute ischemic stroke: A volcano hard to contain. *Chin. J. Contemp. Neurol. Neurosurg.* **2013**, *13*, 964.
37. Pluta, R.; Lossinsky, A.S.; Walski, M.; Wisniewski, H.M.; Mossakowski, M.J. Platelet occlusion phenomenon after short- and long-term survival following complete cerebral ischemia in rats produced by cardiac arrest. *J. Hirnforsch.* **1994**, *35*, 463–471.
38. Tuttolomondo, A.; Maida, C.; Pinto, A. Inflammation and inflammatory cell recruitment in acute cerebrovascular diseases. *Curr. Immunol. Rev.* **2015**, *11*, 24–32. [CrossRef]
39. Davalos, D.; Grutzendler, J.; Yang, G.; Kim, J.V.; Zuo, Y.; Jung, S.; Littman, D.R.; Dustin, M.L.; Gan, W.-B. ATP mediates rapid microglial response to local brain injury in vivo. *Nat. Neurosci.* **2005**, *8*, 752–758. [CrossRef]
40. Geissmann, F.; Gordon, S.; Hume, D.A.; Mowat, A.; Randolph, G.J. Unravelling mononuclear phagocyte heterogeneity. *Nat. Rev. Immunol.* **2010**, *10*, 453–460. [CrossRef] [PubMed]
41. Dabrowska, S.; Andrzejewska, A.; Lukomska, B.; Jankowski, M. Neuroinflammation as a target for treatment of stroke using mesenchymal stem cells and extracellular vesicles. *J. Neuroinflam.* **2019**, *16*, 178. [CrossRef]
42. Nagy, E.E.; Frigy, A.; Szász, J.A.; Horváth, E. Neuroinflammation and microglia/macrophage phenotype modulate the molecular background of post-stroke depression: A literature review. *Exp. Med.* **2020**, *20*, 2510–2523. [CrossRef]
43. Rawlinson, C.; Jenkins, S.; Thei, L.; Dallas, M.L.; Chen, R. Post-ischaemic immunological response in the brain: Targeting microglia in ischaemic stroke therapy. *Brain Sci.* **2020**, *10*, 159. [CrossRef] [PubMed]
44. Xu, S.; Lu, J.; Shao, A.; Zhang, J.H.; Zhang, J. Glial cells: Role of the immune response in ischemic stroke. *Front. Immunol.* **2020**, *11*, 294. [CrossRef]
45. Zhang, W.; Tian, T.; Gong, S.X.; Huang, W.Q.; Zhou, Q.Y.; Wang, A.P.; Tian, Y. Microglia-associated neuroinflammation is a potential therapeutic target for ischemic stroke. *Neural Regen. Res.* **2021**, *16*, 6–11.
46. Fassbender, K.; Rossol, S.; Kammer, T.; Daffertshofer, M.; Wirth, S.; Dollman, M.; Hennerici, M. Proinflammatory cytokines in serum of patients with acute cerebral ischemia: Kinetics of secretion and relation to the extent of brain damage and outcome of disease. *J. Neurol. Sci.* **1994**, *122*, 135–139. [CrossRef]
47. Beamer, N.B.; Coull, B.M.; Clark, W.M.; Briley, D.P.; Wynn, M.; Sexton, G. Persistent inflammatory response in stroke survivors. *Neurology* **1998**, *50*, 1722–1728. [CrossRef] [PubMed]
48. Mazzotta, G.; Sarchielli, P.; Caso, V.; Paciaroni, M.; Floridi, A.; Gallai, V. Different cytokine levels in thrombolysis patients as predictors for clinical outcome. *Eur. J. Neurol.* **2004**, *11*, 377–381. [CrossRef]
49. Beridze, M.; Sanikidze, T.; Shakarishvili, R.; Intskirveli, N.; Bornstein, N.M. Selected acute phase CSF factors in ischemic stroke: Findings and prognostic value. *BMC Neurol.* **2011**, *11*, 41. [CrossRef]
50. Kes, V.B.; Simundic, A.-M.; Nikolac, N.; Topić, E.; Demarin, V. Pro-inflammatory and anti-inflammatory cytokines in acute ischemic stroke and their relation to early neurological deficit and stroke outcome. *Clin. Biochem.* **2008**, *41*, 1330–1334.
51. Guruswamy, R.; ElAli, A. Complex roles of microglial cells in ischemic stroke pathobiology: New insights and future directions. *Int. J. Mol. Sci.* **2017**, *18*, 496. [CrossRef] [PubMed]
52. Emmrich, J.V.; Ejaz, S.; Neher, J.J.; Williamson, D.J.; Baron, J.C. Regional distribution of selective neuronal loss and microglial activation across the MCA territory after transient focal ischemia: Quantitative versus semiquantitative systematic immunohistochemical assessment. *J. Cereb. Blood Flow Metab.* **2015**, *35*, 20–27. [CrossRef] [PubMed]

53. Neumann, J.; Gunzer, M.; Gutzeit, H.O.; Ullrich, O.; Reymann, K.G.; Dinkel, K. Microglia provide neuroprotection after ischemia. *FASEB J.* **2006**, *20*, 714–716. [CrossRef]
54. Denes, A.; Vidyasagar, R.; Feng, J.; Narvainen, J.; McColl, B.W.; Kauppinen, R.A.; Allan, S.M. Proliferating resident microglia after focal cerebral ischaemia in mice. *J. Cereb. Blood Flow Metab.* **2007**, *27*, 1941–1953. [CrossRef] [PubMed]
55. Schilling, M.; Besselmann, M.; Müller, M.; Strecker, J.K.; Ringelstein, E.B.; Kiefer, R. Predominant phagocytic activity of resident microglia over hematogenous macrophages following transient focal cerebral ischemia: An investigation using green fluorescent protein transgenic bone marrow chimeric mice. *Exp. Neurol.* **2005**, *196*, 290–297. [CrossRef]
56. Hu, X.; Li, P.; Guo, Y.; Wang, H.; Leak, R.K.; Chen, S.; Gao, Y.; Chen, J. Microglia/macrophage polarization dynamics reveal novel mechanism of injury expansion after focal cerebral ischemia. *Stroke* **2012**, *43*, 3063–3070. [CrossRef] [PubMed]
57. Jin, W.N.; Shi, S.X.; Li, Z.; Li, M.; Wood, K.; Gonzales, R.J.; Liu, Q. Depletion of microglia exacerbates postischemic inflammation and brain injury. *J. Cereb. Blood Flow Metab.* **2017**, *37*, 2224–2236. [CrossRef]
58. Singhal, G.; Baune, B.T. Microglia: An interface between the loss of neuroplasticity and depression. *Front. Cell. Neurosci.* **2017**, *11*, 270. [CrossRef] [PubMed]
59. Bylicky, M.; Mueller, G.P.; Day, R.M. Mechanisms of endogenous neuroprotective effects of astrocytes in brain injury. *Oxid. Med. Cell. Longev.* **2018**, *2018*, 6501031. [CrossRef] [PubMed]
60. Takano, T.; Oberheim, N.; Cotrina, M.L.; Nedergaard, M. Astrocytes and ischemic injury. *Stroke* **2009**, *40* (Suppl. S1), S8–S12. [CrossRef]
61. Ketheeswaranathan, P.; Turner, N.A.; Spary, E.J.; Batten, T.F.; McColl, B.W.; Saha, S. Changes in glutamate transporter expression in mouse forebrain areas following focal ischemia. *Brain Res.* **2011**, *1418*, 93–103. [CrossRef]
62. Pluta, R.; Kida, E.; Lossinsky, A.S.; Golabek, A.A.; Mossakowski, M.J.; Wisniewski, H.M. Complete cerebral ischemia with short-term survival in rats induced by cardiac arrest. I. Extracellular accumulation of Alzheimer's beta-amyloid protein precursor in the brain. *Brain Res.* **1994**, *649*, 323–328. [CrossRef]
63. Wang, H.; Song, G.; Chuang, H.; Chiu, C.; Abdelmaksoud, A.; Ye, Y.; Zhao, L. Portrait of glial scar in neurological diseases. *Int. J. Immunopathol. Pharmacol.* **2018**, *31*, 2058738418801406. [CrossRef]
64. Pluta, R. Influence of prostacyclin on early morphological changes in the rabbit brain after complete 20-min ischemia. *J. Neurol. Sci.* **1985**, *70*, 305–316. [CrossRef]
65. Pluta, R. Resuscitation of the rabbit brain after acute complete ischemia lasting up to one hour: Pathophysiological and pathomorphological observations. *Resuscitation* **1987**, *15*, 267–287. [CrossRef]
66. Rempe, R.G.; Hartz, A.M.; Bauer, B. Matrix metalloproteinases in the brain and blood–brain barrier: Versatile breakers and makers. *Br. J. Pharmacol.* **2016**, *36*, 1481–1507. [CrossRef]
67. Overman, J.J.; Clarkson, A.N.; Wanner, I.B.; Overman, W.T.; Eckstein, I.; Maguire, J.L.; Dinov, I.D.; Toga, A.W.; Carmichael, S.T. A role for ephrin-A5 in axonal sprouting, recovery, and activity-dependent plasticity after stroke. *Proc. Natl. Acad. Sci. USA* **2012**, *109*, E2230–E2239. [CrossRef] [PubMed]
68. Nowicka, D.; Rogozinska, K.; Aleksy, M.; Witte, O.W.; Skangiel-Kramska, J. Spatiotemporal dynamics of astroglial and microglial responses after photothrombotic stroke in the rat brain. *Acta Neurobiol. Exp.* **2008**, *68*, 155.
69. Endoh, M.; Maiese, K.; Wagner, J. Expression of the inducible form of nitric oxide synthase by reactive astrocytes after transient global ischemia. *Brain Res.* **1994**, *651*, 92–100. [CrossRef]
70. Li, M.; Li, Z.; Yao, Y.; Jin, W.-N.; Wood, K.; Liu, Q.; Shi, F.; Hao, J. Astrocyte-derived interleukin-15 exacerbates ischemic brain injury via propagation of cellular immunity. *Proc. Natl. Acad. Sci. USA* **2017**, *114*, 396–405. [CrossRef] [PubMed]
71. Liu, Z.; Li, Y.; Cui, Y.; Roberts, C.; Lu, M.; Wilhelmsson, U.; Pekny, M.; Chopp, M. Beneficial effects of gfap/vimentin reactive astrocytes for axonal remodeling and motor behavioral recovery in mice after stroke. *Glia* **2014**, *62*, 2022–2033. [CrossRef] [PubMed]
72. Vogelgesang, A.; Becker, K.J.; Dressel, A. Immunological consequences of ischemic stroke. *Acta Neurol. Scand.* **2014**, *129*, 1–12. [CrossRef] [PubMed]
73. Del Zoppo, G.J. The neurovascular unit in the setting of stroke. *J. Intern. Med.* **2010**, *267*, 156–171. [CrossRef]
74. Furlan, J.; Vergouwen, M.; Silver, F. White blood cell count as a marker of stroke severity and clinical outcomes after acute ischemic stroke (P03.011). *Neurology* **2012**, *78* (Suppl. S1), P03.011. [CrossRef]
75. Martynov, M.Y.; Gusev, E.I. Current knowledge on the neuroprotective and neuroregenerative properties of citicoline in acute ischemic stroke. *J. Exp. Pharmacol.* **2015**, *7*, 17–28. [CrossRef]
76. Jickling, G.C.; Liu, D.; Ander, B.P.; Stamova, B.; Zhan, X.; Sharp, F.R. Targeting neutrophils in ischemic stroke: Translational insights from experimental studies. *Br. J. Pharmacol.* **2015**, *35*, 888–901. [CrossRef]
77. Perera, M.N.; Ma, H.K.; Arakawa, S.; Howells, D.W.; Markus, R.; Rowe, C.C.; Donnan, G.A. Inflammation following stroke. *J. Clin. Neurosci.* **2006**, *13*, 1–8. [CrossRef]
78. Wang, Q.; Tang, X.N.; Yenari, M.A. The inflammatory response in stroke. *J. Neuroimmunol.* **2007**, *184*, 53–68. [CrossRef]
79. Weston, R.M.; Jones, N.M.; Jarrott, B.; Callaway, J.K. Inflammatory cell infiltration after endothelin-1-induced cerebral ischemia: Histochemical and myeloperoxidase correlation with temporal changes in brain injury. *J. Cereb. Blood Flow Metab.* **2007**, *27*, 100–114. [CrossRef] [PubMed]

80. Weston, R.M.; Jarrott, B.; Ishizuka, Y.; Callaway, J.K. AM-36 modulates the neutrophil inflammatory response and reduces breakdown of the blood brain barrier after endothelin-1 induced focal brain ischaemia. *Br. J. Pharmacol.* **2006**, *149*, 712–723. [CrossRef]
81. Watcharotayangul, J.; Mao, L.; Xu, H.; Vetri, F.; Baughman, V.L.; Paisansathan, C.; Pelligrino, D.A. Post-ischemic vascular adhesion protein-1 inhibition provides neuroprotection in a rat temporary middle cerebral artery occlusion model. *J. Neurochem.* **2012**, *123*, 116–124. [CrossRef]
82. Pérez-De-Puig, I.; Miró, F.; Ferrer-Ferrer, M.; Gelpi, E.; Pedragosa, J.; Justicia, C.; Urra, X.; Chamorro, Á.; Planas, A.M. Neutrophil recruitment to the brain in mouse and human ischemic stroke. *Acta Neuropathol.* **2014**, *129*, 239–257. [CrossRef]
83. Connolly, E.S.; Winfree, C.J.; Springer, T.A.; Naka, Y.; Liao, H.; Yan, S.D.; Stern, D.M.; Solomon, R.A.; Gutierrez-Ramos, J.C.; Pinsky, D.J. Cerebral protection in homozygous null ICAM-1 mice after middle cerebral artery occlusion. Role of neutrophil adhesion in the pathogenesis of stroke. *J. Clin. Investig.* **1996**, *97*, 209–216. [CrossRef]
84. Funk, J.L.; Frye, J.B.; Davis-Gorman, G.; Spera, A.L.; Bernas, M.J.; Witte, M.H.; Weinand, M.E.; Timmermann, B.N.; McDonagh, P.F.; Ritter, L. Curcuminoids limit neutrophil-mediated reperfusion injury in experimental stroke by targeting the endothelium. *Microcirculation* **2013**, *20*, 544–554. [CrossRef] [PubMed]
85. Kalani, A.; Chaturvedi, P.; Kamat, P.K.; Maldonado, C.; Bauer, P.; Joshua, I.G.; Tyagi, S.C.; Tyagi, N. Curcumin-loaded embryonic stem cell exosomes restored neurovascular unit following ischemia-reperfusion injury. *Int. J. Biochem. Cell Biol.* **2016**, *79*, 360–369. [CrossRef] [PubMed]
86. Castellanos, M.; Leira, R.; Serena, J.; Pumar, J.M.; Lizasoain, I.; Castillo, J.; Dávalos, A. Plasma metalloproteinase-9 concentration predicts hemorrhagic transformation in acute ischemic stroke. *Stroke* **2003**, *34*, 40–46. [CrossRef]
87. Gökhan, S.; Ozhasenekler, A.; Durgun, H.M.; Akil, E.; Ustündag, M.; Orak, M. Neutrophil lymphocyte ratios in stroke subtypes and transient ischemic attack. *Eur. Rev. Med. Pharmacol. Sci.* **2013**, *17*, 653–657. [PubMed]
88. Wisniewski, H.M.; Pluta, R.; Lossinsky, A.S.; Mossakowski, M.J. Ultrastructural studies of cerebral vascular spasm after cardiac arrest-related global cerebral ischemia in rats. *Acta Neuropathol.* **1995**, *90*, 432–440. [CrossRef]
89. Kim, J.Y.; Park, J.; Chang, J.Y.; Kim, S.H.; Lee, J.E. Inflammation after ischemic stroke: The role of leukocytes and glial cells. *Exp. Neurobiol.* **2016**, *25*, 241–251. [CrossRef]
90. Feng, Y.; Liao, S.; Wei, C.; Jia, D.; Wood, K.; Liu, Q.; Wang, X.; Shi, F.-D.; Jin, W.-N. Infiltration and persistence of lymphocytes during late-stage cerebral ischemia in middle cerebral artery occlusion and photothrombotic stroke models. *J. Neuroinflamm.* **2017**, *14*, 248. [CrossRef]
91. Liu, T.; Clark, R.K.; McDonnell, P.C.; Young, P.R.; White, R.F.; Barone, F.C.; Feuerstein, G.Z. Tumor necrosis factor-α expression in ischemic neurons. *Stroke* **1994**, *5*, 1481–1488. [CrossRef]
92. Lambertsen, K.L.; Gregersen, R.; Meldgaard, M.; Clausen, B.H.; Heibøl, E.K.; Ladeby, R.; Knudsen, J.; Frandsen, A.; Owens, T.; Finsen, B. A role for interferon-g in focal cerebral ischemia in mice. *J. Neuropathol. Exp. Neurol.* **2004**, *63*, 942–955. [CrossRef]
93. Yilmaz, G.; Arumugam, T.V.; Stokes, K.Y.; Granger, D.N. Role of T lymphocytes and interferon-g in ischemic stroke. *Circulation* **2006**, *113*, 2105–2112. [CrossRef] [PubMed]
94. Liesz, A.; Suri-Payer, E.; Veltkamp, C.; Doerr, H.; Sommer, C.; Rivest, S.; Giese, T.; Veltkamp, R. Regulatory T cells are key cerebroprotective immunomodulators in acute experimental stroke. *Nat. Med.* **2009**, *15*, 192–199. [CrossRef] [PubMed]
95. Gelderblom, M.; Arunachalam, P.; Magnus, T. $\gamma\delta$ T cells as early sensors of tissue damage and mediators of secondary neurodegeneration. *Front. Cell. Neurosci.* **2014**, *8*, 368. [CrossRef] [PubMed]
96. Nadareishvili, Z.G.; Li, H.; Wright, V.; Maric, D.; Warach, S.; Hallenbeck, J.M.; Dambrosia, J.; Barker, J.L.; Vaird, A.E. Elevated proinflammatory CD4+CD28–lymphocytes and stroke recurrence and death. *Neurology* **2004**, *63*, 1446–1451. [CrossRef]
97. Tuttolomondo, A.; Pecoraro, R.; Casuccio, A.; Di Raimondo, D.; Buttá, C.; Clemente, G.; Della Corte, V.; Guggino, G.; Arnao, V.; Maida, C.; et al. Peripheral frequency of CD4+ CD28− cells in acute ischemic stroke. *Medicine* **2015**, *94*, e813. [CrossRef] [PubMed]
98. Tuttolomondo, A.; Di Raimondo, D.; Pecoraro, R.; Casuccio, A.; Di Bona, D.; Aiello, A.; Accardi, G.; Arnao, V.; Clemente, G.; On behalf of KIRIIND (KIR Infectious and Inflammatory Diseases) Collaborative Group. HLA and killer cell immunoglobulin-like receptor (KIRs) genotyping in patients with acute ischemic stroke. *J. Neuroinflamm.* **2019**, *16*, 1–15. [CrossRef]
99. Xie, L.; Li, W.; Hersh, J.; Liu, R.; Yang, S.H. Experimental ischemic stroke induces long-term T cell activation in the brain. *J. Cereb. Blood Flow Metab.* **2019**, *39*, 2268–2276. [CrossRef]
100. ElAli, A.; Jean LeBlanc, N. The role of monocytes in ischemic stroke pathobiology: New avenues to explore. *Front. Aging Neurosci.* **2016**, *8*, 29. [CrossRef]
101. Buck, B.H.; Liebeskind, D.S.; Saver, J.L.; Bang, O.Y.; Yun, S.W.; Starkman, S.; Ali, L.K.; Kim, O.; Villablanca, J.P.; Salamon, N.; et al. Early neutrophilia is associated with volume of ischemic tissue in acute stroke. *Stroke* **2008**, *39*, 355–360. [CrossRef] [PubMed]
102. Gelderblom, M.; Gallizioli, M.; Ludewig, P.; Thom, V.; Arunachalam, P.; Rissiek, B.; Bernreuther, C.; Glatzel, M.; Korn, T.; Arumugam, T.V.; et al. IL-23 (Interleukin-23)–producing conventional dendritic cells control the detrimental IL-17 (Interleukin-17) response in stroke. *Stroke* **2018**, *49*, 155–164. [CrossRef] [PubMed]
103. Syková, E. Glial diffusion barriers during aging and pathological states. *Prog. Brain Res.* **2001**, *132*, 339–363.
104. Boutin, H.; LeFeuvre, R.A.; Horai, R.A.; Asano, M.; Iwakura, Y.; Rothwell, N.J. Role of IL-1α and IL-1β in ischemic brain damage. *J. Neurosci.* **2001**, *21*, 5528–5534. [CrossRef]

105. Sun, M.; Deng, B.; Zhao, X.; Gao, C.; Yang, L.; Zhao, H.; Yu, D.; Zhang, F.; Xu, L.; Chen, L.; et al. Isoflurane preconditioning provides neuroprotection against stroke by regulating the expression of the TLR4 signalling pathway to alleviate microglial activation. *Sci. Rep.* **2015**, *5*, 11445. [CrossRef]
106. Yuan, Y.; Zha, H.; Rangarajan, P.; Ling, E.A.; Wu, C. Anti-inflammatory effects of Edaravone and Scutellarin in activated microglia in experimentally induced ischemia injury in rats and in BV-2 microglia. *BMC Neurosci.* **2014**, *15*, 125. [CrossRef]
107. Liska, G.M.; Lippert, T.; Russo, E.; Nieves, N.; Borlongan, C.V. A dual role for hyperbaric oxygen in stroke neuroprotection: Preconditioning of the brain and stem cells. *Cond. Med.* **2018**, *1*, 151–166.
108. Dvoriantchikova, G.; Barakat, D.; Brambilla, R.; Agudelo, C.; Hernandez, E.; Bethea, J.R.; Shestopalov, V.I.; Ivanov, D. Inactivation of astroglial NF-kappa B promotes survival of retinal neurons following ischemic injury. *Eur. J. Neurosci.* **2009**, *30*, 175–185. [CrossRef]
109. Martin, D.; Chinookoswong, N.; Miller, G. The interleukin-1 receptor antagonist (rhIL-1ra) protects against cerebral infarction in a rat model of hypoxia-ischemia. *Exp. Neurol.* **1994**, *130*, 362–367. [CrossRef]
110. Mulcahy, N.J.; Ross, J.; Rothwell, N.J.; Loddick, S.A. Delayed administration of interleukin-1 receptor antagonist protects against transient cerebral ischaemia in the rat. *Br. J. Pharmacol.* **2003**, *140*, 471–476. [CrossRef]
111. LaVine, S.D.; Hofman, F.M.; Zlokovic, B.V. Circulating antibody against tumor necrosis factor–alpha protects rat brain from reperfusion injury. *J. Cereb. Blood Flow Metab.* **1998**, *18*, 52–58. [CrossRef]
112. De Bilbao, F.; Arsenijevic, D.; Moll, T.; Gracia-Gabay, I.; Vallet, P.; Langhans, W.; Giannakopoulos, P. In vivo overexpression of interleukin-10 increases resistance to focal brain ischemia in mice. *J. Neurochem.* **2009**, *110*, 12–22. [CrossRef]
113. De Geyter, D.; Stoop, W.; Sarre, S.; De Keyser, J.; Kooijman, R. Neuroprotective efficacy of subcutaneous insulin-like growth factor-I administration in normotensive and hypertensive rats with an ischemic stroke. *Neuroscience* **2013**, *250*, 253–262. [CrossRef]
114. Hyakkoku, K.; Hamanaka, J.; Tsuruma, K.; Shimazawa, M.; Tanaka, H.; Uematsu, S.; Akira, S.; Inagaki, N.; Nagai, H.; Hara, H. Toll-like receptor 4 (TLR4), but not TLR3 or TLR9, knock-out mice have neuroprotective effects against focal cerebral ischemia. *Neuroscience* **2010**, *171*, 258–267. [CrossRef]
115. Patel, A.R.; Ritzel, R.; McCullough, L.D.; Liu, F. Microglia and ischemic stroke: A double-edged sword. *Int. J. Physiol. Pathophysiol. Pharmacol.* **2013**, *5*, 73–90.
116. Taylor, R.A.; Sansing, L. Microglial responses after ischemic stroke and intracerebral hemorrhage. *Clin. Dev. Immunol.* **2013**, *2013*, 1–10. [CrossRef]
117. Hum, P.D.; Subramanian, S.; Parker, S.M.; Afentoulis, M.E.; Kaler, L.J.; Vandenbark, A.A.; Offner, H. T- and B-cell-deficient mice with experimental stroke have reduced lesion size and inflammation. *Br. J. Pharmacol.* **2007**, *27*, 1798–1805. [CrossRef]
118. Liesz, A.; Zhou, W.; Mracskó, É.; Karcher, S.; Bauer, H.; Schwarting, S.; Sun, L.; Bruder, D.; Stegemann, S.; Cerwenka, A. Inhibition of lymphocyte trafficking shields the brain against deleterious neuroinflammation after stroke. *Brain* **2011**, *134*, 704–720. [CrossRef]
119. Panahian, N.; Yoshiura, M.; Maines, M.D. Overexpression of heme oxygenase-1 is neuroprotective in a model of permanent middle cerebral artery occlusion in transgenic mice. *J. Neurochem.* **2008**, *72*, 1187–1203. [CrossRef]
120. Chao, X.D.; Ma, Y.H.; Luo, P.; Cao, L.; Lau, W.B.; Zhao, B.C.; Han, F.; Liu, W.; Ning, W.D.; Su, N.; et al. Up-regulation of Heme oxygenase-1 attenuates brain damage after cerebral ischemia via simultaneous inhibition of superoxide production and preservation of NO bioavailability. *Exp. Neurol.* **2013**, *239*, 163–169. [CrossRef]
121. Shah, Z.A.; Nada, S.E.; Doré, S. Heme oxygenase 1, beneficial role in permanent ischemic stroke and in Gingko biloba (EGb 761) neuroprotection. *Neuroscience* **2011**, *180*, 248–255. [CrossRef]
122. McCombe, P.A.; Read, S.J. Immune and inflammatory responses to stroke: Good or bad? *Int. J. Stroke* **2008**, *3*, 254–265. [CrossRef]
123. Frost, G.R.; Jonas, L.A.; Li, Y.M. Friend, foe or both? Immune activity in Alzheimer's disease. *Front. Aging Neurosci.* **2019**, *11*, 337. [CrossRef] [PubMed]
124. Cappellano, G.; Vecchio, D.; Magistrelli, L.; Clemente, N.; Raineri, D.; Barbero Mazzucca, C.; Virgilio, E.; Dianzani, U.; Chiocchetti, A.; Comi, C. The *Yin-Yang* of osteopontin in nervous system diseases: Damage versus repair. *Neural Regen. Res.* **2021**, *16*, 1131–1137. [PubMed]

International Journal of
Molecular Sciences

Review

Ferroptosis Mechanisms Involved in Neurodegenerative Diseases

Cadiele Oliana Reichert [1], Fábio Alessandro de Freitas [1], Juliana Sampaio-Silva [1], Leonardo Rokita-Rosa [1], Priscila de Lima Barros [1], Debora Levy [1] and Sérgio Paulo Bydlowski [1,2,*]

1 Lipids, Oxidation, and Cell Biology Group, Laboratory of Immunology (LIM19), Heart Institute (InCor), Hospital das Clínicas HCFMUSP, Faculdade de Medicina, Universidade de São Paulo, São Paulo 05403-900, Brazil; kadielli@hotmail.com (C.O.R.); fabio.alessandro@usp.br (F.A.d.F.); jukisbio@gmail.com (J.S.-S.); rokita@usp.br (L.R.-R.); pri_limabarros@hotmail.com (P.d.L.B.); d.levy@hc.fm.usp.br (D.L.)

2 Instituto Nacional de Ciencia e Tecnologia em Medicina Regenerativa (INCT-Regenera), CNPq, Rio de Janeiro 21941-902, Brazil

* Correspondence: spbydlow@usp.br

Received: 6 October 2020; Accepted: 28 October 2020; Published: 20 November 2020

Abstract: Ferroptosis is a type of cell death that was described less than a decade ago. It is caused by the excess of free intracellular iron that leads to lipid (hydro) peroxidation. Iron is essential as a redox metal in several physiological functions. The brain is one of the organs known to be affected by iron homeostatic balance disruption. Since the 1960s, increased concentration of iron in the central nervous system has been associated with oxidative stress, oxidation of proteins and lipids, and cell death. Here, we review the main mechanisms involved in the process of ferroptosis such as lipid peroxidation, glutathione peroxidase 4 enzyme activity, and iron metabolism. Moreover, the association of ferroptosis with the pathophysiology of some neurodegenerative diseases, namely Alzheimer's, Parkinson's, and Huntington's diseases, has also been addressed.

Keywords: ferroptosis; cell death; iron metabolism; neurodegenerative diseases; glutathione peroxidase 4; GSH; system xc^{-}; Alzheimer's disease; Parkinson's disease; Huntington's disease

1. Introduction

The current classification system of cell death has been updated by the Nomenclature Committee on Cell Death (NCCD), according to their guidelines for the definition and interpretation of all aspects of cell death [1,2]. Accidental cell death (ACD) is an instantaneous and catastrophic demise of cells exposed to severe insults of physical or mechanical forces. On the other hand, regulated cell death (RCD) is a dedicated molecular machinery [3]. RCD can occur in two ways: firstly, as programmed cell death that can occur in the absence of any exogenous environmental perturbation. Secondly, RCD can originate from disturbances of the intracellular or extracellular microenvironment that cannot be restored to cellular homeostasis [4–6].

In 2012, Brent R. Stockwell described a unique form of cell death that results from the overwhelming iron-dependent accumulation of lethal amounts of lipid-based reactive oxygen species and named it ferroptosis [7]. Ferroptosis is morphologically and biochemically distinct from other RCDs. It occurs without the chromatin condensation and nuclear reduction seen in apoptosis, cellular and organellar swelling of necrosis, and without the common features of autophagy. Morphologically, only mitochondrial shrinkage distinguishes it from other forms of death [8,9]. Ferroptotic cell death is associated with the iron-dependent mechanism and formation of extremely reactive free radicals, along with severe peroxidation of membrane phospholipids (PLs) rich in polyunsaturated fatty acids (PUFAs), mainly of arachidonic or adrenic acids from phosphatidyl ethanolamine (PE) molecules [10–12].

The complex balance between reactive oxygen species (ROS) and the antioxidant system maintains cell homeostasis by removing dangerous stimuli and controlling oxidative stress by several factors, and is also present in the central nervous system (CNS) [13,14]. Among these factors, system xc$^-$, an amino acid antiporter, maintains the synthesis of glutathione (GSH) and oxidative protection. Inhibition of system xc$^-$ causes a rapid drop of intracellular glutathione levels and cell death caused by the accumulation of lipid-derived ROS. Lipid and protein oxidation lead to inflammation and changes in DNA, and are the trigger for premature aging, loss of function and death of neurons. The increase in oxidative stress generated by free radicals associated with uncontrolled intracellular iron metabolism has been associated with the pathophysiology of neurodegenerative diseases [13–15]. Here we address the main pathways of ferroptosis and its role in the pathophysiology of Alzheimer's disease, Parkinson's disease, and Huntington's disease, the main neurodegenerative diseases in which ferroptosis has been shown to be involved.

2. Ferroptosis

Ferroptosis is a particular form of cell death that is induced by lipid hydroperoxides derived from the oxidation of reactive species generated by free iron. Cell death in ferroptosis involves three main factors: increased free intracellular iron, depletion of the redox glutathione/GPx4/system xc$^-$ and the oxidation of membrane PUFAs [16–18] (Figure 1).

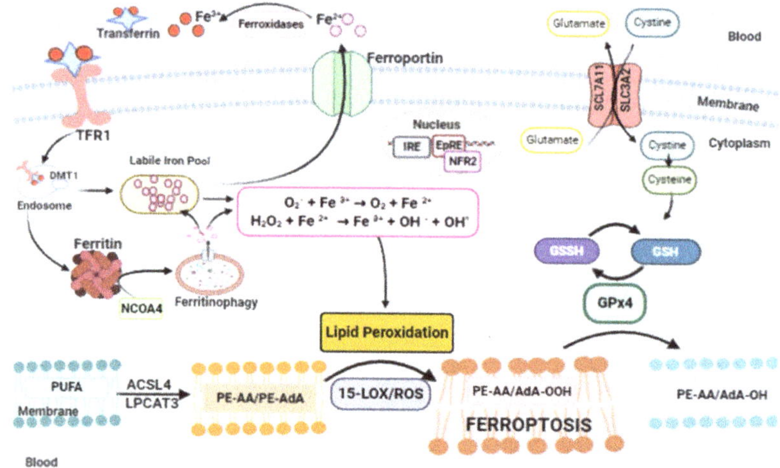

Figure 1. Ferroptosis pathway. Ferroptosis can be initiated through transferrin endocytosis linked to transferrin receptor 1 (TFR1). After endocytosis, ferric iron is released from the Transferrin–TRF1 complex and is reduced to ferrous iron (Fe^{2+}). Fe^{2+} can be stored in ferritin or remain in the cytoplasm as a labile iron Pool (LIP). The LIP is composed mainly of Fe^{2+}, which through Fenton reaction generates species such as: the hydroxyl radical that reacts with membrane lipids, providing the lipid peroxidation of arachidonic acid (AA) or adrenic acid (AdA). Lipid peroxidation can also occur via enzyme. However, it is necessary for the free polyunsaturated fatty acids (PUFAs) to be esterified as membrane PUFA by the enzymes ACSL4 and LPCAT3, forming arachidonic or adrenic acids esterified in phosphatidyl ethanolamine (PE-AA/PE-AdA). Dioxigenation by 15-LOX generates PE-AA/AdA-OOH, which reacts with other membrane lipids, forming pores in the lipid bilayer, destabilizing it and then rupturing the membrane. Ferroptosis is inhibited by GPx4, which converts PE-AA/AdA-OOH to alcohol and water. This reaction occurs through the use of glutathione (GSH) as a substrate. GSH synthesis occurs via the entry of cystine into the cell by system xc$^-$.

2.1. Lipid Peroxidation and Ferroptosis

Lipid peroxidation is the trigger for the activation of ferroptosis [13,18]. Lipid peroxides (PL-OOH), mainly lipid hydroperoxides (L-OOH), have the ability to cause damage to the lipid bilayer of the plasma membrane due to the accelerated oxidation of the membrane lipids which leads to ferroptosis. The increase in the concentration of lipid peroxides can alter the structure and function of nucleic acids and proteins, as well as the Michael acceptors and aldehydes. In fact, it can generate additional toxicity due to its degradation products [19–22]. The cellular lipids include thousands of lipid species that vary in quantity, intra- and extracellular distribution, functions and cell type [8]. Thus, the higher the concentration of free polyunsaturated fatty acids (PUFAs) in the cell, the greater the damage caused by lipid hydroperoxidation and the extent of ferroptosis, which can vary among diseases and organs/tissues [4,8,17].

PUFAs are good substrates for autoxidation because the C–H bonds of the methylene groups flanked by C-C double bonds are among the weakest C–H bonds known [19–22]. The structure of the PUFA molecule contains bis-allyl hydrogen atoms that can be abstracted. Then, there is a rearrangement of the resonance radical structure, with subsequent addition of molecular oxygen, giving rise to the peroxyl radical and the formation of the primary molecular product, lipid hydroperoxide (L-OOH). Soon after, the cleavage of the L-OOH molecule occurs, giving rise to highly electrophilic secondary oxidation products, including epoxy, oxo- or aldehyde groups, which are highly reactive and toxic to membranes and cells [8,23].

First, PUFAs are esterified with membrane phospholipids, such as phosphatidyl ethanolamine (PE). The esterification reaction is catalyzed by acyl-CoA synthetase long-chain family member 4 (ACSL4), which binds coenzyme A to long-chain PUFAs, which can then be used for esterification of lysophospholipids by lysophosphatidylcholine acyltransferase 3 (LPCAT3); the substrates can undergo peroxidation resulting in the formation of arachidonoyl (AA) and adrenoyl (AdA) acids, which can lead to ferroptosis. Suppression of the ACSL4 enzyme inhibits ferroptosis by depleting the substrates for lipid peroxidation [24,25].

The PUFA oxidation process that leads to ferroptosis can occur enzymatically or non-enzymatically [26]. The non-enzymatic oxidation process occurs through ROS and hydroxyl radical, from the Fenton reaction. This process is both non-selective and non-specific. Thus, oxidation rates are proportional to the number of readily abstractable bis-allyl hydrogens in the PUFA molecule, resulting in the accumulation of a highly diversified pattern of oxidation products with the predominance of oxygenated PUFA-PLs with 6, 5, 4, 3 and 2 double bonds [8,19]. Enzymatic oxidation of PUFAs occurs through lipoxygenases (LOXs) [27]. LOXs are dioxigenases containing iron in their catalytic region that promote the dioxigenation of polyunsaturated fatty acids containing at least two isolated cis-double bonds. In humans, there are different isoforms of LOX (5-LOX, 12S-LOX, 12R-LOX, 15-LOX-1, 15-LOX-2 and eLOX3) [19,27].

Membrane ester lipids are cleaved by cytosolic phospholipase A2 in different fatty acids: arachidonic acid (AA), eicosapentaenoic acid (EPA) and docosahexaenoic acid (DHA). Oxygenation by cyclooxygenases (COXs) generates prostanglandins-G (PGG2, PGG3 and PGG4, respectively). However, oxygenation by LOX generates doubly and triply oxygenated (15-hydroperoxy)-diacylated PE species [28–31]. Oxidation induced by 15-LOX is selective and specific, occurring preferably in arachidonic acid-phosphatidylethanolamine (AA-PE) or adrenoyl acid (AdA)-PE. The product of this oxidation is 15-hydroperoxy-arachidonic acid-phosphatidylethanolamines (15-HOO-AA-PEs) or 15-hydroperoxy-adrenoyl acid-phosphatidylethanolamines (15-HOO-AdA-PEs) (Figure 1) [28–31]. The catalytic activity 15-LOX is dependent on the pro-ferroptotic PEBP1 protein [32].

Stoyanovsky et al. [33] showed that the ferroptosis process includes two stages: (i) selective and specific enzymatic production of 15-HOO-AA-PE by 15-LOX; (ii) oxidative cleavage of these initial HOO derivatives to proximate electrophiles capable of interacting with protein targets to cause the formation of pores in plasma membranes, or to rupture them. The two types of oxidatively truncated products can be formed from 15-HOO-AA-PE with the carbonyl function either on the shortened AA-residue esterified into PE, or on the leaving aldehyde.

In addition, tocopherols and tocotrienols suppress LOX and protect against ferroptosis [24]. On the other hand, ferrostatins inhibit ferroptosis by efficiently scavenging free radicals in lipid bilayers [34].

Recently, Zou et al. [35] have shown that cytochrome P450 oxidoreductase (POR) is a key mediator in the induction of ferroptosis in cells that exhibit intrinsic and induced susceptibility to ferroptosis by enabling membrane polyunsaturated phospholipid peroxidation. POR depletion suppressed arachidonic acid-induced sensitivity to ML210/RSL3 in a dose-dependent manner. In addition to suppressing PUFA-induced ferroptosis susceptibility, POR depletion by constitutive or inducible knockout also compromised the intrinsic ferroptosis sensitivity in ccRCC cells 786-O and 769-P.

2.2. Glutathione Peroxidase 4 and Ferroptosis

Cells have several escape mechanisms against cell death [36,37]. In the ferroptotic process, one of the most important and most studied so far is the enzyme glutathione peroxidase 4 (GPx4), (also called Phospholipid Hydroperoxide Glutathione Peroxidase (PHGPx)) [38,39]. In the human organism, there are several isozymes of glutathione peroxidase, which vary in cell location and substrate specificity [40]. The GPx4 enzyme is a selenoprotein, with approximately 20–21 kDa, composed of 197 amino acids, and encoded by the GPx4 gene in chromosome 19 localization [41]. GPx4 has in its active site the amino acid selenocysteine, which is necessary for protection against ferroptosis [42]. The catalytic site of selenocysteine involves three different redox states: selenol, selenenic acid and seleninic acid. These different forms of the redox state allow the regulation of the catalytic efficiency of the peroxide reduction, which is dependent on the cellular redox state [43]. The enzymatic activity of GPx4 is vital to cells, since the enzyme can reduce H_2O_2 and is the only enzyme that can reduce phospholipid hydroperoxides [44].

In addition, by structural similarity, GPx4 can reduce both peroxidized fatty acids and esterified cholesterol hydroperoxides, as well as thymine hydroperoxide, a product of free radical attack on DNA. The reduction reaction can occur in membranes, in the cytoplasm and/or in lipoproteins [45,46]. In the antiferroptotic process, the GPx4 enzyme directly reduces toxic lipid peroxides (PL-OOH) to non-toxic lipid alcohols (PL-OH) using reduced glutathione (GSH) as a substrate [47–49]. The synthesis of GSH through the cystine/glutamate antiporter system xc^- is a limiting step for the function of detoxification of lipid peroxides by GPx4 [50]. The rate-limiting compound of GSH synthesis is the non-essential amino acid cysteine. Cysteine can be imported into cells directly or in its oxidized form, cystine, through system xc^-. Within the cell, cystine is reduced to cysteine by biosynthesis of GSH [51]. Figure 2 shows the complete GSH biosynthesis pathway.

GPx4 inhibitors, including ML210, ML162 and (1S), (3R)-RSL3 (RSL3), are used as specific ferroptosis inducers [52–55]. Moreover, the overexpression or silencing of the gene coding for 14-3-3 proteins controls the inactivation of GPx4 by RSL3 [56]. In addition, liproxstatin-1 is able to suppress ferroptosis in cells, inhibits mitochondrial lipid peroxidation, and restores the expression of GSH, GPX4 and ferroptosis suppressor protein 1 [57,58]. A variety of ferroptosis inducers can inhibit cystine absorption by inhibiting system xc^-, such as: erastin, sulfasalazine and sorafenib, resulting in reduced GPx4 activity in different cells lines. Thus, there is no synthesis of GSH and the activity of GPx4 decreases. As a consequence, there is a reduction in the cell antioxidant capacity and hence increased L-ROS, ultimately leading to ferroptosis [59–67].

GSH biosynthesis is regulated by the ubiquitously expressed transcription factor nuclear factor erythroid-2 related factor 2 (Nrf2). In baseline conditions, Nrf2-dependent transcription is suppressed due to proteasomal degradation in the cytosol by Keap1 (kelch-like ECH-associated protein 1). However, due to the exposure to a variety of different stimuli, including oxidative stress, the ubiquitination and degradation of Nrf2 are blocked, leading to the stabilization and nuclear accumulation of Nrf2, where it induces dependent electrophilic response element (EpRE) gene expression to restore cellular redox homeostasis [72]. Nrf2 regulates several steps of GSH biosynthesis transcription enzymes, such as catalytic and regulatory subunits of glutamate cysteine ligase (GCL), GSH synthase, GPx2, GSH S-transferases (GSTs) and GSH reductase (GR) as well as the light-chain subunit of the xc- [72–75]. Nrf2 is also associated with the regulation of antioxidant enzymes, NADPH: quinine oxidoreductase-1 (NQO-1 and NQO-2) and nicotinamide adenine dinucleotide phosphate oxidase 2 (NOX2). In addition,

Nrf2 can regulate iron metabolism enzymes [76,77] and proteins associated with multiple drug resistance (ABCG2, MRP3, MRP4, glutathione S-transferase P (GSTP)) [72,78].

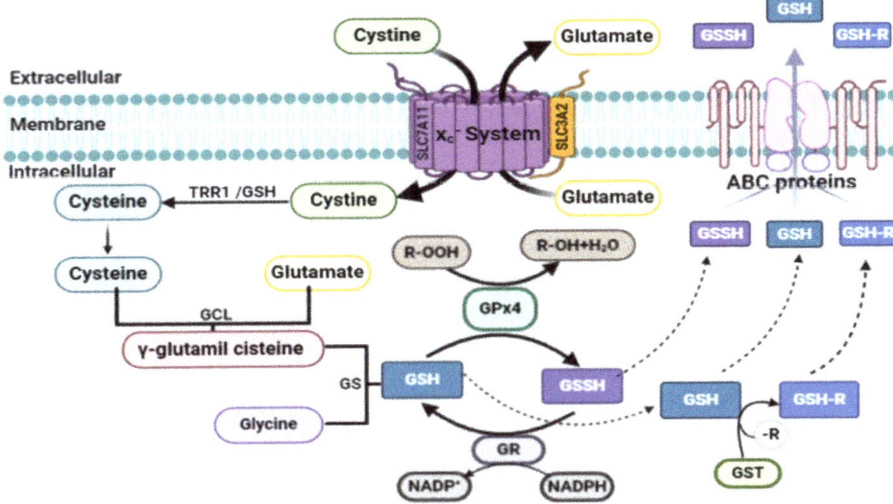

Figure 2. Glutathione (GSH) biosynthesis pathway. GSH is known as one of the small-molecule water-soluble antioxidants, the most important of somatic cells. GSH is a linear tripeptide formed by three amino acids: glutamic acid, cysteine and glycine. The thiol group present in the amino acid cysteine is considered the active site responsible for the antioxidant biochemical properties of glutathione. In biological systems, glutathione can be found in reduced form (GSH) or in oxidized form (GSSG). The oxidized form is a heterodimerization of the reduced form. The GSH/GSSG ratio is used to estimate the redox state of biological systems [51]. The rate-limiting compound of GSH synthesis is the non-essential amino acid cysteine. Cysteine can be imported into cells directly or in its oxidized form, cystine, through the cystine/glutamate antiporter system xc^-. In humans, on chromosome 4, the SLC7A11 gene (solute carrier family 7A11) encodes the SLCA11 antiporter, which is part of a system called system xc^-. The structure of this protein is heterodimeric and includes two chains: a specific light chain, xCT (SLCA11), and a heavy chain, 4F2hc (SLC3A2), which are linked by a disulfide bridge. The xCT chain has 12 transmembrane domains consisting of 501 amino acids, with the N and C terminal regions located intracellularly; it is not glycosylated and has a molecular mass of approximately 55 kDa. The heavy chain, 4F2hc, is a type II glycoprotein with a single transmembrane domain, an intracellular NH 2 terminal and a molecular weight of approximately 85 kDa. The 4F2hc chain is a subunit common to amino acid transport systems, while the xCT chain is unique to cystine/glutamate exchange. System xc^- transports amino acids, independently of sodium and dependent on chloride, which are specific to import cystine and export glutamate at the same time through the plasma membrane. Both amino acids are transported in anionic form. The ratio of counter transport between cystine and glutamate is 1:1. Currently, it is known that system xc^- is involved in (a) cystine uptake to maintain the extracellular balance of cysteine/redox cystine, (b) cysteine/cystine uptake for GSH synthesis and (c) non-vesicular glutamate export [68]. Within the cell, cystine is reduced to cysteine. This reduction reaction can be performed by intracellular GSH or by the enzyme thioredoxin reductase 1 (TRR1) [69]. The beginning of GSH synthesis is the formation of the γ-glutamylcysteine molecule, which is catalyzed by the enzyme glutamate cysteine ligase (GCL). GCL catalyzes the binding of glutamate and cysteine in the presence of adenosine triphosphate (ATP). Then, the enzyme GSH synthase (GS) catalyzes the formation of GSH through the link between γ-glutamylcysteine and glycine [69]. GSH reduces radicals (R•) non-enzymatically and organic hydroperoxides catalyzed by GSH peroxidase (GPx) and is thus

converted to GSH disulfide (GSSG). GSSG is recycled to GSH by GSH reductase (GR), a reaction that uses reduced nicotinamide adenine dinucleotide phosphate (NADPH) as a cofactor [69]. GSH S-transferase (GST) forms GSH (GS-R) adducts from organic molecules (R) and GSH, which together with GSH and GSSG are exported from the cell by ABC transporters, mainly ABC-1 and ABC-G2 [70,71]. Extracellular GSH is metabolized by the γ-glutamyl transferase (GGT) ectoenzyme, which transfers the γ-glutamyl residue to different acceptor amino acids, leading to the formation of a dipeptide containing γ-glutamyl and the cysteine glycine dipeptide, which is cleaved by extracellular dipeptides to generate cysteine and glycine that can be taken up by cells, starting the glutathione biosynthesis cycle [69].

Recently, Doll et al. [79] described an in vitro model, a parallel pathway that included FSP1-CoQ10-NADPH, which cooperates with GPx4 and the glutathione system to suppress lipid peroxidation of phospholipids. Ferroptosis suppressor protein 1 (FSP1) provides protection against ferroptosis induced by the deletion of the GPx4 enzyme via RSL3. This effect is mediated by coenzyme Q10 (CoQ10). The reduced form of the enzyme, ubiquinol, captures lipid peroxyl radicals that mediate lipid peroxidation, while FSP1 catalyzes the regeneration of CoQ10 using NADPH as a cofactor. Moreover, the authors described that the antiferroptotic function of FSP1 is independent of cellular glutathione concentration, GPx4 activity, ACSL4 expression and oxidizable fatty acid content [79]. Coenzyme Q10 is an endogenous lipophilic antioxidant produced in the mevalonate pathway, as well as a part of the mitochondrial respiratory chain, and from the metabolism of fatty acid and pyrimidine [80,81]. Indeed, the homologous proteins MDM2 and MDMX, negative regulators of the tumor suppressor p53, promote ferroptosis by regulating lipid peroxidation by altering PPARα activity. MDM2–MDMX complex inhibition increased the levels of both FSP1 proteins and coenzyme Q10 [82].

2.3. *Iron and Ferroptosis*

In the human body, iron metabolism is regulated by means of a perfectly adjusted balance between plasma proteins. They are associated with the transport, absorption and recycling of iron, in order to avoid the accumulation of iron, which is highly harmful and reactive in tissues. Figure 3 shows several aspects of human iron metabolism. Biochemically, iron is capable of accepting and donating electrons, interconverting between the ferric (Fe^{3+}) and ferrous (Fe^{2+}) forms, which are both found in the human body. The Fe^{3+}/Fe^{2+} redox potential participates in a large number of protein complexes, especially those that involve oxygen reduction for adenosine triphosphate (ATP) synthesis and the reduction of DNA precursors. Iron is also a necessary component in the formation of molecules that bind and transport oxygen (hemoglobin and myoglobin) and for the activities of cytochrome enzymes, as well as in many enzymes that perform the redox process, functioning as electron carriers [83–85].

Physiologically, iron that will be distributed to tissues needs to bind to transferrin (apotransferrin), giving rise to holotransferrin [87]. The distribution of iron to the tissues occurs through the endocytosis of holotransferrin, mediated by binding to transferrin receptor type 1 (TFR1) and type 2 (TFR2) [88]. After endocytosis, the holotransferrin–TFR1 complex is mobilized to the endosomes. In the acid environment of the endosome, Fe^{3+} is released from TF and converted to Fe^{2+} by oxidation-reduction, by six-transmembrane epithelial antigen of the prostate 3 (STEAP3) and then exported into cytosol by divalent metal transporter 1 (DMT-1). Iron can be stored in ferritin/hemosiderin or remain labile [89].

The labile iron pool (LIP), composed mostly of Fe^{2+}, is a pool of chelable iron and active redox present in the cell; it can be present in mitochondria, lysosomes, cytosol and in the nucleus [90]. The concentration of LIP is essentially regulated by the absorption, use, distribution and export of iron in the cell and in the body. Labile iron has high chemical reactivity and exhibits high cytotoxic potential. Cytotoxicity is associated with the fact that labile iron catalyzes the formation of hydroxyl radicals (OH·) derived from hydrogen peroxide (H_2O_2) through the Fenton reaction (Figure 1). Moreover, H_2O_2 has a lower capacity to react with molecules, while OH· generated from the iron-dependent Fenton reaction has high reactivity with biological molecules, such as proteins and DNA, and generates lipid (hydro) peroxidation in ferroptosis [12–17].

Indeed, glutathione has a high affinity with Fe^{2+} and the major component of LIP in cytosol is presented as the glutathione-Fe^{2+} conjugates [90]. This is important because the decrease in intracellular glutathione increases the concentration of Fe^{2+}, facilitating the Fenton reaction [90–93]. The storage of labile iron in ferritin serves to prevent its high reactivity, avoiding the generation of reactive species [94]. The ferritin structure is formed by 24 subunit-composed chains both light (L) and heavy (H) with a spherical "shell" shape, which accommodates about 4500 iron atoms. H-ferritin contains a ferroxidase, which oxidizes Fe^{2+} to Fe^{3+}, to store iron inside the nucleus. When necessary, iron stocks are mobilized and exported by ferroportin (FPN). This process is downregulated by hepcidin [83–85].

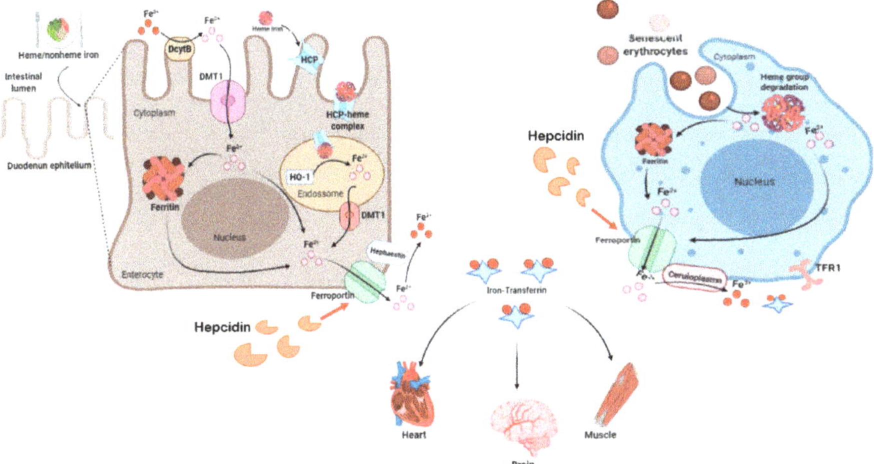

Figure 3. Human Iron metabolism. Iron concentration in the body is maintained through diet and the recycling of senescent erythrocytes. The daily diet provides approximately 1–2 mg of iron. Enterocytes, present in the duodenum and in the proximal portion of the jejunum, can absorb both ferrous iron (heme iron) and ferric iron (non-heme). However, it is necessary to reduce ferric iron to ferrous iron, by apical ferric reductase enzymes, such as enzyme duodenal cytochrome b (Dcytb), for absorption to occur. Then, iron is transported by the divalent metal type transporter-1 (DMT-1) and stored inside the cell [86]. Ferrous iron from the diet is internalized by the heme-1 carrier protein (HCP) in cells, where it is stored as hemosiderin and/or ferritin, to prevent the Fenton reaction. Physiologically, iron stores are mobilized from intracellular to the extracellular by ferroportin (FPN) when the serum iron is low. Iron released in its ferrous state is oxidized to ferric iron and binds to serum apotransferrin to be transported through the body, giving rise to holotransferrin. This oxidation reaction occurs through the action of oxidase enzymes: hephestine present in enterocytes, ceruloplasmin present in hepatocytes macrophages. The distribution of iron to the tissues occurs through the endocytosis of holotransferrin, mediated by the binding to transferrin receptors type 1 (TFR1) and type 2 (TFR2). Ferroportin mediates the efflux of iron within cells, maintaining systemic iron homeostasis. This process is negatively regulated by hepcidin, which promotes ferroportin endocytosis and then proteolysis in lysosomes by induced ubiquitination [83]. The recycling of iron by macrophages occurs through the phagocytosis of senescent erythrocytes and hemoglobin and the heme group of intravascular hemolysis. Once internalized in the macrophage, the heme group releases ferrous iron through the activity of the enzyme heme oxygenase, which can be exported to the extracellular medium by ferroportin or stored as ferritin [83,84].

The intracellular iron content is regulated by the iron regulatory proteins (IRP1 and IRP2) and the iron-responsive element (IRE) [95,96]. IRPs can bind to RNA stem-loops containing an IRE in the untranslated region (UTR), affecting the translation of target Mrna: 3′ UTR of H-ferritin mRNA and in the 5′ UTR of TFR1 mRNA. In response to cellular iron demand IRE/IRP interaction promotes TFR1 mRNA stability and inhibits H-ferritin translation, thus modulating cellular iron uptake and storage. Overexpression of both TF and TFR1 sensitizes cells to ferroptosis by enhancing iron uptake; on the other hand, silencing TFR1 can inhibit erastin-induced ferroptosis [16,97–99]. In addition, anti-TfR1 antibodies identify tumor ferroptotic cells from different tissues [100].

Autophagy in fibroblasts leads to erastin-induced ferroptosis through the degradation of ferritin and induction of TfR1 expression [101]. On the other hand, cellular senescence has been associated with intracellular iron accumulation and impaired ferritinophagy [102]. Ferritinophagy induces ferroptosis by increasing the nuclear receptor coactivator 4 (NCOA4), followed by an increase in ferritin degradation in the phagolysosome and release of Fe^{2+} (labile iron pool) in the cytoplasm. In fact, some authors consider ferroptosis to be a type of autophagy [103–107].

Another molecule involved in ferroptosis is the sigma-1 receptor (S1R), which protects hepatocellular carcinoma cells against ferroptosis. S1R regulates ROS accumulation via Nrf2. Knockdown of S1R blocks the expression of GPx4 and HO-1. Moreover, knockdown of S1R significantly increases Fe^{2+} levels and MDA production in HCC cells treated with erastin and sorafenib, as well as the upregulation of H-ferritin chain and TRF1 [108]. In addition, it has been shown that heat shock protein β-1 (HSPβ1) is a negative regulator of ferroptotic cancer cell death, and erastin stimulates heat shock factor 1 (HSF1)-dependent HSPβ1. Knockdown of HSF1 and HSPβ1 enhances erastin-induced ferroptosis, whereas heat shock pretreatment and overexpression of HSPβ1 inhibits erastin-induced ferroptosis by protein kinase C. Moreover, the increase in cellular iron in HSPβ1 knockdown cells has been associated with increased expression of TFR1 and a mild decrease in the expression of the H-ferritin chain [109]. HSPA5, an endoplasmic reticulum (ER)-sessile chaperone, was shown to bind and stabilize GPx4, thus indirectly counteracting lipid peroxidation in ferroptosis [97].

3. Ferroptosis in Neurodegenerative Diseases

Iron is vital to the physiology of all human tissues. However, under certain conditions it can be harmful, especially for the brain. Although the cellular metabolism of the CNS requires iron as a redox metal for energy generation, mainly the production of ATP, nervous tissue is vulnerable to oxidative damage generated by excess iron and decreased antioxidant systems [110–113]. The explicit identification of ferroptosis in vivo is hampered by the lack of specific biomarkers, due to several factors that may be associated with the ferroptotic process. However, there is considerable evidence that implicates ferroptosis in the pathophysiology of neurodegeneration. Ferroptosis involves the accumulation of brain iron, glutathione depletion and lipid peroxidation simultaneously, which triggers a cascade of events including activation of inflammation, neurotransmitter oxidation, neuronal communication failure, myelin sheath degeneration, astrocyte dysregulation, dementia and cell death. Iron or free iron overload can initiate lipid peroxidation in neurons, astrocytes, oligodendrocytes, microglia and Schwann cells. In addition, the low activity of GPx4 and the glutathione system have been shown to be associated with ferroptosis in motor neurodegeneration [114–117].

Recently, it has been proposed that the modulation of ferroptosis may be beneficial for neurodegenerative diseases and that inhibition of ferroptosis by GPx4 could provide protective mechanisms against neurodegeneration [118]. First, it was demonstrated that the non-oxidative form of dopamine is a strong inhibitor of ferroptotic cell death. Dopamine reduced erastin-induced ferrous iron accumulation, glutathione depletion, and malondialdehyde production. Moreover, dopamine increased the stability of GPx4 [119]. The GPx4 enzyme is essential for the survival of parvalbumin-positive interneurons and prevention of seizures, as well as protection against ferroptosis in animal models [42,120]. Next, in a study with PC12 cell line (a model system for neurobiological and neurochemical studies), cell death was induced by *tert*-butylhydroperoxide (t-BHP), a widespread inducer of oxidative stress; it was observed that t-BHP increased the generation of lipid ROS, decreased the expression of GPx4 and the ratio of GSH/GSSG. All these effects could be reversed by the ferroptosis inhibitor, ferrostatin-1 and deferoxamine, iron chelator. In addition, JNK1/2 and ERK1/2 were activated upstream from the ferroptosis and mitochondrial dysfunction [121].

In addition, in C57BL/6 J male mice treated with arsenite for 6 months, it was observed that arsenite induced ferroptotic cell death in neurons by the accumulation of reactive oxygen species and lipid peroxidation products, disruption of Fe^{2+} homeostasis, depletion of glutathione and adenosine triphosphate, inhibition of system xc^-, activation of mitogen-activated protein kinases and mitochondrial voltage-dependent anion channel pathways, and upregulation of endoplasmic reticulum stress [122]. This is an important issue because arsenite (inorganic arsenic) has been associated with neural loss and Alzheimer's and Parkinson's diseases as well as amyotrophic lateral sclerosis (ALS) [123–130]. An in vitro study showed that exposure to paraquat and maneb induced ferroptosis in dopaminergic SHSY5Y cells, associated with the activation of NADPH oxidase. The activation of NADPH oxidase contributed to the dopaminergic neurodegeneration associated with lipid peroxidation and neuroinflammation [131].

In a multiple sclerosis model and in an experimental autoimmune encephalomyelitis (EAE) animal model [124], it has been observed that mRNA expression of the cytoplasmic, mitochondrial and nuclear GPx4 enzyme decreased in multiple sclerosis gray matter and in the spinal cord of EAE. Neuronal GPx4 was lower in EAE spinal cords than in controls. Moreover, γ-glutamylcysteine ligase and cysteine/glutamate antiporter were diminished in EAE, which is associated with high accumulation of lipid peroxidation products and the reduction in the proportion of the docosahexaenoic acid in non-myelin lipids. These results, together with the presence of abnormal neuronal mitochondrial morphology, which includes an irregular matrix, ruptured external membrane and reduced/absent ridges, are consistent with the occurrence of ferroptotic damage in inflammatory demyelinating disorders [132].

In fact, iron can bind to IRPs, leading to the dissociation of IRPs from the IRE and altered translation of the target transcripts. Recently, an IRE was found in the 5′-UTR of the amyloid precursor protein (APP) and α-synuclein transcripts (α-Syn). The levels of α-Syn, APP and amyloid β (Aβ) peptide, as well as protein aggregation, can be negatively regulated by IRPs, but are regulated positively in the presence of iron accumulation. Therefore, it has been suggested that the inhibition of the IRE-modulated expression of APP and α-Syn or iron chelation in patient brains has therapeutic significance for human neurodegenerative diseases [133].

3.1. *Ferroptosis in Alzheimer's Disease*

Alzheimer's disease (AD) is considered a neurodegenerative disease associated with multiple brain complications. It was initially described by the German Alois Alzheimer in 1907 [134,135]. AD is characterized by progressive disorder in the cortical and hippocampal neuronal areas which leads to both loss of neuronal function and cell death, and is the most common type of dementia (Figure 4). The hallmark of AD is the histopathological presence of an extracellular β-amyloid (Aβ) deposition in senile plaques (SPs) and intracellular neurofibrillary tangles (NFTs) formed from the hyperphosphorylation of the tau protein. Neurocognitive decline is associated with synapse decrease and neurotransmitter oxidation. These changes are due to the increase in oxidative stress, mainly an increase in ROS and intra- and extracellular hydrogen peroxides [136–138]. Moreover, genetic changes include alterations in amyloid precursor protein (APP), apolipoprotein E (APOE), presenilin 1 (PSEN1) and presenilin 2 (PSEN 2) genes [139].

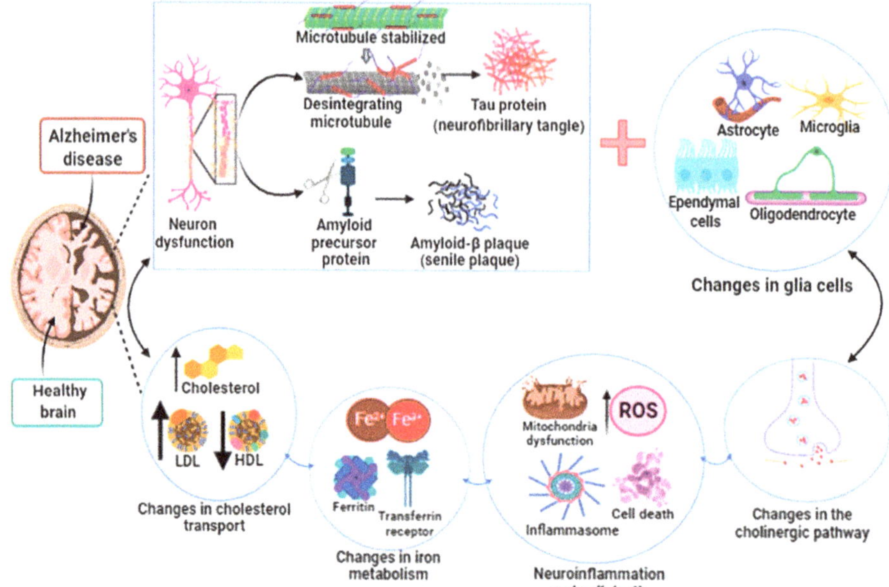

Figure 4. Alzheimer's disease. The development and progression of Alzheimer's disease (AD) lead to atrophy, loss and dysfunction of both neurons and glial cells. AD begins in the dorsal raphe nucleus with subsequent progression to the cortex, which is the center of information processing and memory storage. The factors that promote the development of AD are still unknown. However, it seems that the intracellular accumulation in neurons of the phosphorylated Tau protein (neurofibrillary tangle) and the formation of amyloid-B plaque (senile plaque) in the extracellular environment and brain tissue both lead to neuron loss and dysfunction. In addition, the formation of neurofibrillary tangle and senile plaque alters the functions of glial cells, such as oligodendrocytes (responsible for the myelination of neurons), microglia cells (phagocytic cells) and astrocytes (responsible for the absorption and exchange of nutrients between neurons and blood vessels). Dysregulation of cholesterol transport and iron metabolism in the central nervous system contributes to poor prognosis of Alzheimer's disease. All these associated factors lead to an increase in neuroinflammation and oxidative stress associated with mitochondrial dysfunction, compromising the production of ATP, altering the concentration of neurotransmitters in the synaptic cleft, finally promoting cell death.

Evidence of the association between the accumulation of iron in the cerebral cortex and the development of Alzheimer's disease emerged in the early 1960s [140,141]. Since then, several studies have demonstrated the direct association between free iron, oxidative stress, lipid peroxidation and cell death of neurons, usually associated with apoptosis and/or necrosis, due to increased neuroinflammation. The iron dyshomeostasis is associated with ROS production and neurodegeneration in AD [138]. Furthermore, aging and changes in iron metabolism are associated with the development of Aβ plaques and NFTs. Svobodová et al. [142] demonstrated in an APP/PS1 transgenic mice model that free iron and ferritin accumulation follows amyloid plaque formation in the cerebral cortex area. In fact, iron deposition has been involved in the misfolding process of the Aβ plaques and NFTs [143].

Iron is related to the development of Tau protein and, consequently, NFTs. In fact, iron is present through the induction and regulation of tau phosphorylation [143,144]. The association of NFT with neurodegenerative dysfunctions is termed tauopathy [145–147]. The oxidation process slows down or excludes the regular action of the Aβ and tau protein [148]. In animal models of tauopathies, increased iron associated with aging and neurodegeneration has been observed [149]. Indeed, animals with tauopathies treated with the iron chelator deferiprone showed a trend toward improved cognitive function associated with the decrease in brain iron levels and sarkosyl-insoluble tau [150].

APP is a type 1 transmembrane protein and its function in heathy individuals appears to be associated with the development of synaptic activity [145]. Proteolytic cleavage of the β-amyloid precursor protein (APP) to form the β-amyloid peptide (Aβ) is related to the pathogenesis of AD because APP mutations that influence this process induce familial AD or decrease the risk of AD [145]. The amyloid cascade hypothesis states that the agglomeration and production of Aβ plaques in the brain would occur, resulting in cell death. Presenilins 1 (PSEN1) and presenilins 2 (PSEN2) precisely cleave the APP and other proteins as they are part of the catalytic protease compounds [151]. Acetylcholinesterase participates in the aggregation of Aβ plaques [152]. Moreover, the Aβ plaques in the presence of free iron participate efficiently in the generation of ROS resulting in increased lipid peroxidation, protein oxidation and DNA damage [153]. Deferiprone derivatives act as acetylcholinesterase inhibitors and in iron chelation [154].

The proteolytic cleavage in APP occurs by enzymatic complexes involving α-secretase or β-secretase and γ-secretase. The proteolytic cleavage in APP by β-secretase produces a neurotoxic 40 to 42 amino acid amyloid [155]. Tsatsanis et al. [156] showed that APP promotes neuronal iron efflux by stabilizing the cell-surface presentation of ferroportin, and that β-cleveage of APP depletes surface ferroportin, leading to intracellular iron retention independently on the generation of Aβ. Furthermore, these findings indicate how β-secretase's processing of APP might indirectly promote ferroptosis. Iron overload alters the neuronal sAPPα distribution and directly inhibits β-secretase activity [157]. Cortical iron has been strongly associated with the rate of cognitive decline [158]. Iron in the brain increases lipid peroxidation, oxidative stress, and neuroinflammation due to the depletion of neuronal antioxidant systems—mainly the glutathione system [143]. In addition, increased hepcidin expression in APP/PS1 mice astrocytes improves cognitive decline and partially decreases the formation of Aβ plaques in the cortex and hippocampus. Indeed, decreased iron levels in neurons led to a reduction in oxidative stress (induced by iron accumulation), decrease in neuroinflammation and decreased neuronal cell death in the cortex and the hippocampus. [159]. As mentioned before, the hepcidin peptide binds ferroportin, which is followed by cell internalization and further degradation [160].

In order to investigate whether neurons of the cerebral cortex and hippocampus severely affected in patients with AD may be vulnerable to ferroptosis, Hambright et al. [161] have shown in GPx4BIKO mice (a mice model with a conditional deletion in neurons of the forebrain of GPx4) that tamoxifen led to the deletion of GPx4 mainly in neurons of the forebrain. GPx4BIKO mice exhibited significant deficits in spatial learning and memory function, as well as hippocampal neurodegeneration. These results were associated with ferroptosis markers, such as increased lipid peroxidation, ERK activation and neuroinflammation. In addition, GPx4BIKO mice fed a vitamin E-deficient diet had an accelerated rate of hippocampal neurodegeneration and behavioral dysfunction. On the other hand, treatment with Liproxstatin-1, a ferroptosis inhibitor, improved neurodegeneration in these mice. Moreover, in an in vitro model, iron increased nerve cell death in conditions where GSH levels were reduced, by decreasing the activity of glutamate cysteine ligase [162].

The HT22 cell line has high concentrations of extracellular glutamate, which inhibit the glutamate-cystine antiport, leading to the depletion of intracellular GSH and resulting in excessive ROS production. In a study with these cells, Hirata et al. [163] found that an oxindole compound, GIF-0726-r, prevented cell death induced by oxidative stress, including oxytosis induced by glutamate and ferroptosis induced by erastin. Moreover, an excess of extracellular glutamate associated with high levels of extracellular iron cause the overactivation of glutamate receptors. As a consequence, there was an increase in iron uptake in neurons and astrocytes, increasing the production of membrane peroxides. Glutamate-induced neuronal death can be mitigated by iron chelating compounds or free radical scavenging molecules. Ferroptosis is induced by reactive oxygen species in the excitotoxicity of glutamate [110,164,165]. In addition, the sterubin compound maintained GSH levels in HT22 cell lines treated with erastin and RSL3, suggesting protection against ferroptosis [166]. 7-*O*-esters of taxifolin 1 and 2 were described as having neuroprotective action against ferroptosis induced by RSL3 in HT22 cells [167].

Chalcones 14a–c were shown to inhibit β-amyloid aggregation, and in addition, protect neural cells against toxicity induced by Aβ aggregation and from erastin and RSL3-induced ferroptosis in human neuroblastoma SH-SY5Y cells [160]. The authors suggested that the inhibition of toxicity induced by Aβ plaques' aggregation and of ferroptosis occurs due to the presence of hydroxyl groups in the chalcone derivatives. Chalcone 14a-c can react with lipid peroxyl radicals by transferring the hydrogen (H) atom, thus inhibiting lipid peroxidation [168].

After treatment with high dietary iron (HDI), WT (wild type) mice and the APP/PS1 double Tg mouse model of ADon (HDI) showed upregulation of divalent metal transporter 1 (DMT1) and ferroportin expression, and downregulation of TFR1 expression, with fewer NeuN-positive neurons in both animal models. Moreover, the iron-induced neuron loss may involve increased ROS production and oxidative mitochondria dysfunction, decreased DNA repair, and exacerbated apoptosis and autophagy [169]. Using X-ray spectromicroscopy and electron microscopy it was found that the coaggregation of Aβ and ferritin resulted in the conversion of the ferritin inert ferric core into more reactive low oxidation states [170].

Ates et al. [171] showed in an animal model that inhibition of fatty acid synthase (FASN) by CMS121 decreased lipid peroxidation. CMS121 treatment reduced the levels of 15LOX2 in the hippocampus compared to those of untreated WT mice. Relative levels of endocannabinoids, fatty acids, and PUFAs were significantly higher in untreated AD mice as compared to CMS121-treated AD mice, suggesting that other enzymes may be involved in the process of ferroptosis in Alzheimer's disease.

It is important to highlight the heterogeneity of Alzheimer's disease and the involvement of multiple metabolic pathways which contribute to the poor prognosis of this disease (Figure 4). In fact, multiple patterns of cell death are involved in the neurodegeneration process, such as apoptosis, necrosis, and autophagy associated with disturbed BBB (brain blood barrier) permeability. In vitro experiments are the main evidence of ferroptosis in human neurodegenerative processes. The identification of ferroptosis in in vivo models of Alzheimer's disease is difficult since specific markers for ferroptotic cells, such as specific antibodies, are not available. In addition, other metal ions, such as copper, can also regulate ferroptosis and lipid peroxidation [172,173]. Taking all under consideration, we still do not know whether ferroptosis is the cause or consequence of neurodegeneration processes such as Alzheimer's disease.

3.2. Ferroptosis in Parkinson's Disease

Parkinson's disease (PD) is one of the most common and best-known diseases of the nervous system, affecting roughly 0.1–0.2% of the general population and 1% of the population above 60 years [174]. PD is characterized as a slowly progressing neurodegenerative ailment with motor and non-motor clinical manifestations, due to an intense decrease in dopamine production [175]. Classic hallmarks in PD are still related to the motor manifestation such as bradykinesia, resting tremor and rigidity [176]. However, non-motor symptoms associated with PD have recently gained more attention due to their relevance and impact on the patient's quality of life. Non-motor symptoms of PD include anosmia, constipation, pain, anxiety, depression, psychosis and cognitive disorders that can progress to dementia [177–179].

The pathophysiological characteristics of PD include the slow and progressive degeneration of dopaminergic neurons in the pars compacta of the substantia nigra (SNpc), which is associated with a systematic and progressive accumulation of iron, leading to striatum dopamine depletion, disappearance of neuromelanin and the appearance of intracellular Lewy bodies having aggregated α-synuclein as the main component [180,181]. During PD progression there is an increase in oxidative stress, lipid peroxidation, and mitochondrial dysfunction associated with the depletion of antioxidant enzymes in the glutathione systems. All of these associated factors lead to neuronal death and the functional disability of the organism (Figure 5). Currently, the pharmacological treatment of PD aims to increase dopamine levels in the synaptic cleft. Levodopa is the drug of choice, being associated with dopamine agonists, dopamine metabolism inhibitors and decarboxylase inhibitors. Treatment is stable for a period of 5–6 years. Then, however, the disease progresses with marked neurodegeneration and development of dementia [182–184].

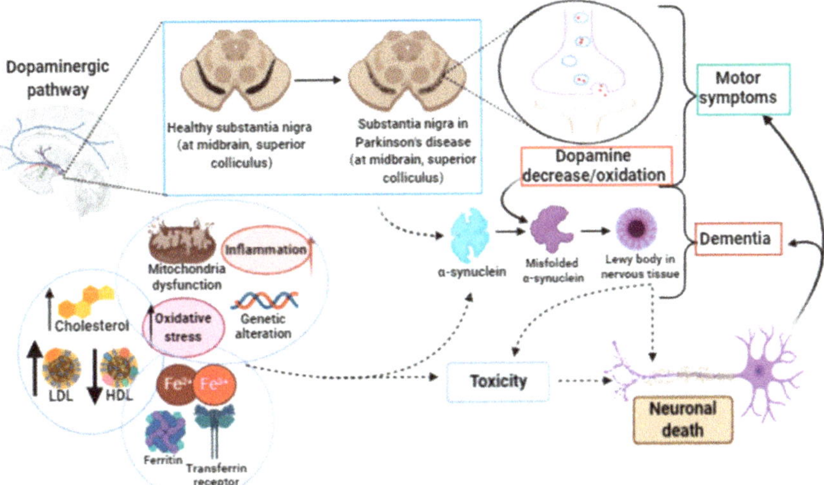

Figure 5. Parkinson's disease. Parkinson's disease (PD) occurs due to the decrease in and/or oxidation of dopamine in the substantia nigra, involving the motor system. The incorrect folding of α-synuclein leads to the accumulation of protein (Lewy body) in nervous tissue. The formation of Lewy bodies may be due to a highly pro-oxidative environment, due to dysfunction in the transport of lipids, iron, inflammation and mitochondrial changes. The increase in Lewy bodies is the trigger for the development of dementia, neurotoxicity and neuronal death.

The association between iron and PD is long-standing [185,186]. Daily exposure to elevated iron levels is a risk factor for the development of PD [187]. In addition, an increase in the iron content in substantia nigra and globus pallidus of PD patients was observed by magnetic resonance imaging (MRI). This increase was associated with time of disease, neurodegeneration and severity of motor impairment [188,189]. In mice, treatment with deferiprone (DFP) was shown to significantly reduce labile iron and biological damage in oxidation-stressed cells, improving motor functions while raising striatal dopamine levels. Furthermore, in patients with PD, a decrease in iron overload has been described in the substantia nigra after 6 months of deferiprone treatment, as has an increase in ceruloplasmin activity in cerebrospinal fluid (CSF) [190–193].

Ferroptosis is also proving to be a mechanism of immeasurable importance for the pathogenesis of PD [194]. In fact, since the early 2000s it has been known that elevated levels of iron could be found in the brain of patients with PD, although an iron-dependent cell death mechanism had not yet been proposed at that time. Additionally, several genes and proteins related to iron metabolism of brain cells have been found to be mutated in PD patients, strengthening the correlation between iron metabolism and Parkinson's disease [195,196].

Some previous works on Parkinson's disease described the presence of PUFA peroxidation, a decrease in GPx4 activity and exhaustion of the glutathione system, associated with increased oxidative stress. The first evidence of ferroptosis in Parkinson's disease was described by Do Van et al. [197]. PD models, both in vitro and in vivo, have shown that the characteristic features of ferroptosis were present in differentiated Lund human mesencephalic (LUHMES) cells intoxicated with erastin. The characteristics of ferroptosis in LUHMES cells were different from those reported for other cell lines. Moreover, the calcium chelator 1.2-bis(o-aminophenoxy)ethane-N,N,N′,N′-tetraacetic acid (BAPTA) and protein kinase C (PKC) inhibitors (the bisindolylmaleimide analog Bis-III, and small interfering RNA (siRNA)) were very effective in counteracting erastin-induced cell death. In LUHMES cells, ferroptosis requires activated mitogen-activated protein kinase (MEK) signaling but is independent of Ras activation. Moreover, ferroptosis involvement in dopaminergic cell death was observed in a mouse model in which toxicity was inhibited by the specific ferroptosis inhibitor ferrostatin 1. Lastly, the regulation of dopaminergic cell death by ferroptosis and its inhibition by PKC were also confirmed ex vivo by studying organotypic slice cultures (OSCs). It is important to note that the ferroptosis activation pathways were initially described in cancer models, in which there was elevated metabolic activity and cellular proliferation due to uncontrolled cellular repair pathways. These pathways are not activated in neurodegeneration models. Therefore, ferroptosis can be triggered by different mechanisms in different cells and different tissues.

A plethora of new evidence is clarifying the molecular mechanisms involved in the interaction of PD and ferroptosis cell death. α-Synuclein, a protein abundantly expressed in the nervous system and a main component of Lewy bodies, has been widely studied in PD as its pathogenic effects are strongly correlated with PD's pathophysiology [198]. Additionally, it has been recently shown that α-synuclein aggregation (a common feature in PD) is responsible for the production of ROS followed by lipid peroxidation in an iron-dependent manner, resulting in increased calcium influx and consequent cell death [199]. In this way, the use of ferroptosis inhibitors such as ferrostatin or iron chelators [200] has been sufficient to suppress cell death, supporting the hypothesis that ferroptosis is a major player in this process and may harbor therapeutic potential. Several studies suggest the modulation in ferroptosis as a therapeutic target in neurodegenerative diseases [201,202].

Another molecule that may link PD to ferroptosis is the transcription factor Nrf2. As Nrf2 is involved in the regulation of processes such as the metabolism of iron, lipids and glutathione, several works have focused on understanding how the modulation of this transcription factor can intervene in the ferroptosis pathway [203]. For instance, it was shown that Nrf2 overexpression in brain tumor cells was an indication of poor survival outcomes since Nrf2 bestowed these tumor cells with resistance to cell death mechanisms such as ferroptosis [204]. In addition, the activation of the Nrf2 pathway (p62-Keap1-Nrf2) has also been shown to prevent 6-hydroxydopamine (6-OHDA)-induced ferroptosis in a human dopaminergic cell line (SH-SY5Y) [205]. Taken together, these results demonstrate that the modulation of Nrf2 expression can provide new therapeutic approaches for neurodegenerative diseases such as PD [206]. In addition, studies performed on a monkey model of PD has shown that clioquinol (CQ), a drug primarily used as an antiparasitic agent that also presents iron chelation properties [207], was able to improve both, motor and non-motor manifestations in treated monkeys. Shi et al. [208] observed that CQ not only led to a decrease in iron levels in the substantia nigra but also managed to suppress the well-known oxidative stress present in PD. At a molecular level, CQ was shown to be able to reduce p53-mediated cell apoptosis and to diminish the iron content and oxidative stress by activating the AKT/mTOR pathway, which was found downregulated in the PD monkey model [208]. These results, taken together, point out once again to the involvement of ferroptosis in PD and demonstrate how pharmacological interventions could be useful to revert this outcome. Unfortunately, available therapies for PD patients are only capable of mitigating their symptoms and cannot reverse the loss of dopaminergic cells [209]. Therefore, seeking new therapeutic options that may intervene in this primary process of cell death could potentially change PD treatment.

3.3. Ferroptosis in Huntington's Disease

Huntington's disease (HD) is an autosomal dominant late-onset neurodegenerative disorder (age of onset: 30–50 years). HD is caused by a polymorphic sequence of three CAG nucleotides in exon 1 of the IT15 gene (Huntingtin (HTT)), which is located at 4p16.3. HD was described by George Huntington in 1872, after he observed in Long Island a rare disease present in some families in the region. He called this disease "hereditary chorea" (Huntington, 1872). The main observed clinical signs are motor disorders (such as involuntary movements), cognitive, emotional, and psychiatric disorders (such as personality change and dementia). Carriers of this disease may also have dysphagia, which leads to weight loss. In patients affected with juvenile HD under the age of 20, the most observed disorders are behavioral disorders, learning difficulties, and often seizures [210–214].

The mutation occurs in exon 1 of the HTT gene; this region is polymorphic and encodes a polyglutamine segment (polyQ)—in this fragment, expansion and generation of mutant proteins that can lead to the development of HD may occur [215]. The huntingtin protein has a molecular weight of ~348 kDa, and the expression level is different according to the cell types in which it is found: neuronal cell bodies, dendritic cells, and axons. Inside the cell, the huntingtin protein is located in the cytoplasm and partially in the nucleus and can move between these compartments [216–218].

Such evidence of expression and location suggests that this protein plays an important role in the nervous system, suggesting that changes in its conformation can lead to an imbalance in the performance of its functions, which in turn can result in the development of HD. The pathophysiological mechanism associated with HD includes loss of glial cells (astrocytes and oligodendrocytes), neuronal death, and atrophy of brain tissues, which may start in the striatum, followed by the cerebral cortex (Figure 6). Although mutated huntingtin protein (mHTT) is found in the brain, its expression has been evidenced in cells other than the ones in the central nervous system [216].

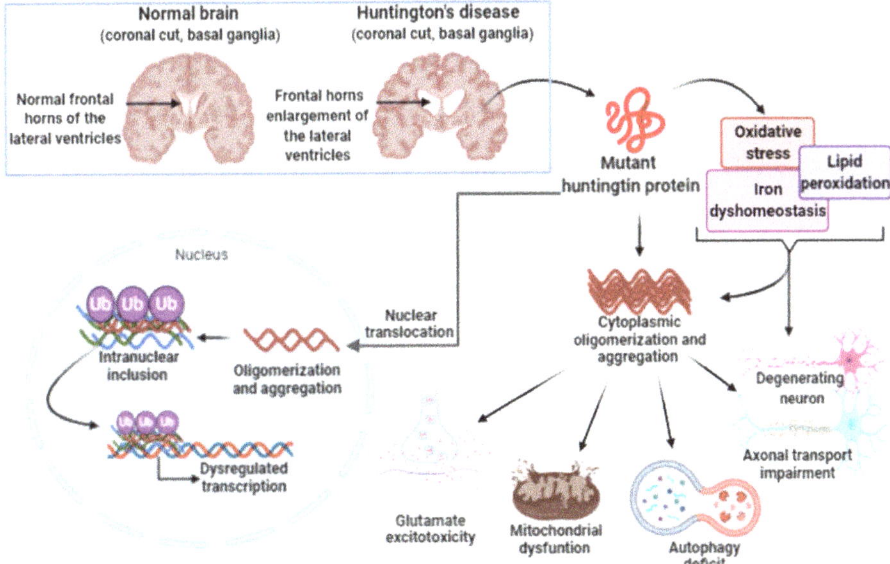

Figure 6. Huntington's disease. Huntington's disease is caused by the repetition of autosomal dominant CAG trinucleotide in the Huntingtin gene (HTT gene) on chromosome 4, giving rise to the mutant huntingtin protein. The mutated protein translocates to the nucleus and remains in the cytoplasm. In the nucleus, association, oligomerization and aggregation with other proteins occurs, leading to the formation of inclusions. Protein inclusions disrupt the transcriptional process in nerve tissue cells. In the cytoplasm, the oligomerization, aggregation and precipitation of the huntingtin protein occurs. This process alters the metabolism and both intra- and extra-cellular signaling pathways. The increase in oxidative stress, lipid peroxidation and iron dyshomeostasis contribute to the aggregation and oligomerization of huntingtin protein with other cytoplasmic proteins. Aberrant protein aggregation increases the excitotoxicity of glutamate. Mitochondrial dysfunction changes autophagy mechanisms, and transport in the neuronal axon, leading to nerve cell degeneration.

One of the cellular processes involved in the development of HD is ferroptosis. A study in mouse models with HD showed an accumulation of toxic iron in neurons compared to the wild model, which suggests that iron accumulation may contribute to the neurodegenerative process [219]. Another study with mice that had GPx4 ablation showed the importance of inhibiting ferroptosis to prevent spinal motor neuron degeneration since mice with this characteristic showed motor disorders. The tests carried out to confirm ferroptosis included the absence of apoptotic markers (caspase-3 and TUNEL) and activation of the ERK pathway [220]. Magnetic resonance images showed an accumulation of iron in the brain of patients with HD [221]. However, the pathway that induces ferroptosis in the brain has not been fully elucidated [222]. The mutated huntingtin protein (mHTT) is cleaved at different points since it has different cleavage sites than those present in the normal protein, which result in fragments with different sizes inside neurons (small oligomers or monomers) [223]. It has been shown that mHTT and the wild version are associated with the outer mitochondrial membrane [224].

Inside the cells, there must be a balance between the state of mitochondrial fission and fusion for the proper functioning of this organelle [225]. When an imbalance occurs, cellular respiration is affected and, consequently, there is an increase in ROS inside cells [225]. Mitochondria can generate significant amounts of ROS, resulting from normal organelle metabolism and the electron transport chain that contributes to oxidative stress. Additionally, an mHTT leads to increased oxidative stress, which consequently increases ROS levels in the cell [226]. Under normal conditions, GSH regulates the activity of GPx4, and its function is to inhibit ferroptosis and eliminate the excess of lipid peroxides [225]. However, an increase in ROS levels and, consequently, an increase in lipid peroxides, which leads to a depletion of GSH, decreases GPx4 [225]. There is a series of intracellular signals that culminate in an imbalance in cell homeostasis and leads to the process of ferroptosis. In patients with HD, there is a deregulation of GSH which interferes with their functions and enzymes dependent on its action [117,225].

In patients with HD and asymptomatic carriers, high lipid peroxidation and low GSH plasma levels have been found, showing that oxidative stress may be linked to the pathophysiological mechanism of HD [227]. Iron chelators could be an alternative for treatment [228].

Nfr2 is a transcriptional regulator of genes involved with ferroptosis, such as GPx4, which can be found in cytosol and modulates mitochondrial function [206]. These are essential components for the process of ferroptosis and demonstrate the importance of Nfr2 in protecting against ferroptosis. Therefore, Nfr2 can be an alternative therapy for reducing ferroptosis [206]. Currently there is no specific treatment available for HD, only palliative care [212]. In general, the multidisciplinary treatment available at the moment is focused on the palliative treatment of symptoms and neuroprotection of patients [218,229]. Although there is no specific treatment for HD, several studies are being developed focusing on silencing the DNA or mRNA of the mutated allele whose gene varies in the number of copies [215,229]. Another promising alternative is the creation of mouse models with HD and the involvement of GSH activity as sources of possible target treatments [117].

4. Conclusions and Perspectives

The description of the ferroptosis mode of cell death is still recent. Although its participation in several diseases such as cancer has been increasingly established, in neurodegenerative diseases information is still lacking. Most data have been obtained from experimental studies and more clinical approaches are necessary. For instance, the experimental use of iron chelators and antioxidants has shown to be effective as a possible alternative to interrupt the process of ferroptosis in neurodegenerative diseases. However, when tested in humans, they showed reduced efficacy, which shows that research involving signaling molecules from other pathways that trigger the process of ferroptosis are needed. Nevertheless, there is a growing body of evidence that ferroptosis plays a central role in several neurodegenerative diseases, at least in those described here. This field of research may still yield promising results and disruptive therapeutic alternatives for patients with neurodegenerative diseases.

Author Contributions: Conceptualization, C.O.R., D.L., and S.P.B.; writing—original draft preparation C.O.R., F.A.d.F., J.S.-S., L.R.-R., P.d.L.B., D.L., and S.P.B.; writing—review and editing, C.O.R., D.L., and S.P.B.; supervision, S.P.B; funding acquisition, S.P.B. The figures were created by authors using bioBender.com. All authors have read and agreed to the published version of the manuscript.

Funding: This work was supported by grants from Conselho Nacional de Desenvolvimento Científico e Tecnológico (CNPq); Coordenação de Aperfeiçoamento de Pessoal de Nível Superior (CAPES); Instituto Nacional de Ciência e Tecnologia–Fluidos Complexos (INCT-FCx); Instituto Nacional de Ciência e Tecnologia em Medicina Regenerativa (INCT-Regenera), all from Brazil.

Conflicts of Interest: The authors declare no conflict of interest.

References

1. Tang, D.; Kang, R.; Berghe, T.V.; Vandenabeele, P.; Kroemer, G. The molecular machinery of regulated cell death. *Cell Res.* **2019**, *29*, 347–364. [CrossRef]
2. Wimmer, K.; Sachet, M.; Oehler, R. Circulating biomarkers of cell death. *Clin. Chim. Acta* **2020**, *500*, 87–97. [CrossRef]
3. Galluzzi, L.; Vitale, I.; Aaronson, S.A.; Abrams, J.M.; Adam, D.; Agostinis, P.; Alnemri, E.S.; Altucci, L.; Amelio, I.; Andrews, D.W.; et al. Molecular mechanisms of cell death: Recommendations of the Nomenclature Committee on Cell Death 2018. *Cell Death Differ.* **2018**, *25*, 486–541. [CrossRef]
4. Galluzzi, L.; Pedro, J.M.B.-S.; Kepp, O.; Kroemer, G. Regulated cell death and adaptive stress responses. *Cell. Mol. Life Sci.* **2016**, *73*, 2405–2410. [CrossRef]
5. Levy, D.; de Melo, T.C.; Oliveira, B.A.; Paz, J.L.; de Freitas, F.A.; Reichert, C.O.; Rodrigues, A.; Bydlowski, S.P. 7-Ketocholesterol and cholestane-triol increase expression of SMO and LXRα signaling pathways in a human breast cancer cell line. *Biochem. Biophys. Rep.* **2018**, *19*. [CrossRef]
6. Paz, J.L.; Levy, D.; Oliveira, B.A.; de Melo, T.C.; de Freitas, F.A.; Reichert, C.O.; Rodrigues, A.; Pereira, J.; Bydlowski, S.P. 7-Ketocholesterol Promotes Oxiapoptophagy in Bone Marrow Mesenchymal Stem Cell from Patients with Acute Myeloid Leukemia. *Cells* **2019**, *8*, 482. [CrossRef]
7. Dixon, S.J.; Lemberg, K.M.; Lamprecht, M.R.; Skouta, R.; Zaitsev, E.M.; Gleason, C.E.; Patel, D.N.; Bauer, A.J.; Cantley, A.M.; Yang, W.S.; et al. Ferroptosis: An iron-dependent form of nonapoptotic cell death. *Cell* **2012**. [CrossRef]
8. Bayır, H.; Anthonymuthu, T.S.; Tyurina, Y.Y.; Patel, S.J.; Amoscato, A.A.; Lamade, A.M.; Yang, Q.; Vladimirov, G.K.; Philpott, C.C.; Kagan, V.E. Achieving Life through Death: Redox Biology of Lipid Peroxidation in Ferroptosis. *Cell Chem. Biol.* **2020**, *27*, 387–408. [CrossRef]
9. Fricker, M.; Tolkovsky, A.M.; Borutaite, V.; Coleman, M.; Brown, G.C. Neuronal cell death. *Physiol. Rev.* **2018**, *98*, 813–880. [CrossRef]
10. Sun, Y.; Chen, P.; Zhai, B.; Zhang, M.; Xiang, Y.; Fang, J.; Xu, S.; Gao, Y.; Chen, X.; Sui, X.; et al. The emerging role of ferroptosis in inflammation. *Biomed. Pharmacother.* **2020**, *127*, 11010. [CrossRef] [PubMed]
11. Zhong, S.; Li, L.; Shen, X.; Li, Q.; Xu, W.; Wang, X.; Tao, Y.; Yin, H. An update on lipid oxidation and inflammation in cardiovascular diseases. *Free Radic. Biol. Med.* **2019**, *144*, 266–278. [CrossRef] [PubMed]
12. Galaris, D.; Barbouti, A.; Pantopoulos, K. Iron homeostasis and oxidative stress: An intimate relationship. *Biochim. Biophys. Acta Mol. Cell Res.* **2019**, *1866*, 118535. [CrossRef] [PubMed]
13. Stockwell, B.R.; Angeli, J.P.F.; Bayir, H.; Bush, A.I.; Conrad, M.; Dixon, S.J.; Fulda, S.; Gascón, S.; Hatzios, S.K.; Kagan, V.E.; et al. Ferroptosis: A Regulated Cell Death Nexus Linking Metabolism, Redox Biology, and Disease. *Cell* **2017**, *171*, 273–285. [CrossRef] [PubMed]
14. Mao, H.; Zhao, Y.; Li, H.; Lei, L. Ferroptosis as an emerging target in inflammatory diseases. *Prog. Biophys. Mol. Biol.* **2020**, *155*, 20–28. [CrossRef]
15. Bridges, R.J.; Natale, N.R.; Patel, S.A. System x c- cystine/glutamate antiporter: An update on molecular pharmacology and roles within the CNS. *Br. J. Pharmacol.* **2012**, *165*, 20–34. [CrossRef]
16. Hirschhorn, T.; Stockwell, B.R. The development of the concept of ferroptosis. *Free Radic. Biol. Med.* **2019**, *133*, 130–143. [CrossRef]
17. Yang, W.S.; Stockwell, B.R. Ferroptosis: Death by Lipid Peroxidation. *Trends Cell Biol.* **2016**, *26*, 165–176. [CrossRef]
18. Stockwell, B.R.; Jiang, X. The Chemistry and Biology of Ferroptosis. *Cell Chem. Biol.* **2020**, *27*, 365–375. [CrossRef]
19. Conrad, M.; Pratt, D.A. The chemical basis of ferroptosis. *Nat. Chem. Biol.* **2019**. [CrossRef]
20. Wang, Y.; Wei, Z.; Pan, K.; Li, J.; Chen, Q. The function and mechanism of ferroptosis in cancer. *Apoptosis* **2020**, *25*, 1–13. [CrossRef]
21. Gaschler, M.M.; Stockwell, B.R. Lipid peroxidation in cell death. *Biochem. Biophys. Res. Commun.* **2017**, *482*, 419–425. [CrossRef] [PubMed]
22. Desai, S.N.; Farris, F.F.; Ray, S.D. Lipid Peroxidation. In *Encyclopedia of Toxicology*, 3rd ed.; Elsevier: Amsterdam, The Netherland, 2014; ISBN 9780123864543.

23. Tyurina, Y.Y.; Croix, C.M.S.; Watkins, S.C.; Watson, A.M.; Epperly, M.W.; Anthonymuthu, T.S.; Kisin, E.R.; Vlasova, I.I.; Krysko, O.; Krysko, D.V.; et al. Redox (phospho)lipidomics of signaling in inflammation and programmed cell death. *J. Leukoc. Biol.* **2019**, *106*, 57–81. [CrossRef]
24. Kagan, V.E.; Mao, G.; Qu, F.; Angeli, J.P.F.; Doll, S.; Croix, C.S.; Dar, H.H.; Liu, B.; Tyurin, V.A.; Ritov, V.B.; et al. Oxidized arachidonic and adrenic PEs navigate cells to ferroptosis. *Nat. Chem. Biol.* **2017**, *13*, 81–90. [CrossRef]
25. Doll, S.; Proneth, B.; Tyurina, Y.Y.; Panzilius, E.; Kobayashi, S.; Ingold, I.; Irmler, M.; Beckers, J.; Aichler, M.; Walch, A.; et al. ACSL4 dictates ferroptosis sensitivity by shaping cellular lipid composition. *Nat. Chem. Biol.* **2017**, *13*, 91–98. [CrossRef] [PubMed]
26. Konstorum, A.; Tesfay, L.; Paul, B.T.; Torti, F.M.; Laubenbacher, R.C.; Torti, S.V. Systems biology of ferroptosis: A modeling approach. *J. Theor. Biol.* **2020**, *493*, 110222. [CrossRef] [PubMed]
27. Yang, W.S.; Kim, K.J.; Gaschler, M.M.; Patel, M.; Shchepinov, M.S.; Stockwell, B.R. Peroxidation of polyunsaturated fatty acids by lipoxygenases drives ferroptosis. *Proc. Natl. Acad. Sci. USA* **2016**, *113*, E4966–E4975. [CrossRef]
28. Anthonymuthu, T.S.; Kenny, E.M.; Shrivastava, I.; Tyurina, Y.Y.; Hier, Z.E.; Ting, H.C.; Dar, H.H.; Tyurin, V.A.; Nesterova, A.; Amoscato, A.A.; et al. Empowerment of 15-Lipoxygenase Catalytic Competence in Selective Oxidation of Membrane ETE-PE to Ferroptotic Death Signals, HpETE-PE. *J. Am. Chem. Soc.* **2018**, *140*, 17835–17839. [CrossRef]
29. Dar, H.H.; Tyurina, Y.Y.; Mikulska-Ruminska, K.; Shrivastava, I.; Ting, H.C.; Tyurin, V.A.; Krieger, J.; Croix, C.M.S.; Watkins, S.; Bayir, E.; et al. Pseudomonas aeruginosa utilizes host polyunsaturated phosphatidylethanolamines to trigger theft-ferroptosis in bronchial epithelium. *J. Clin. Investig.* **2018**, *128*, 4639–4653. [CrossRef]
30. Kuhn, H.; Banthiya, S.; Van Leyen, K. Mammalian lipoxygenases and their biological relevance. *Biochim. Biophys. Acta Mol. Cell Biol. Lipids* **2015**, *1851*, 308–330. [CrossRef]
31. Kuhn, H.; Walther, M.; Kuban, R.J. Mammalian arachidonate 15-lipoxygenases: Structure, function, and biological implications. *Prostaglandins Other Lipid Mediat.* **2002**, *68*, 263–290. [CrossRef]
32. Wenzel, S.E.; Tyurina, Y.Y.; Zhao, J.; Croix, C.M.S.; Dar, H.H.; Mao, G.; Tyurin, V.A.; Anthonymuthu, T.S.; Kapralov, A.A.; Amoscato, A.A.; et al. PEBP1 Wardens Ferroptosis by Enabling Lipoxygenase Generation of Lipid Death Signals. *Cell* **2017**, *171*, 628–641.e26. [CrossRef] [PubMed]
33. Stoyanovsky, D.A.; Tyurina, Y.Y.; Shrivastava, I.; Bahar, I.; Tyurin, V.A.; Protchenko, O.; Jadhav, S.; Bolevich, S.B.; Kozlov, A.V.; Vladimirov, Y.A.; et al. Iron catalysis of lipid peroxidation in ferroptosis: Regulated enzymatic or random free radical reaction? *Free Radic. Biol. Med.* **2019**, *133*, 153–161. [CrossRef]
34. Sheng, X.H.; Cui, C.C.; Shan, C.; Li, Y.Z.; Sheng, D.H.; Sun, B.; Chen, D.Z. O-Phenylenediamine: A privileged pharmacophore of ferrostatins for radical-trapping reactivity in blocking ferroptosis. *Org. Biomol. Chem.* **2018**, *16*, 3952–3960. [CrossRef]
35. Zou, Y.; Li, H.; Graham, E.T.; Deik, A.A.; Eaton, J.K.; Wang, W.; Sandoval-Gomez, G.; Clish, C.B.; Doench, J.G.; Schreiber, S.L. Cytochrome P450 oxidoreductase contributes to phospholipid peroxidation in ferroptosis. *Nat. Chem. Biol.* **2020**, *16*, 302–309. [CrossRef] [PubMed]
36. Green, D.R.; Ferguson, T.; Zitvogel, L.; Kroemer, G. Immunogenic and tolerogenic cell death. *Nat. Rev. Immunol.* **2009**, *9*, 353–363. [CrossRef] [PubMed]
37. Green, D.R.; Galluzzi, L.; Kroemer, G. Metabolic control of cell death. *Science* **2014**, *345*, 1250256. [CrossRef] [PubMed]
38. Conrad, M.; Angeli, J.P.F. Glutathione peroxidase 4 (Gpx4) and ferroptosis: What's so special about it? *Mol. Cell. Oncol.* **2015**, *2*. [CrossRef]
39. Conrad, M.; Kagan, V.E.; Bayir, H.; Pagnussat, G.C.; Head, B.; Traber, M.G.; Stockwell, B.R. Regulation of lipid peroxidation and ferroptosis in diverse species. *Genes Dev.* **2018**, *32*, 602–619. [CrossRef]
40. Margis, R.; Dunand, C.; Teixeira, F.K.; Margis-Pinheiro, M. Glutathione peroxidase family—An evolutionary overview. *FEBS J.* **2008**, *275*, 3959–3970. [CrossRef]
41. Chu, F.F. The human glutathione peroxidase genes GPX2, GPX3, and GPX4 map to chromosomes 14, 5, and 19, respectively. *Cytogenet. Genome Res.* **1994**, *66*, 96–98. [CrossRef]
42. Ingold, I.; Berndt, C.; Schmitt, S.; Doll, S.; Poschmann, G.; Buday, K.; Roveri, A.; Peng, X.; Porto Freitas, F.; Seibt, T.; et al. Selenium Utilization by GPX4 Is Required to Prevent Hydroperoxide-Induced Ferroptosis. *Cell* **2018**, *172*, 409–422.e21. [CrossRef] [PubMed]

43. Borchert, A.; Kalms, J.; Roth, S.R.; Rademacher, M.; Schmidt, A.; Holzhutter, H.G.; Kuhn, H.; Scheerer, P. Crystal structure and functional characterization of selenocysteine-containing glutathione peroxidase 4 suggests an alternative mechanism of peroxide reduction. *Biochim. Biophys. Acta Mol. Cell Biol. Lipids* **2018**, *1863*, 1095–1107. [CrossRef] [PubMed]
44. Angeli, J.P.F.; Conrad, M. Selenium and GPX4, a vital symbiosis. *Free Radic. Biol. Med.* **2018**, *127*, 153–159. [CrossRef]
45. Imai, H.; Nakagawa, Y. Biological significance of phospholipid hydroperoxide glutathione peroxidase (PHGPx, GPx4) in mammalian cells. *Free Radic. Biol. Med.* **2003**, *34*, 145–169. [CrossRef]
46. Conrad, M.; Schneider, M.; Seiler, A.; Bornkamm, G.W. Physiological role of phospholipid hydroperoxide glutathione peroxidase in mammals. *Biol. Chem.* **2007**, *388*, 1019–1025. [CrossRef]
47. Toppo, S.; Vanin, S.; Bosello, V.; Tosatto, S.C.E. Evolutionary and structural insights into the multifaceted glutathione peroxidase (Gpx) superfamily. *Antioxid. Redox Signal.* **2008**, *10*, 1501–1514. [CrossRef] [PubMed]
48. Kumar, S.; Bhadhadhara, K.; Govil, S.; Mathur, N.; Pathak, A.N. In-silico studies for comparative analysis of glutathione peroxidaes isozymes in Homo sapiens. *Int. J. Pharma Bio Sci.* **2014**, *5*, B352–B363.
49. Imai, H.; Matsuoka, M.; Kumagai, T.; Sakamoto, T.; Koumura, T. Lipid peroxidation-dependent cell death regulated by GPx4 and ferroptosis. In *Current Topics in Microbiology and Immunology*; Springer: Cham, Switzerland, 2017.
50. Meister, A. Glutathione metabolism. *Methods Enzymol.* **1995**, *251*, 3–7.
51. Kalinina, E.V.; Chernov, N.N.; Novichkova, M.D. Role of glutathione, glutathione transferase, and glutaredoxin in regulation of redox-dependent processes. *Biochemistry* **2014**, *79*, 1562–1583. [CrossRef]
52. Sui, X.; Zhang, R.; Liu, S.; Duan, T.; Zhai, L.; Zhang, M.; Han, X.; Xiang, Y.; Huang, X.; Lin, H.; et al. RSL3 drives ferroptosis through GPX4 inactivation and ros production in colorectal cancer. *Front. Pharmacol.* **2018**, *9*, 1371. [CrossRef]
53. Eaton, J.; Furst, L.; Ruberto, R.; Moosmayer, D.; Hillig, R.; Hilpmann, A.; Zimmermann, K.; Ryan, M.; Niehues, M.; Badock, V.; et al. Targeting a Therapy-Resistant Cancer Cell State Using Masked Electrophiles as GPX4 Inhibitors. *bioRxiv* **2018**. [CrossRef]
54. Eaton, J.K.; Furst, L.; Cai, L.L.; Viswanathan, V.S.; Schreiber, S.L. Structure-activity relationships of GPX4 inhibitor warheads. *Bioorganic Med. Chem. Lett.* **2020**, *30*, 127538. [CrossRef] [PubMed]
55. Eaton, J.K.; Furst, L.; Ruberto, R.A.; Moosmayer, D.; Hilpmann, A.; Ryan, M.J.; Zimmermann, K.; Cai, L.L.; Niehues, M.; Badock, V.; et al. Selective covalent targeting of GPX4 using masked nitrile-oxide electrophiles. *Nat. Chem. Biol.* **2020**, *16*, 497–506. [CrossRef]
56. Vučković, A.M.; Travain, V.B.; Bordin, L.; Cozza, G.; Miotto, G.; Rossetto, M.; Toppo, S.; Venerando, R.; Zaccarin, M.; Maiorino, M.; et al. Inactivation of the glutathione peroxidase GPx4 by the ferroptosis-inducing molecule RSL3 requires the adaptor protein 14-3-3ε. *FEBS Lett.* **2020**, *594*, 611–624. [CrossRef] [PubMed]
57. Angeli, J.P.F.; Schneider, M.; Proneth, B.; Tyurina, Y.Y.; Tyurin, V.A.; Hammond, V.J.; Herbach, N.; Aichler, M.; Walch, A.; Eggenhofer, E.; et al. Inactivation of the ferroptosis regulator Gpx4 triggers acute renal failure in mice. *Nat. Cell Biol.* **2014**, *16*, 1180–1191. [CrossRef]
58. Fan, B.-Y.; Pang, Y.-L.; Li, W.-X.; Zhao, C.-X.; Zhang, Y.; Wang, X.; Ning, G.-Z.; Kong, X.-H.; Liu, C.; Yao, X.; et al. Liproxstatin-1 is an effective inhibitor of oligodendrocyte ferroptosis induced by inhibition of glutathione peroxidase 4. *Neural Regen. Res.* **2020**, *16*, 561–566. [CrossRef] [PubMed]
59. Li, Y.; Yan, H.; Xu, X.; Liu, H.; Wu, C.; Zhao, L. Erastin/sorafenib induces cisplatin-resistant non-small cell lung cancer cell ferroptosis through inhibition of the Nrf2/xCT pathway. *Oncol. Lett.* **2020**, *19*, 323–333. [CrossRef]
60. Hasegawa, M.; Takahashi, H.; Rajabi, H.; Alam, M.; Suzuki, Y.; Yin, L.; Tagde, A.; Maeda, T.; Hiraki, M.; Sukhatme, V.P.; et al. Functional interactions of the cystine/glutamate antiporter, CD44V and MUC1-C oncoprotein in triple-negative breast cancer cells. *Oncotarget* **2016**, *7*, 11756–11769. [CrossRef] [PubMed]
61. Chen, L.; Li, X.; Liu, L.; Yu, B.; Xue, Y.; Liu, Y. Erastin sensitizes Glioblastoma cells to temozolomide by restraining xCT and cystathionine-γ-lyase function. *Oncol. Rep.* **2015**, *33*, 1465–1474. [CrossRef]
62. Wang, S.F.; Chen, M.S.; Chou, Y.C.; Ueng, Y.F.; Yin, P.H.; Yeh, T.S.; Lee, H.C. Mitochondrial dysfunction enhances cisplatin resistance in human gastric cancer cells via the ROS-activated GCN2-eIF2α-ATF4-xCT pathway. *Oncotarget* **2016**, *7*, 74132–74151. [CrossRef]

63. Zhang, Z.; Guo, M.; Li, Y.; Shen, M.; Kong, D.; Shao, J.; Ding, H.; Tan, S.; Chen, A.; Zhang, F.; et al. RNA-binding protein ZFP36/TTP protects against ferroptosis by regulating autophagy signaling pathway in hepatic stellate cells. *Autophagy* **2020**, *16*, 1482–1505. [CrossRef] [PubMed]
64. Yu, H.; Yang, C.; Jian, L.; Guo, S.; Chen, R.; Li, K.; Qu, F.; Tao, K.; Fu, Y.; Luo, F.; et al. Sulfasalazine-induced ferroptosis in breast cancer cells is reduced by the inhibitory effect of estrogen receptor on the transferrin receptor. *Oncol. Rep.* **2019**, *42*, 826–838. [CrossRef] [PubMed]
65. Kim, E.H.; Shin, D.; Lee, J.; Jung, A.R.; Roh, J.L. CISD2 inhibition overcomes resistance to sulfasalazine-induced ferroptotic cell death in head and neck cancer. *Cancer Lett.* **2018**, *432*, 180–190. [CrossRef] [PubMed]
66. Roh, J.L.; Kim, E.H.; Jang, H.J.; Park, J.Y.; Shin, D. Induction of ferroptotic cell death for overcoming cisplatin resistance of head and neck cancer. *Cancer Lett.* **2016**, *381*, 96–103. [CrossRef] [PubMed]
67. Zhao, Y.; Li, Y.; Zhang, R.; Wang, F.; Wang, T.; Jiao, Y. The role of Erastin in ferroptosis and its prospects in cancer therapy. *Onco. Targets. Ther.* **2020**, *13*, 5429–5441. [CrossRef]
68. Armada-Moreira, A.; Gomes, J.I.; Pina, C.C.; Savchak, O.K.; Gonçalves-Ribeiro, J.; Rei, N.; Pinto, S.; Morais, T.P.; Martins, R.S.; Ribeiro, F.F.; et al. Going the Extra (Synaptic) Mile: Excitotoxicity as the Road Toward Neurodegenerative Diseases. *Front. Cell. Neurosci.* **2020**, *14*, 1–27. [CrossRef] [PubMed]
69. Lewerenz, J.; Hewett, S.J.; Huang, Y.; Lambros, M.; Gout, P.W.; Kalivas, P.W.; Massie, A.; Smolders, I.; Methner, A.; Pergande, M.; et al. The cystine/glutamate antiporter system xc- in health and disease: From molecular mechanisms to novel therapeutic opportunities. *Antioxid. Redox Signal.* **2013**, *18*, 522–555. [CrossRef] [PubMed]
70. Ballatori, N.; Hammond, C.L.; Cunningham, J.B.; Krance, S.M.; Marchan, R. Molecular mechanisms of reduced glutathione transport: Role of the MRP/CFTR/ABCC and OATP/SLC21A families of membrane proteins. *Toxicol. Appl. Pharmacol.* **2005**, *204*, 238–255. [CrossRef]
71. Lorendeau, D.; Dury, L.; Nasr, R.; Boumendjel, A.; Teodori, E.; Gutschow, M.; Falson, P.; Di Pietro, A.; Baubichon-Cortay, H. MRP1-dependent Collateral Sensitivity of Multidrug-resistant Cancer Cells: Identifying Selective Modulators Inducing Cellular Glutathione Depletion. *Curr. Med. Chem.* **2017**. [CrossRef]
72. Shelton, P.; Jaiswal, A.K. The transcription factor NF-E2-related factor 2 (nrf2): A protooncogene? *FASEB J.* **2013**, *27*, 414–423. [CrossRef]
73. Reisman, S.A.; Csanaky, I.I.; Yeager, R.I.; Klaassen, C.D. Nrf2 activation enhances biliary excretion of sulfobromophthalein by inducing glutathione-S-Transferase activity. *Toxicol. Sci.* **2009**, *109*, 24–30. [CrossRef] [PubMed]
74. Wild, A.C.; Mulcahy, R.T. Regulation of γ-Glutamylcysteine synthetase subunit gene expression: Insights into transcriptional control of antioxidant defenses. *Free Radic. Res.* **2000**, *32*, 281–301. [CrossRef] [PubMed]
75. Habib, E.; Linher-Melville, K.; Lin, H.X.; Singh, G. Expression of xCT and activity of system xc- are regulated by NRF2 in human breast cancer cells in response to oxidative stress. *Redox Biol.* **2015**, *5*, 33–42. [CrossRef]
76. Zhang, L.; Wang, H. Targeting the NF-E2-Related Factor 2 Pathway: A Novel Strategy for Traumatic Brain Injury. *Mol. Neurobiol.* **2018**, *55*, 1773–1785. [CrossRef]
77. Kerins, M.J.; Ooi, A. The Roles of NRF2 in Modulating Cellular Iron Homeostasis. *Antioxid. Redox Signal.* **2018**, *29*, 1756–1773. [CrossRef]
78. Bartolini, D.; Giustarini, D.; Pietrella, D.; Rossi, R.; Galli, F. Glutathione S-transferase P influences the Nrf2-dependent response of cellular thiols to seleno-compounds. *Cell Biol. Toxicol.* **2020**, *36*, 379–386. [CrossRef]
79. Doll, S.; Freitas, F.P.; Shah, R.; Aldrovandi, M.; da Silva, M.C.; Ingold, I.; Grocin, A.G.; da Silva, T.N.X.; Panzilius, E.; Scheel, C.H.; et al. FSP1 is a glutathione-independent ferroptosis suppressor. *Nature* **2019**, *575*, 693–698. [CrossRef]
80. Hargreaves, I.; Heaton, R.A.; Mantle, D. Disorders of human coenzyme q10 metabolism: An overview. *Int. J. Mol. Sci.* **2020**, *21*, 6695. [CrossRef] [PubMed]
81. Suresh, P.K.; Sah, A.K.; Daharwal, S.J. Role of free radicals in ocular diseases: An overview. *Res. J. Pharm. Technol.* **2014**, *7*, 1330–1344.
82. Venkatesh, D.; O'Brien, N.A.; Zandkarimi, F.; Tong, D.R.; Stokes, M.E.; Dunn, D.E.; Kengmana, E.S.; Aron, A.T.; Klein, A.M.; Csuka, J.M.; et al. MDM2 and MDMX promote ferroptosis by PPARα-mediated lipid remodeling. *Genes Dev.* **2020**, *34*, 526–543. [CrossRef]
83. Reichert, C.O.; Da Cunha, J.; Levy, D.; Maselli, L.M.F.; Bydlowski, S.P.; Spada, C. Hepcidin: Homeostasis and Diseases Related to Iron Metabolism. *Acta Haematol.* **2017**, *137*. [CrossRef]

84. Reichert, C.O.; Da Cunha, J.; Levy, D.; Maselli, L.M.F.; Bydlowski, S.P.; Spada, C. Hepcidin: SNP-Like Polymorphisms Present in Iron Metabolism and Clinical Complications of Iron Accumulation and Deficiency. In *Genetic Polymorphisms*; Intechopen: London, UK, 2017.
85. Reichert, C.O.; Marafon, F.; Levy, D.; Maselli, L.M.F.; Bagatini, M.D.; Blatt, S.L.; Bydlowski, S.P.; Spada, C. Influence of Hepcidin in the Development of Anemia. In *Current Topics in Anemia*; Intechopen: London, UK, 2018.
86. Lane, D.J.R.; Bae, D.H.; Merlot, A.M.; Sahni, S.; Richardson, D.R. Duodenal cytochrome b (DCYTB) in Iron metabolism: An update on function and regulation. *Nutrients* **2015**, *7*, 2274–2296. [CrossRef] [PubMed]
87. Ganz, T. Iron Homeostasis: Fitting the Puzzle Pieces Together. *Cell Metab.* **2008**, *7*, 288–290. [CrossRef] [PubMed]
88. Andrews, N.C.; Schmidt, P.J. Iron homeostasis. *Annu. Rev. Physiol.* **2007**, *69*, 69–85. [CrossRef] [PubMed]
89. Knutson, M.D. Steap proteins: Implications for iron and copper metabolism. *Nutr. Rev.* **2007**, *65*, 335–340.
90. Lv, H.; Shang, P. The significance, trafficking and determination of labile iron in cytosol, mitochondria and lysosomes. *Metallomics* **2018**, *10*, 899–916. [CrossRef]
91. Xie, Y.; Hou, W.; Song, X.; Yu, Y.; Huang, J.; Sun, X.; Kang, R.; Tang, D. Ferroptosis: Process and function. *Cell Death Differ.* **2016**, *23*, 369–379. [CrossRef]
92. Lei, P.; Bai, T.; Sun, Y. Mechanisms of ferroptosis and relations with regulated cell death: A review. *Front. Physiol.* **2019**, *10*, 139. [CrossRef]
93. Dixon, S.J.; Stockwell, B.R. The hallmarks of ferroptosis. *Annu. Rev. Cancer Biol.* **2019**, *3*, 35–54. [CrossRef]
94. Cabantchik, Z.I. Labile iron in cells and body fluids: Physiology, pathology, and pharmacology. *Front. Pharmacol.* **2014**, *5*, 45. [CrossRef]
95. Muckenthaler, M.U.; Galy, B.; Hentze, M.W. Systemic iron homeostasis and the iron-responsive element/iron-regulatory protein (IRE/IRP) regulatory network. *Annu. Rev. Nutr.* **2008**, *28*, 197–213. [CrossRef]
96. Piccinelli, P.; Samuelsson, T. Evolution of the iron-responsive element. *RNA* **2007**, *13*, 952–966. [CrossRef]
97. Battaglia, A.M.; Chirillo, R.; Aversa, I.; Sacco, A.; Costanzo, F.; Biamonte, F. Ferroptosis and Cancer: Mitochondria Meet the "Iron Maiden" Cell Death. *Cells* **2020**, *9*, 1505. [CrossRef] [PubMed]
98. Hintze, K.J.; Theil, E.C. Cellular regulation and molecular interactions of the ferritins. *Cell. Mol. Life Sci.* **2006**, *63*, 591–600. [CrossRef]
99. Anderson, G.J.; Frazer, D.M. Current understanding of iron homeostasis. *Am. J. Clin. Nutr.* **2017**, *106*, 1559S–1566S. [CrossRef] [PubMed]
100. Feng, H.; Schorpp, K.; Jin, J.; Yozwiak, C.E.; Hoffstrom, B.G.; Decker, A.M.; Rajbhandari, P.; Stokes, M.E.; Bender, H.G.; Csuka, J.M.; et al. Transferrin Receptor Is a Specific Ferroptosis Marker. *Cell Rep.* **2020**, *30*, 3411–3423.e7. [CrossRef]
101. Park, E.; Chung, S.W. ROS-mediated autophagy increases intracellular iron levels and ferroptosis by ferritin and transferrin receptor regulation. *Cell Death Dis.* **2019**, *10*, 1–10. [CrossRef]
102. Masaldan, S.; Clatworthy, S.A.S.; Gamell, C.; Meggyesy, P.M.; Rigopoulos, A.T.; Haupt, S.; Haupt, Y.; Denoyer, D.; Adlard, P.A.; Bush, A.I.; et al. Iron accumulation in senescent cells is coupled with impaired ferritinophagy and inhibition of ferroptosis. *Redox Biol.* **2018**, *14*, 100–115. [CrossRef] [PubMed]
103. Zhou, B.; Liu, J.; Kang, R.; Klionsky, D.J.; Kroemer, G.; Tang, D. Ferroptosis is a type of autophagy-dependent cell death. *Semin. Cancer Biol.* **2019**, *66*, 89–100. [CrossRef]
104. Hamaï, A.; Mehrpour, M. Autophagy and iron homeostasis. *Medecine/Sciences* **2017**, *33*, 260–267. [CrossRef]
105. Li, N.; Wang, W.; Zhou, H.; Wu, Q.; Duan, M.; Liu, C.; Wu, H.; Deng, W.; Shen, D.; Tang, Q. Ferritinophagy-mediated ferroptosis is involved in sepsis-induced cardiac injury. *Free Radic. Biol. Med.* **2020**, *160*, 303–318. [CrossRef]
106. Fuhrmann, D.C.; Mondorf, A.; Beifuß, J.; Jung, M.; Brüne, B. Hypoxia inhibits ferritinophagy, increases mitochondrial ferritin, and protects from ferroptosis. *Redox Biol.* **2020**, *36*, 101670. [CrossRef] [PubMed]
107. Zhang, Z.; Yao, Z.; Wang, L.; Ding, H.; Shao, J.; Chen, A.; Zhang, F.; Zheng, S. Activation of ferritinophagy is required for the RNA-binding protein ELAVL1/HuR to regulate ferroptosis in hepatic stellate cells. *Autophagy* **2018**, *14*, 2083–2103. [CrossRef]
108. Bai, T.; Lei, P.; Zhou, H.; Liang, R.; Zhu, R.; Wang, W.; Zhou, L.; Sun, Y. Sigma-1 receptor protects against ferroptosis in hepatocellular carcinoma cells. *J. Cell. Mol. Med.* **2019**, *23*, 7349–7359. [CrossRef]
109. Sun, X.; Ou, Z.; Xie, M.; Kang, R.; Fan, Y.; Niu, X.; Wang, H.; Cao, L.; Tang, D. HSPB1 as a novel regulator of ferroptotic cancer cell death. *Oncogene* **2015**, *34*, 5617–5625. [CrossRef]

110. DeGregorio-Rocasolano, N.; Martí-Sistac, O.; Gasull, T. Deciphering the iron side of stroke: Neurodegeneration at the crossroads between iron dyshomeostasis, excitotoxicity, and ferroptosis. *Front. Neurosci.* **2019**, *13*, 85. [CrossRef] [PubMed]
111. Hare, D.; Ayton, S.; Bush, A.; Lei, P. A delicate balance: Iron metabolism and diseases of the brain. *Front. Aging Neurosci.* **2013**, *5*, 34. [CrossRef] [PubMed]
112. Reichert, C.O.; de Macedo, C.G.; Levy, D.; Sini, B.C.; Monteiro, A.M.; Gidlund, M.; Maselli, L.M.F.; Gualandro, S.F.M.; Bydlowski, S.P. Paraoxonases (PON) 1, 2, and 3 polymorphisms and PON-1 activities in patients with sickle cell disease. *Antioxidants* **2019**, *8*, 252. [CrossRef]
113. Levy, D.; Reichert, C.O.; Bydlowski, S.P. Paraoxonases activities and polymorphisms in elderly and old-age diseases: An overview. *Antioxidants* **2019**, *8*, 118. [CrossRef]
114. Blesa, J.; Trigo-Damas, I.; Quiroga-Varela, A.; Jackson-Lewis, V.R. Oxidative stress and Parkinson's disease. *Front. Neuroanat.* **2015**, *9*, 91. [CrossRef]
115. Dias, V.; Junn, E.; Mouradian, M.M. The role of oxidative stress in parkinson's disease. *J. Parkinson Dis.* **2013**, *3*, 461–491. [CrossRef]
116. Gu, F.; Chauhan, V.; Chauhan, A. Glutathione redox imbalance in brain disorders. *Curr. Opin. Clin. Nutr. Metab. Care* **2015**, *18*, 89–95. [CrossRef]
117. Johnson, W.M.; Wilson-Delfosse, A.L.; Mieyal, J.J. Dysregulation of glutathione homeostasis in neurodegenerative diseases. *Nutrients* **2012**, *4*, 1399–1440. [CrossRef]
118. Cardoso, B.R.; Hare, D.J.; Bush, A.I.; Roberts, B.R. Glutathione peroxidase 4: A new player in neurodegeneration? *Mol. Psychiatry* **2017**, *22*, 328–335. [CrossRef]
119. Wang, D.; Peng, Y.; Xie, Y.; Zhou, B.; Sun, X.; Kang, R.; Tang, D. Antiferroptotic activity of non-oxidative dopamine. *Biochem. Biophys. Res. Commun.* **2016**, *480*, 602–607. [CrossRef]
120. Ingold, I.; Berndt, C.; Schmitt, S.; Doll, S.; Poschmann, G.; Roveri, A.; Peng, X.; Freitas, F.P.; Aichler, M.; Jastroch, M.; et al. Selenium utilization by GPX4 was an evolutionary requirement to prevent hydroperoxide-induced ferroptosis. *Free Radic. Biol. Med.* **2017**, *112*, 24. [CrossRef]
121. Wu, C.; Zhao, W.; Yu, J.; Li, S.; Lin, L.; Chen, X. Induction of ferroptosis and mitochondrial dysfunction by oxidative stress in PC12 cells. *Sci. Rep.* **2018**, *8*, 574. [CrossRef]
122. Tang, Q.; Bai, L.L.; Zou, Z.; Meng, P.; Xia, Y.; Cheng, S.; Mu, S.; Zhou, J.; Wang, X.; Qin, X.; et al. Ferroptosis is newly characterized form of neuronal cell death in response to arsenite exposure. *Neurotoxicology* **2018**, *67*, 27–36. [CrossRef]
123. Yang, Y.W.; Liou, S.H.; Hsueh, Y.M.; Lyu, W.S.; Liu, C.S.; Liu, H.J.; Chung, M.C.; Hung, P.H.; Chung, C.J. Risk of Alzheimer's disease with metal concentrations in whole blood and urine: A case-control study using propensity score matching. *Toxicol. Appl. Pharmacol.* **2018**, *356*, 8–14. [CrossRef]
124. García-Chávez, E.; Segura, B.; Merchant, H.; Jiménez, I.; Del Razo, L.M. Functional and morphological effects of repeated sodium arsenite exposure on rat peripheral sensory nerves. *J. Neurol. Sci.* **2007**, *258*, 104–110. [CrossRef]
125. Muhammad, A.; Odunola, O.A.; Gbadegesin, M.A.; Sallau, A.B.; Ndidi, U.S.; Ibrahim, M.A. Inhibitory effects of sodium arsenite and acacia honey on acetylcholinesterase in rats. *Int. J. Alzheimers Dis.* **2015**, *2015*, 1–7. [CrossRef]
126. Chin-Chan, M.; Navarro-Yepes, J.; Quintanilla-Vega, B. Environmental pollutants as risk factors for neurodegenerative disorders: Alzheimer and Parkinson diseases. *Front. Cell. Neurosci.* **2015**, *9*, 1–22. [CrossRef] [PubMed]
127. Sampayo-Reyes, A.; Zakharyan, R.A. Inhibition of human glutathione S-transferase omega by tocopherol succinate. *Biomed. Pharmacother.* **2006**, *60*, 238–244. [CrossRef]
128. Patti, F.; Fiore, M.; Chisari, C.G.; D'Amico, E.; Lo Fermo, S.; Toscano, S.; Copat, C.; Ferrante, M.; Zappia, M. CSF neurotoxic metals/metalloids levels in amyotrophic lateral sclerosis patients: Comparison between bulbar and spinal onset. *Environ. Res.* **2020**, *188*, 109820. [CrossRef]
129. Bello, A.; Woskie, S.R.; Gore, R.; Sandler, D.P.; Schmidt, S.; Kamel, F. Retrospective assessment of occupational exposures for the GENEVA study of ALS among military veterans. *Ann. Work Expo. Heal.* **2017**, *61*, 299–310. [CrossRef]
130. De Benedetti, S.; Lucchini, G.; Del Bò, C.; Deon, V.; Marocchi, A.; Penco, S.; Lunetta, C.; Gianazza, E.; Bonomi, F.; Iametti, S. Blood trace metals in a sporadic amyotrophic lateral sclerosis geographical cluster. *BioMetals* **2017**, *30*, 355–365. [CrossRef] [PubMed]

131. Hou, L.; Huang, R.; Sun, F.; Zhang, L.; Wang, Q. NADPH oxidase regulates paraquat and maneb-induced dopaminergic neurodegeneration through ferroptosis. *Toxicology* **2019**, *417*, 64–73. [CrossRef]
132. Hu, C.L.; Nydes, M.; Shanley, K.L.; Pantoja, I.E.M.; Howard, T.A.; Bizzozero, O.A. Reduced expression of the ferroptosis inhibitor glutathione peroxidase-4 in multiple sclerosis and experimental autoimmune encephalomyelitis. *J. Neurochem.* **2019**, *148*, 426–439. [CrossRef]
133. Zhou, Z.D.; Tan, E.K. Iron regulatory protein (IRP)-iron responsive element (IRE) signaling pathway in human neurodegenerative diseases. *Mol. Neurodegener.* **2017**, *12*, 1–12.
134. Koukoulitsa, C.; Villalonga-Barber, C.; Csonka, R.; Alexi, X.; Leonis, G.; Dellis, D.; Hamelink, E.; Belda, O.; Steele, B.R.; Micha-Screttas, M.; et al. Biological and computational evaluation of resveratrol inhibitors against Alzheimers disease. *J. Enzym. Inhib. Med. Chem.* **2016**, *31*, 67–77. [CrossRef]
135. Zeitschrift, A. About a peculiar disease of the cerebral cortex. By Alois Alzheimer, 1907 (Translated by L. Jarvik and H. Greenson). *Alzheimer Dis. Assoc. Disord.* **1987**, *1*, 3–8.
136. Butterfield, D.A.; Lauderback, M.C. Serial Review: Causes and Consequences of Oxidative Stress in Alzheimer's Disease. *Free Radic. Biol. Med.* **2002**, *32*, 1050–1060. [CrossRef]
137. Markesbery, W.R.; Carney, J.M. Oxidative alterations in Alzheimer's disease. *Brain Pathol.* **1999**, *9*, 133–146. [CrossRef] [PubMed]
138. Querfurth, H.W.; Laferla, F.M. Mechanisms of disease Alzheimer's Disease. *N. Engl. J. Med.* **2010**, *326*, 329–344. [CrossRef] [PubMed]
139. Lane, C.A.; Hardy, J.; Schott, J.M. Alzheimer's disease. *Eur. J. Neurol.* **2018**, *25*, 59–70. [CrossRef]
140. Hallgren, B.; Sourander, P. The Non-Haemin Iron in the Cerebral Cortex in Alzheimer's Disease. *J. Neurochem.* **1960**, *5*, 307–310. [CrossRef]
141. Dedman, D.J.; Treffry, A.; Candy, J.M.; Taylor, G.A.A.; Morris, C.M.; Bloxham, C.A.; Perry, R.H.; Edwardson, J.A.; Harrison, P.M. Iron and aluminium in relation to brain ferritin in normal individuals and Alzheimer's-disease and chronic renal-dialysis patients. *Biochem. J.* **1992**, *287*, 509–514. [CrossRef]
142. Svobodová, H.; Kosnáč, D.; Balázsiová, Z.; Tanila, H.; Miettinen, P.O.; Sierra, A.; Vitovič, P.; Wagner, A.; Polák, S.; Kopáni, M. Elevated age-related cortical iron, ferritin and amyloid plaques in APPswe/PS1ΔE9 transgenic mouse model of Alzheimer's disease. *Physiol. Res.* **2019**. [CrossRef]
143. Yan, N.; Zhang, J.J. Iron Metabolism, Ferroptosis, and the Links With Alzheimer's Disease. *Front. Neurosci.* **2020**, *13*, 1443. [CrossRef]
144. Nikseresht, S.; Bush, A.I.; Ayton, S. Treating Alzheimer's disease by targeting iron. *Br. J. Pharmacol.* **2019**, *176*, 3622–3635. [CrossRef]
145. Serrano-Pozo, A.; Frosch, M.P.; Masliah, E.; Hyman, B.T. Neuropathological alterations in Alzheimer disease. *Cold Spring Harb. Perspect. Med.* **2011**, *1*, 1–23. [CrossRef]
146. Bonda, D.J.; Castellani, R.J.; Zhu, X.; Nunomura, A.; Lee, H.-G.; Perry, G.; Smith, M.A. A Novel Perspective on Tau in Alzheimers Disease. *Curr. Alzheimer Res.* **2011**, *8*, 639–642. [CrossRef] [PubMed]
147. Council, M.R.; Road, H.; Kingdom, U. N Eurodegenerative auopathies. *Genetics* **2001**, *24*, 1121–1161.
148. Pohanka, M. Alzheimer's Disease and Oxidative Stress: A Review. *Curr. Med. Chem.* **2013**, *21*, 356–364. [CrossRef]
149. Rao, S.S.; Lago, L.; De Vega, R.G.; Bray, L.; Hare, D.J.; Clases, D.; Doble, P.A.; Adlard, P.A. Characterising the spatial and temporal brain metal profile in a mouse model of tauopathy. *Metallomics* **2020**, *12*, 301–313. [CrossRef]
150. Rao, S.S.; Portbury, S.D.; Lago, L.; Bush, A.I.; Adlard, P.A. The Iron Chelator Deferiprone Improves the Phenotype in a Mouse Model of Tauopathy. *J. Alzheimers Dis.* **2020**. [CrossRef] [PubMed]
151. Walter, J. Twenty years of presenilins—Important proteins in health and disease. *Mol. Med.* **2015**, *21*, S41–S48. [CrossRef] [PubMed]
152. Rees, T.; Hammond, P.I.; Soreq, H.; Younkin, S.; Brimijoin, S. Acetylcholinesterase promotes beta-amyloid plaques in cerebral cortex. *Neurobiol. Aging* **2003**, *24*, 777–787. [CrossRef]
153. Smith, D.G.; Cappai, R.; Barnham, K.J. The redox chemistry of the Alzheimer's disease amyloid β peptide. *Biochim. Biophys. Acta Biomembr.* **2007**, *1768*, 1976–1990. [CrossRef]
154. Bortolami, M.; Pandolfi, F.; De Vita, D.; Carafa, C.; Messore, A.; Di Santo, R.; Feroci, M.; Costi, R.; Chiarotto, I.; Bagetta, D.; et al. New deferiprone derivatives as multi-functional cholinesterase inhibitors: Design, synthesis and in vitro evaluation. *Eur. J. Med. Chem.* **2020**, *198*, 112350. [CrossRef]

155. Neuner, S.M.; Tcw, J.; Goate, A.M. Neurobiology of Disease Genetic architecture of Alzheimer's disease. *Neurobiol. Dis.* **2020**, *143*, 104976. [CrossRef] [PubMed]
156. Tsatsanis, A.; Wong, B.X.; Gunn, A.P.; Ayton, S.; Bush, A.I.; Devos, D.; Duce, J.A. Amyloidogenic processing of Alzheimer's disease β-amyloid precursor protein induces cellular iron retention. *Mol. Psychiatry* **2020**. [CrossRef] [PubMed]
157. Chen, Y.T.; Chen, W.Y.; Huang, X.T.; Xu, Y.C.; Zhang, H.Y. Iron dysregulates APP processing accompanying with sAPPα cellular retention and β-secretase inhibition in rat cortical neurons. *Acta Pharmacol. Sin.* **2017**, *39*, 177–183. [CrossRef]
158. Ayton, S.; Wang, Y.; Diouf, I.; Schneider, J.A.; Brockman, J.; Morris, M.C.; Bush, A.I. Brain iron is associated with accelerated cognitive decline in people with Alzheimer pathology. *Mol. Psychiatry* **2019**, *25*, 2932–2941. [CrossRef] [PubMed]
159. Xu, Y.; Zhang, Y.; Zhang, J.H.; Han, K.; Zhang, X.; Bai, X.; You, L.H.; Yu, P.; Shi, Z.; Chang, Y.Z.; et al. Astrocyte hepcidin ameliorates neuronal loss through attenuating brain iron deposition and oxidative stress in APP/PS1 mice. *Free Radic. Biol. Med.* **2020**, *158*, 84–95. [CrossRef]
160. Nemeth, E.; Tuttle, M.S.; Powelson, J.; Vaughn, M.D.; Donovan, A.; Ward, D.M.V.; Ganz, T.; Kaplan, J. Hepcidin regulates cellular iron efflux by binding to ferroportin and inducing its internalization. *Science.* **2004**, *360*, 2090–2093. [CrossRef]
161. Hambright, W.S.; Fonseca, R.S.; Chen, L.; Na, R.; Ran, Q. Ablation of ferroptosis regulator glutathione peroxidase 4 in forebrain neurons promotes cognitive impairment and neurodegeneration. *Redox Biol.* **2017**, *12*, 8–17. [CrossRef]
162. Maher, P. Potentiation of glutathione loss and nerve cell death by the transition metals iron and copper: Implications for age-related neurodegenerative diseases. *Free Radic. Biol. Med.* **2018**, *115*, 92–104. [CrossRef]
163. Hirata, Y.; Yamada, C.; Ito, Y.; Yamamoto, S.; Nagase, H.; Oh-hashi, K.; Kiuchi, K.; Suzuki, H.; Sawada, M.; Furuta, K. Novel oxindole derivatives prevent oxidative stress-induced cell death in mouse hippocampal HT22 cells. *Neuropharmacology* **2018**, *135*, 242–252. [CrossRef]
164. Yagami, T.; Yamamoto, Y.; Koma, H. Pathophysiological Roles of Intracellular Proteases in Neuronal Development and Neurological Diseases. *Mol. Neurobiol.* **2019**, *56*, 3090–3112. [CrossRef]
165. Kostandy, B.B. The role of glutamate in neuronal ischemic injury: The role of spark in fire. *Neurol. Sci.* **2012**, *33*, 223–237. [CrossRef] [PubMed]
166. Fischer, W.; Currais, A.; Liang, Z.; Pinto, A.; Maher, P. Old age-associated phenotypic screening for Alzheimer's disease drug candidates identifies sterubin as a potent neuroprotective compound from Yerba santa. *Redox Biol.* **2019**, *21*, 101089. [CrossRef]
167. Gunesch, S.; Hoffmann, M.; Kiermeier, C.; Fischer, W.; Pinto, A.F.M.; Maurice, T.; Maher, P.; Decker, M. 7-O-Esters of taxifolin with pronounced and overadditive effects in neuroprotection, anti-neuroinflammation, and amelioration of short-term memory impairment in vivo. *Redox Biol.* **2020**, *29*, 101378. [CrossRef]
168. Cong, L.; Dong, X.; Wang, Y.; Deng, Y.; Li, B.; Dai, R. On the role of synthesized hydroxylated chalcones as dual functional amyloid-β aggregation and ferroptosis inhibitors for potential treatment of Alzheimer's disease. *Eur. J. Med. Chem.* **2019**, *166*, 11–21. [CrossRef] [PubMed]
169. Li, L.-B.; Chai, R.; Zhang, S.; Xu, S.-F.; Zhang, Y.-H.; Li, H.-L.; Fan, Y.-G.; Guo, C. Iron Exposure and the Cellular Mechanisms Linked to Neuron Degeneration in Adult Mice. *Cells* **2019**, *8*, 198. [CrossRef] [PubMed]
170. Everett, J.; Brooks, J.; Lermyte, F.; O'Connor, P.B.; Sadler, P.J.; Dobson, J.; Collingwood, J.F.; Telling, N.D. Iron stored in ferritin is chemically reduced in the presence of aggregating Aβ(1-42). *Sci. Rep.* **2020**. [CrossRef]
171. Ates, G.; Goldberg, J.; Currais, A.; Maher, P. CMS121, a fatty acid synthase inhibitor, protects against excess lipid peroxidation and inflammation and alleviates cognitive loss in a transgenic mouse model of Alzheimer's disease. *Redox Biol.* **2020**. [CrossRef]
172. Ashraf, A.; So, P.W. Spotlight on Ferroptosis: Iron-Dependent Cell Death in Alzheimer's Disease. *Front. Aging Neurosci.* **2020**, *12*, 196. [CrossRef]
173. Li, J.; Cao, F.; Yin, H.-L.; Huang, Z.-J.; Lin, Z.-T.; Mao, N.; Sun, B.; Wang, G. Ferroptosis: Past, present and future. *Cell Death Dis.* **2020**, *11*, 1–13. [CrossRef]
174. Tysnes, O.B.; Storstein, A. Epidemiology of Parkinson's disease. *J. Neural Transm.* **2017**, *124*, 901–905. [CrossRef]

175. Schneider, R.B.; Iourinets, J.; Richard, I.H. Parkinson's disease psychosis: Presentation, diagnosis and management. *Neurodegener. Dis. Manag.* **2017**, *7*, 365–376. [CrossRef] [PubMed]
176. Hayes, M.T. Parkinson's Disease and Parkinsonism. *Am. J. Med.* **2019**, *132*, 802–807. [CrossRef] [PubMed]
177. Bayulkem, K.; Lopez, G. Nonmotor fluctuations in Parkinson's disease: Clinical spectrum and classification. *J. Neurol. Sci.* **2010**, *289*, 89–92. [CrossRef] [PubMed]
178. Witjas, T.; Kaphan, E.; Azulay, J.P.; Blin, O.; Ceccaldi, M.; Pouget, J.; Poncet, M.; Ali Chérif, A. Nonmotor fluctuations in Parkinson's disease: Frequent and disabling. *Neurology* **2002**, *59*, 408–413. [CrossRef] [PubMed]
179. Chaudhuri, K.R.; Healy, D.G.; Schapira, A.H.V. Non-motor symptoms of Parkinson's disease: Diagnosis and management. *Lancet Neurol.* **2006**, *5*, 235–245. [CrossRef]
180. Dauer, W.; Przedborski, S. Parkinson's disease: Mechanisms and models. *Neuron* **2003**, *39*, 889–909. [CrossRef]
181. Lees, A.J.; Hardy, J.; Revesz, T. Parkinson's disease. *Lancet* **2009**, *373*, 2055–2066. [CrossRef]
182. Kalia, L.V.; Lang, A.E. Parkinson's disease. *Lancet* **2015**, *386*, 896–912. [CrossRef]
183. Samii, A.; Nutt, J.G.; Ransom, B.R. Parkinson's disease. *Lancet* **2004**, *363*, 1783–1793. [CrossRef]
184. Schapira, A.H.V.; Bezard, E.; Brotchie, J.; Calon, F.; Collingridge, G.L.; Ferger, B.; Hengerer, B.; Hirsch, E.; Jenner, P.; Le Novère, N.; et al. Novel pharmacological targets for the treatment of Parkinson's disease. *Nat. Rev. Drug Discov.* **2006**, *5*, 845–854. [CrossRef]
185. Dexter, D.T.; Wells, F.R.; Lee, A.J.; Agid, F.; Agid, Y.; Jenner, P.; Marsden, C.D. Increased Nigral Iron Content and Alterations in Other Metal Ions Occurring in Brain in Parkinson's Disease. *J. Neurochem.* **1989**, *52*, 1830–1836. [CrossRef] [PubMed]
186. Dexter, D.T.; Jenner, P.; Schapira, A.H.V.; Marsden, C.D. Alterations in levels of iron, ferritin, and other trace metals in neurodegenerative diseases affecting the basal ganglia. *Ann. Neurol.* **1992**, *32*, S94–S100. [CrossRef] [PubMed]
187. Gorell, J.M.; Johnson, C.C.; Rybicki, B.A.; Peterson, E.L.; Kortsha, G.X.; Brown, G.G.; Richardson, R.J. Occupational exposures to metals as risk factors for Parkinson's disease. *Neurology* **1997**, *48*, 650–658. [CrossRef]
188. Graham, J.M.; Paley, M.N.J.; Grünewald, R.A.; Hoggard, N.; Griffiths, P.D. Brain iron deposition in Parkinson's disease imaged using the PRIME magnetic resonance sequence. *Brain* **2000**, *123*, 2423–2431. [CrossRef] [PubMed]
189. Rossi, M.; Ruottinen, H.; Soimakallio, S.; Elovaara, I.; Dastidar, P. Clinical MRI for iron detection in Parkinson's disease. *Clin. Imaging* **2013**, *37*, 631–636. [CrossRef] [PubMed]
190. Devos, D.; Moreau, C.; Devedjian, J.C.; Kluza, J.; Petrault, M.; Laloux, C.; Jonneaux, A.; Ryckewaert, G.; Garçon, G.; Rouaix, N.; et al. Targeting chelatable iron as a therapeutic modality in Parkinson's disease. *Antioxid. Redox Signal.* **2014**, *21*, 195–210. [CrossRef]
191. Grolez, G.; Moreau, C.; Sablonnière, B.; Garçon, G.; Devedjian, J.C.; Meguig, S.; Gelé, P.; Delmaire, C.; Bordet, R.; Defebvre, L.; et al. Ceruloplasmin activity and iron chelation treatment of patients with Parkinson's disease. *BMC Neurol.* **2015**, *15*, 1–6. [CrossRef]
192. Martin-Bastida, A.; Ward, R.J.; Newbould, R.; Piccini, P.; Sharp, D.; Kabba, C.; Patel, M.C.; Spino, M.; Connelly, J.; Tricta, F.; et al. Brain iron chelation by deferiprone in a phase 2 randomised double-blinded placebo controlled clinical trial in Parkinson's disease. *Sci. Rep.* **2017**, *7*, 1398. [CrossRef]
193. Vassiliev, V.; Harris, Z.L.; Zatta, P. Ceruloplasmin in neurodegenerative diseases. *Brain Res. Rev.* **2005**, *49*, 633–640. [CrossRef]
194. Zhao, Z.; Bao, X.Q.; Zhang, D. Mechanisms of ferroptosis and its involvement in Parkinson's disease. *Yaoxue Xuebao* **2019**, *54*, 399–406.
195. Rhodes, S.L.; Ritz, B. Genetics of iron regulation and the possible role of iron in Parkinson's disease. *Neurobiol. Dis.* **2008**, *32*, 183–195. [CrossRef]
196. Valko, M.; Jomova, K.; Rhodes, C.J.; Kuča, K.; Musílek, K. Redox- and non-redox-metal-induced formation of free radicals and their role in human disease. *Arch. Toxicol.* **2016**, *90*, 1–37. [CrossRef] [PubMed]
197. Do Van, B.; Gouel, F.; Jonneaux, A.; Timmerman, K.; Gelé, P.; Pétrault, M.; Bastide, M.; Laloux, C.; Moreau, C.; Bordet, R.; et al. Ferroptosis, a newly characterized form of cell death in Parkinson's disease that is regulated by PKC. *Neurobiol. Dis.* **2016**, *94*, 169–178. [CrossRef] [PubMed]
198. Stefanis, L. α-Synuclein in Parkinson's disease. *Cold Spring Harb. Perspect. Med.* **2012**, *2*, 1–23. [CrossRef] [PubMed]

199. Angelova, P.R.; Choi, M.L.; Berezhnov, A.V.; Horrocks, M.H.; Hughes, C.D.; De, S.; Rodrigues, M.; Yapom, R.; Little, D.; Dolt, K.S.; et al. Alpha synuclein aggregation drives ferroptosis: An interplay of iron, calcium and lipid peroxidation. *Cell Death Differ.* **2020**, *27*, 2781–2796. [CrossRef]
200. Miotto, G.; Rossetto, M.; Di Paolo, M.L.; Orian, L.; Venerando, R.; Roveri, A.; Vučković, A.M.; Bosello Travain, V.; Zaccarin, M.; Zennaro, L.; et al. Insight into the mechanism of ferroptosis inhibition by ferrostatin-1. *Redox Biol.* **2020**, *28*, 101328. [CrossRef]
201. Guiney, S.J.; Adlard, P.A.; Bush, A.I.; Finkelstein, D.I.; Ayton, S. Ferroptosis and cell death mechanisms in Parkinson's disease. *Neurochem. Int.* **2017**, *104*, 34–48. [CrossRef]
202. Weiland, A.; Wang, Y.; Wu, W.; Lan, X.; Han, X.; Li, Q.; Wang, J. Ferroptosis and Its Role in Diverse Brain Diseases. *Mol. Neurobiol.* **2019**, *56*, 4880–4893. [CrossRef]
203. Kaidery, N.A.; Ahuja, M.; Thomas, B. Crosstalk between Nrf2 signaling and mitochondrial function in Parkinson's disease. *Mol. Cell. Neurosci.* **2019**, *101*, 103413. [CrossRef]
204. Fan, Z.; Wirth, A.-K.; Chen, D.; Wruck, C.J.; Rauh, M.; Buchfelder, M.; Savaskan, N. Nrf2-Keap1 pathway promotes cell proliferation and diminishes ferroptosis. *Oncogenesis* **2017**, *6*, e371. [CrossRef]
205. Sun, Y.; He, L.; Wang, T.; Hua, W.; Qin, H.; Wang, J.; Wang, L.; Gu, W.; Li, T.; Li, N.; et al. Activation of p62-Keap1-Nrf2 Pathway Protects 6-Hydroxydopamine-Induced Ferroptosis in Dopaminergic Cells. *Mol. Neurobiol.* **2020**, *57*, 4628–4641. [CrossRef]
206. Abdalkader, M.; Lampinen, R.; Kanninen, K.M.; Malm, T.M.; Liddell, J.R. Targeting Nrf2 to suppress ferroptosis and mitochondrial dysfunction in neurodegeneration. *Front. Neurosci.* **2018**, *12*, 1–9. [CrossRef] [PubMed]
207. Kaur, D.; Andersen, J.K. Ironing out Parkinson's disease: Is therapeutic treatment with iron chelators a real possibility? *Aging Cell* **2002**, *1*, 17–21. [CrossRef] [PubMed]
208. Shi, L.; Huang, C.; Luo, Q.; Xia, Y.; Liu, W.; Zeng, W.; Cheng, A.; Shi, R.; Zhengli, C. Clioquinol improves motor and non-motor deficits in MPTP-induced monkey model of Parkinson's disease through AKT/mTOR pathway. *Aging (Albany N. Y.)* **2020**, *12*, 9515–9533. [CrossRef] [PubMed]
209. Levy, O.A.; Malagelada, C.; Greene, L.A. Cell death pathways in Parkinson's disease: Proximal triggers, distal effectors, and final steps. *Apoptosis* **2009**, *14*, 478–500. [CrossRef]
210. Gellera, C.; Meoni, C.; Castellotti, B.; Zappacosta, B.; Girotti, F.; Taroni, F.; DiDonato, S. Errors in Huntington disease diagnostic test caused by trinucleotide deletion in the IT15 gene. *Am. J. Hum. Genet.* **1996**, *59*, 475–477. [PubMed]
211. Roos, R.A.C. Huntington's disease: A clinical review. *Orphanet J. Rare Dis.* **2010**, *5*, 40. [CrossRef] [PubMed]
212. Wexler, N.S. Huntington's disease: Advocacy driving science. *Annu. Rev. Med.* **2012**, *63*, 1–22. [CrossRef]
213. Wexler, A.; Wild, E.J.; Tabrizi, S.J. George Huntington: A legacy of inquiry, empathy and hope. *Brain* **2016**, *139*, 2326–2333. [CrossRef]
214. Nopoulos, P.C. Huntington disease: A single-gene degenerative disorder of the striatum. *Dialogues Clin. Neurosci.* **2016**, *18*, 91–98.
215. Caterino, M.; Squillaro, T.; Montesarchio, D.; Giordano, A.; Giancola, C.; Melone, M.A.B. Huntingtin protein: A new option for fixing the Huntington's disease countdown clock. *Neuropharmacology* **2018**, *135*, 126–138. [CrossRef]
216. Bates, G.P.; Dorsey, R.; Gusella, J.F.; Hayden, M.R.; Kay, C.; Leavitt, B.R.; Nance, M.; Ross, C.A.; Scahill, R.I.; Wetzel, R.; et al. Huntington disease. *Nat. Rev. Dis. Prim.* **2015**, *1*, 1–21. [CrossRef]
217. DiFiglia, M.; Sapp, E.; Chase, K.; Schwarz, C.; Meloni, A.; Young, C.; Martin, E.; Vonsattel, J.P.; Carraway, R.; Reeves, S.A.; et al. Huntingtin is a cytoplasmic protein associated with vesicles in human and rat brain neurons. *Neuron* **1995**, *14*, 1075–1081. [CrossRef]
218. Jimenez-Sanchez, M.; Licitra, F.; Underwood, B.R.; Rubinsztein, D.C. Huntington's disease: Mechanisms of pathogenesis and therapeutic strategies. *Cold Spring Harb. Perspect. Med.* **2017**, *7*, 1–22. [CrossRef]
219. Chen, J.; Marks, E.; Lai, B.; Zhang, Z.; Duce, J.A.; Lam, L.Q.; Volitakis, I.; Bush, A.I.; Hersch, S.; Fox, J.H. Iron Accumulates in Huntington's Disease Neurons: Protection by Deferoxamine. *PLoS ONE* **2013**, *8*, e77023. [CrossRef]
220. Chen, L.; Hambright, W.S.; Na, R.; Ran, Q. Ablation of the ferroptosis inhibitor glutathione peroxidase 4 in neurons results in rapid motor neuron degeneration and paralysis. *J. Biol. Chem.* **2015**, *290*, 28097–28106. [CrossRef]

221. Rosas, H.D.; Chen, Y.I.; Doros, G.; Salat, D.H.; Chen, N.K.; Kwong, K.K.; Bush, A.; Fox, J.; Hersch, S.M. Alterations in brain transition metals in Huntington disease: An evolving and intricate story. *Arch. Neurol.* **2012**, *69*, 887–893. [CrossRef]
222. Magtanong, L.; Dixon, S.J. Ferroptosis and Brain Injury. *Dev. Neurosci.* **2019**, *40*, 382–395. [CrossRef]
223. Ross, C.A.; Tabrizi, S.J. Huntington's disease: From molecular pathogenesis to clinical treatment. *Lancet Neurol.* **2011**, *10*, 83–98. [CrossRef]
224. Choo, Y.S.; Johnson, G.V.W.; MacDonald, M.; Detloff, P.J.; Lesort, M. Mutant huntingtin directly increases susceptibility of mitochondria to the calcium-induced permeability transition and cytochrome c release. *Hum. Mol. Genet.* **2004**, *13*, 1407–1420. [CrossRef] [PubMed]
225. Mi, Y.; Gao, X.; Xu, H.; Cui, Y.; Zhang, Y.; Gou, X. The Emerging Roles of Ferroptosis in Huntington's Disease. *NeuroMol. Med.* **2019**, *21*, 110–119. [CrossRef] [PubMed]
226. Wyttenbach, A.; Sauvageot, O.; Carmichael, J.; Diaz-Latoud, C.; Arrigo, A.P.; Rubinsztein, D.C. Heat shock protein 27 prevents cellular polyglutamine toxicity and suppresses the increase of reactive oxygen species caused by huntingtin. *Hum. Mol. Genet.* **2002**, *11*, 1137–1151. [CrossRef]
227. Klepac, N.; Relja, M.; Klepac, R.; Hećimović, S.; Babić, T.; Trkulja, V. Oxidative stress parameters in plasma of Huntington's disease patients, asymptomatic Huntington's disease gene carriers and healthy subjects: A cross-sectional study. *J. Neurol.* **2007**, *254*, 1676–1683. [CrossRef] [PubMed]
228. Han, C.; Liu, Y.; Dai, R.; Ismail, N.; Su, W.; Li, B. Ferroptosis and Its Potential Role in Human Diseases. *Front. Pharmacol.* **2020**, *11*, 239. [CrossRef] [PubMed]
229. Csobonyeiova, M.; Polak, S.; Danisovic, L. Recent overview of the use of iPSCs huntington's disease modeling and therapy. *Int. J. Mol. Sci.* **2020**, *21*, 2239. [CrossRef]

International Journal of
Molecular Sciences

Review

Involvement of Microglia in Neurodegenerative Diseases: Beneficial Effects of Docosahexahenoic Acid (DHA) Supplied by Food or Combined with Nanoparticles

Karine Charrière [1], Imen Ghzaiel [2], Gérard Lizard [2] and Anne Vejux [2,*]

1 Centre Hospitalier Universitaire de Besançon, Centre d'Investigation Clinique, INSERM CIC 1431, 25030 Besançon, France; karine.charriere@gmail.com
2 Team Bio-PeroxIL, "Biochemistry of the Peroxisome, Inflammation and Lipid Metabolism" (EA7270), Université de Bourgogne Franche-Comté, INSERM, UFR Sciences Vie Terre et Environnement, 21000 Dijon, France; imenghzaiel93@gmail.com (I.G.); gerard.lizard@u-bourgogne.fr (G.L.)
* Correspondence: anne.vejux@u-bourgogne.fr; Tel.: +33-3-8039-3701; Fax: +33-3-8039-6250

Abstract: Neurodegenerative diseases represent a major public health issue and require better therapeutic management. The treatments developed mainly target neuronal activity. However, an inflammatory component must be considered, and microglia may constitute an important therapeutic target. Given the difficulty in developing molecules that can cross the blood–brain barrier, the use of food-derived molecules may be an interesting therapeutic avenue. Docosahexaenoic acid (DHA), an omega-3 polyunsaturated fatty acid (22:6 omega-3), has an inhibitory action on cell death and oxidative stress induced in the microglia. It also acts on the inflammatory activity of microglia. These data obtained in vitro or on animal models are corroborated by clinical trials showing a protective effect of DHA. Whereas DHA crosses the blood–brain barrier, nutritional intake lacks specificity at both the tissue and cellular level. Nanomedicine offers new tools which favor the delivery of DHA at the cerebral level, especially in microglial cells. Because of the biological activities of DHA and the associated nanotargeting techniques, DHA represents a therapeutic molecule of interest for the treatment of neurodegenerative diseases.

Keywords: docosahexaenoic acid; microglia; neurodegenerative disease; inflammation; nanomedicine

Citation: Charrière, K.; Ghzaiel, I.; Lizard, G.; Vejux, A. Involvement of Microglia in Neurodegenerative Diseases: Beneficial Effects of Docosahexahenoic Acid (DHA) Supplied by Food or Combined with Nanoparticles. *Int. J. Mol. Sci.* **2021**, 22, 10639. https://doi.org/10.3390/ijms221910639

Academic Editor: Fabrizio Michetti

Received: 2 September 2021
Accepted: 27 September 2021
Published: 30 September 2021

1. Introduction

Neurodegenerative diseases represent a major public health issue in the world. Indeed, due to the progressive aging of the population and the lack of curative treatments, the number of people suffering from neurodegenerative diseases has increased considerably in recent years and is expected to continue to grow steadily in the years to come. These pathologies are chronic progressive diseases that affect the central nervous system, mainly the neurons that are often the target of therapies. The causes of these pathologies are to be found in genetics or risk factors such as the presence of chemical molecules in food, air, water, houses, and everyday objects. These different risk factors can contribute to oxidative stress, inflammation, and peroxisomal and mitochondrial dysfunctions, ultimately leading to neuronal death [1]. Most therapies target neurons and their function. However, there is an inflammatory component that must also be considered, and thus the involvement of microglia, in these pathologies [2]. Microglial cells are the major resident immune cells in the brain. Microglia activation is often classified into two opposite states: M1 and M2 [3]. The M1 state corresponds to a "classical activation" and is considered to be proinflammatory with a high capacity to present antigens, and high production of nitric oxide (NO) and reactive oxygen species (ROS) as well as pro-inflammatory cytokines. The M2 state includes both "alternative activation" and "acquired deactivation" and expresses the phenotypic markers arginase-1 (Arg1), CD206, interleukin (IL)-10, transforming growth factor β (TGF-β), and IL-1. This M2 state is considered to be an anti-inflammatory state [4] with the capacity to fine-tune inflammation, debris removal, promotion of angiogenesis, and tissue remodeling and repair. This separation into two opposite states does not reflect all the microglia phenotypes which will depend on the brain injury, its stage, and its location [2,5]. Nevertheless, microglial cells can be neuroprotective or neurotoxic depending on their activation status and the M1/M2 terminology remains useful to describe these two properties, while keeping in mind that it does not allow all reactive states of microglia to be described.

In physiological conditions, microglial cells work as sentinels. When they are activated by injurious stimuli, they can turn into several phenotypes, the two main ones being M1 and M2.

M1 produces pro-inflammatory cytokines that allow neuroprotection by removing pathological agents or recruiting additional cells. In this case, neuro-inflammation caused by microglia is neuroprotective. In contrast, a prolonged neuro-inflammation induces neurotoxicity and leads to neurodegeneration [6]. Briefly, M1 cells act as potent effectors that drive the inflammatory response, can have detrimental effects on neural cells, and participate in neuronal cell death if the switch to the M2 state does not occur in an appropriate time frame [7].

Therefore, modulating the activation state of microglia and their tendency toward the M2 state could be a promising therapeutic approach for central nervous system repair and regeneration.

Given the lack of effective treatment and the difficulty of developing a molecule capable of crossing the blood–brain barrier (BBB), the use of food-derived molecules has been raised as a possible therapeutic option to target the current inflammatory state or to improve the phagocytic activity of microglia. Among these molecules, docosahexaenoic acid (DHA; C22:6 omega-3), an omega-3 polyunsaturated fatty acid, has many advantages.

In this review, we will focus on the potential cytoprotective effects of DHA on oxidative stress, cell death affecting microglia, and on microglia-controlled inflammation. We will also present some human clinical trials that show the benefits of using DHA to improve the therapeutic management of patients with neurodegenerative diseases along with certain limits to the use of DHA in capsules or from food. Finally, we will present different approaches using nanoparticles that could allow for a better availability of DHA at the cerebral level.

2. DHA Biochemistry

DHA belongs to the family of fatty acids (FA) which are divided into distinct families according to the amount of carbon they contain and the presence or absence of unsaturations in their hydrocarbon chain (Figure 1) (see review [8]). Thus, we find saturated fatty acids (SFA) that have no double bond, the most common of which are palmitic acid (C16:0) and stearic acid (C18:0). Then, the presence of a single double bond in the carbon chain qualifies the fatty acid as monounsaturated (MUFA); this is the case for oleic acid (C18:1 omega-9). Finally, from two double bonds, we have polyunsaturated fatty acids (PUFA). This last family is divided into two subfamilies: the omega-6 family and the omega-3 family, which are distinguished by the position of the first double bond carried, respectively, by the sixth (n-6 or ω6) or third carbon (n-3 or ω3) from the terminal methyl end. PUFAs 6 and 3 (C20 and C22) are derived from two indispensable precursors, respectively: linoleic acid and α-linolenic acid (consisting of 18 carbon atoms and 3 double bonds).

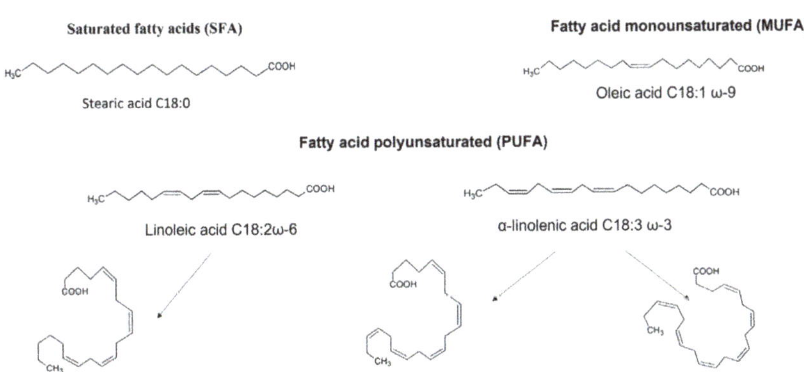

Figure 1. The different families of fatty acids and their main members. Fatty acids are classified according to the saturation or unsaturation of the carbon chain. There are saturated fatty acids (SFA), monounsaturated fatty acids (MUFA), and polyunsaturated fatty acids (PUFA). Among the latter, we distinguish between omega-3 and omega-6. Docosahexaenoic acid belongs to the omega-3 family.

DHA is a major component of PUFAs in all cell membranes. It is necessary for proper development of the retina and the central nervous system (CNS) [9]. DHA deficiency disrupts the composition membrane lipid and thus the functions of astrocytes at the CNS level [10]. DHA is synthesized in the organism from essential precursors (α-linolenic acid C18:3 ω-3) provided by food [11]. The formation of DHA involves a succession of desaturations and elongations, which take place mainly in the liver, muscles, or even adipose tissue (Figure 2) [12]. DHA is synthesized from α-linolenic acid by two steps: conversion of C18:3 ω-3 to C20:5 ω-3 (eicosapentaenoic acid) then to C22:5 ω-3 (docosapentaenoic acid) and finally to C24:5 ω-3 (tetracosapentaenoic acid) in the endoplasmic reticulum. The second step consists of a single-ring peroxisomal β-oxidation of C24:6 ω-3 to C22:6 ω-3 (DHA) [13]. This step of peroxisomal β-oxidation requires the intervention of straight-chain acyl-CoA oxidase (SCOX), D-bifunctional protein (DBP), 3-ketoacyl-CoA thiolase, and sterol carrier protein (SCPx).

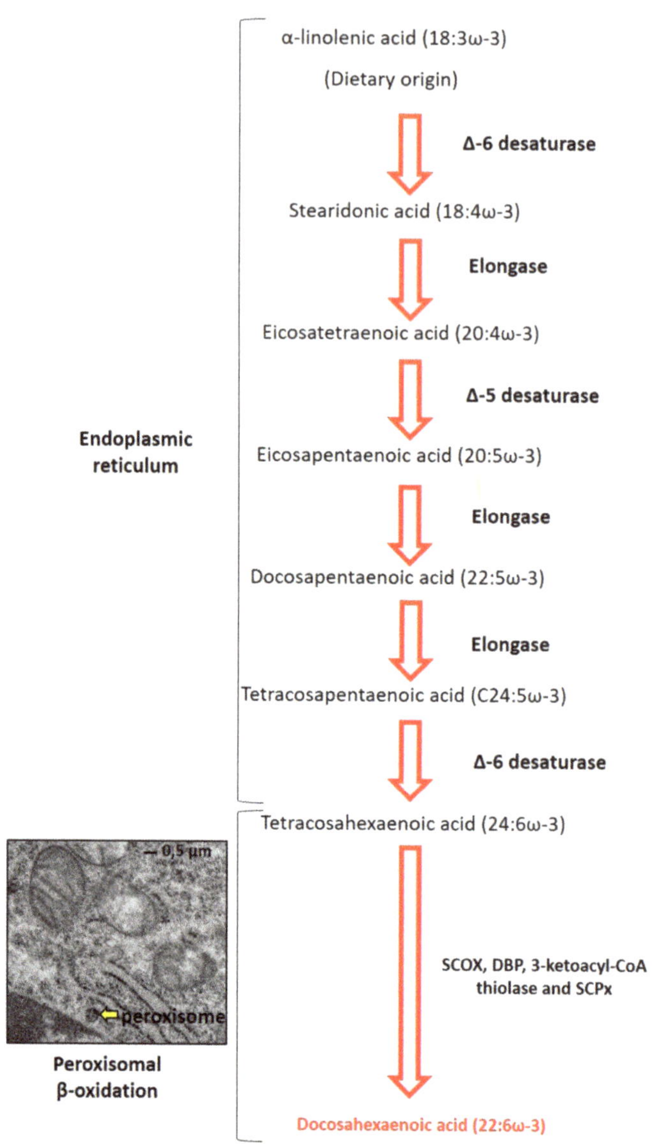

Figure 2. Biochemical pathway of DHA biosynthesis. In the endoplasmic reticulum, α-linolenic acid is converted to tetracosahexaenoic acid by the action of elongases and desaturases. Then, in the peroxisome (yellow arrow: electron microscopy of C2C12 myoblasts; provided by Imen Ghzaiel, PhD student), β-oxidation occurs and leads to the synthesis of DHA.

3. Impact of DHA on Cell Death and Oxidative Stress Induced in Microglial: In Vitro Studies

In the pathophysiology of neurodegenerative diseases, a process of cell death is described that is most often accompanied by oxidative stress. This cell death process mainly affects neurons, but other cells of the nervous system, such as oligodendrocytes or microglia, can also be affected. This raises the question of whether DHA can prevent microglia cell death. Indeed, few studies have targeted the action of DHA on microglial cell death.

Among the studies carried out, it was shown that DHA could protect microglial cells from death induced by oxidized derivatives of cholesterol and oxysterols, identified in the cerebrospinal fluid, plasma, and tissue of patients with different neurodegenerative diseases (multiple sclerosis (MS), Alzheimer's disease (AD), X-linked adrenoleukodystrophy). In this study, 7-ketocholesterol induces cell death characterized by an apoptotic process associated with oxidative stress but also with autophagy in a microglial murine cell line, the BV-2 line. This process is called oxiapoptophagy (OXIdative stress + APOPTOsis + autoPHAGY) [14]. DHA used at 12 μM is capable of inhibiting the oxiapoptophagy process [15]. Another study has shown that at too high concentrations, DHA can, on the contrary, induce cell death. In a BV-2 cell model, DHA used at 200 μM is capable of inducing pyroptosis [16]. This pyroptosis process can be inhibited by 12-lipoxygenase (12-LOX, Alox12e) inhibitors [17].

Concerning oxidative stress, a team showed that DHA was able to inhibit oxidative stress and increase antioxidant defenses in microglial cells. The use of oligomeric amyloid-β peptide (oAβ) induces oxidative stress via the production of ROS in primary mouse microglia and in immortalized mouse microglia: the BV-2 line. This oxidative stress can be inhibited by the use of DHA (10 μM), which can also positively regulate antioxidant pathways by involving the nuclear factor (erythroid-derived 2)-like 2/heme oxygenase-1 (Nrf2/HO-1) [18]. In another model where BV-2 cells were stimulated with lipopolysaccharide (LPS), it was shown that DHA was indeed able to decrease oxidative stress production and increase antioxidant responses via Nrf2/HO-1 [19]. In this model, the use of quercetin coupled with DHA increases the beneficial effects of these two molecules and reduces the concentrations of DHA (IC50 in the range of 60-80 μM used alone) necessary to observe a positive effect. The most effective combination seems to be the following: quercetin (2.5 μM) in combination with DHA (10 μM) [20].

Few data are available on the effect of DHA on cell death of microglia or oxidative stress; the main investigations concern the relationship between microglia and inflammation, which will be developed in the following paragraph. The different results presented are summarized in Table 1.

Table 1. In vitro data on the effect of DHA on cell death, oxidative stress, and inflammation.

Experimentation Type	Target	DHA Forms	Concentration or Dose	Effects	References
BV-2 cells treated with 7-ketocholesterol	Cell death oxidative stress	Chemical form	12 μM	Protection against cell death and oxidative stress	[15]
BV-2 cells	Cell death		200 μM	Induction of cell death (pyroptosis)	[16]
BV-2 cells treated with oligomeric amyloid-β peptide	Oxidative stress		10 μM	Reduce oxidative stress (involving Nrf2/HO-1)	[18]
BV-2 cells treated with LPS			1.25–10 μM	Enhance Nrf-2/HO-1 signaling	[19]
			10 μM DHA + quercetin (2.5 μM)		[20]
BV-2 cells treated with interferon-γ	Inflammation		30 μM	Reduce expression of proteins implicated in inflammation	[21]
BV-2 cells and MG6 cells stimulated with LPS			100 μM DHA and 100 μM EPA	Inhibition of inflammation by involving SIRT1	[22]
BV-2 cells stimulated with LPS			30 μM	Decrease cytokine expression	[17,23]
Primary cultures of mice microglial cells stimulated with LPS			20 μM to 80 μM	Reduce NO and TNF-α release, modulation of phenotypic polarization of microglia	[24]

Table 1. *Cont.*

Experimentation Type	Target	DHA Forms	Concentration or Dose	Effects	References
Human CHME3 microglial cells exposed to Amyloid-β42	Inflammation	Chemical form	0.1 to 1 μM	Decrease pro-inflammatory cytokine production, induction of a shift in phenotype away from pro-inflammatory M1 activation	[25]
Rat glial primary cell cultures, LPS/ IFN-γ stimulation			100 μM	Regulation of the pro-inflammatory response	[26]
EOC20 microglia cells treated by Polyinosinic-Polycytidylic acid or 10 μg/mL Imiquimod			50 μM	Reduction of the production of cytokines TNF-α and IL-6	[27]
Primary neuron/glia rat primary cultures infected with Japanese Encephalitis virus			25 or 50 μM	Increase of neurotoxin cytokines production	[28]
BV-2 cells activated with LPS and IFN-γ		Triglyceride forms	20 μM	Reduce IL-6 and TNF-α production	[29]
BV-2 cells treated with LPS		N-docosahexaenoyl dopamine (DHDA)	2 μM	Decrease IL-6 and CCL-20 production	[30]
Primary cultures of rat microglia and BV-2 cells treated with LPS		Synaptamide	1–100 nM	Suppression of LPS-induced TNF-α and iNOS mRNA expression	[31,32]

4. Impact of DHA on Inflammation Induced by Microglia in Neurodegenerative Diseases: In Vitro and Animal Studies

Immune activation of the central nervous system is present in neurodegenerative pathologies and involves microglia, major players in neuroinflammation [33]. This inflammatory response, necessary to respond to aggression, can lead to the production of neurotoxic factors when prolonged. It is also characteristic of many neurodegenerative diseases, such as AD, Parkinson's disease (PD), amyotrophic lateral sclerosis (ALS), and MS [3]. For several years, the anti-inflammatory capacity of DHA or derivatives forms of DHA has been demonstrated.

In the first part, we present the results obtained in vitro (summarized in Table 1) and in the second part, we present the data obtained in vivo with neurodegenerative models (summarized in Table 2).

Table 2. In vivo data on the effect of DHA.

Experimentation Type	Target	DHA Forms	Concentration or Dose	Effects	References
Injection of LPS intraperitoneally (i.p.) or in brain of C57Bl/6J mice	Inflammation	Synaptamide, endogenous metabolite derived from docosahexaenoic acid	5 mg/kg, i.p	Decrease of TNF-α, IL-1β, IL-6, iNOS, and CCL2 mRNA involving orphan adhesion G-protein-coupled receptor 110 (GPR110)	[32]

Table 2. *Cont.*

Experimentation Type	Target	DHA Forms	Concentration or Dose	Effects	References
C57Bl/6J mice intraperitoneally injected with LPS	Inflammation	Fish hydrolysate supplement	143 µg in 150 µL of fish hydrolysate supplement/day	Reduce expression of TLR4, cytokines (IL-6, TNF-α, IL-1β,), CCL2, and IκB	[34]
		Chemical form	10 mg/day	Reduce IL-6 expression	[34]
C57Bl6/J mice injected with LPS (i.p.).		1-palmitoyl,2-docosahexaenoyl-PC (PC-DHA)	4.33 µg/g of mouse	Decrease IL-6 production	[23]
Piglets		Herring oil	32.30% W/W total fatty acids	No attenuation of the LPS induced inflammation	[35]
C57Bl/6 male mice with intracerebroventricular LPS injections		Fish oil	1.4% of total fatty acids	No change in the expression of pro-inflammatory genes	[36]
Triple-transgenic mouse model of AD		Neuroprotectin D1 (organic synthesis)	50 nM	Downregulation of Aβ42-induced expression of COX-2, TNF-α and B-94	[37]
Male C57BL/6 mice with intracerebroventricular amyloid-β infusion (AD model)		Fish oil	1.5% of total fatty acid	Decrease some elements of the inflammatory response	[38]
Male albino Swiss mice, administration of $AlCl_3$ (20 mg/kg) intragastrically (i.g.) then an intraperitoneal injection with D-gal (120 mg/kg) (AD model)		Chemical form	200 mg/kg	Downregulation of TNF-α expression	[39]
TgCRND8 mice (AD model)		Whole-food diet contained skinless freeze-dried Atlantic salmon and a proprietary mixture of powdered, freeze-dried vegetables and fruits	0.246% of DHA (wt/wt) in a whole-food diet	Increase TNF-α expression	[40]
Male Tg2576 (APPswe) transgenic mice (AD model)		Chemical form	50 mg/kg body weight	Microglial activation	[41]
Male C57/BL6 mice fed for 5 weeks with a diet containing 0.2% cuprizone (MS model)		Regular diet supplemented with n-3 PUFAs	DHA + EPA, 15 g/kg	Suppression of the increase of M1-associated genes and increase of the expression of M2-associated genes	[24]

Table 2. *Cont.*

Experimentation Type	Target	DHA Forms	Concentration or Dose	Effects	References
Female C57BL/6J mice immunization with >95% pure synthetic MOG35-55 peptide (MS model)	Development of the pathology	TG-DHA obtained by enzymatic synthesis	250 mg/kg/day	Improve clinical score	[29]
C57BL/6J female/EAE model (MS model)		Phospholipid-DHA 0.3% or 1% and triacylglycerol-DHA	0.3 or 1% either 0.48 or 1.6 mg DHA/g body weight/day respectively	Reduce EAE onset and severity	[42]

4.1. *In Vitro Results*

Using the BV-2 microglial cell line, Lu et al. have shown that DHA (30 µM) reduced expression of inducible nitric oxide synthase (iNOS), cyclooxygenase-2 (COX-2), and tumor necrosis factor α (TNF-α) induced by interferon-γ (INF-γ), and antagonized IFN-γ induced NO production [21]. Inoue et al. investigated the implication of the sirtuin (SIRT) signaling in the anti-inflammatory response mediated by microglia, with BV-2 cells and with the MG6 line. They also found that DHA inhibited production of TNF-α and IL-6 induced by stimulation in two cellular models (BV-2 cells and MG6 microglia), but had no effect on IL-10 production induced by LPS. Results obtained with the MG6 microglia cells and a treatment with DHA (100 µM) + eicopentaenoic acid (EPA, 100 µM) suggest that the anti-inflammatory properties of DHA and/or EPA could be due to a SIRT1-mediated NF-κB (nuclear factor-kappa B) p65 deacetylation, through a positive feedback regulation of SIRT1 gene expression [22]. Others studies have shown that DHA (30 µM) decreased IL-1β [17,23] and IL-6 [23] expression in BV-2 cells stimulated with LPS. In primary cultures of mice microglial cells, treatment by DHA (20 µM to 80 µM) prior to LPS induction significantly attenuated LPS-induced NO and TNF-α release in a dose dependent manner. The inhibitory effect of DHA (20 µM) on TNF-α and NO release was also observed when cells were treated with myelin + IFNγ. In the same study, the authors observed that DHA probably modulates phenotypic polarization of microglia, with upregulation of M2-associated genes (including chemokine ligands (*CCL), CCL2, CCL17, Arg1*, and *IL-5*) and downregulation of M1-associated genes (including *IL-6, CCL5, TNF-α*, and *IL-1α*) [24]. With the human CHME3 microglial cells treated by DHA (0.1 to 1 µM) and exposed to amyloid-β42 (Aβ42), Hjorth et al. have shown a decrease in the levels of TNF-α and cluster of differentiation (CD) CD40 and CD86, as well as an increase in CD206 [25].

Furthermore, in rat glial primary cell cultures, DHA (100 µM) seems to have an active role in the regulation of the pro-inflammatory response. Indeed, pre-incubation of rat glial primary cell cultures with DHA before LPS/IFN-γ stimulation led to a decrease in the DNA binding activity of the activating protein-1 (AP1) and phosphorylation of c-Jun N-terminal kinase (JNK) and c-Jun. This pre-incubation also led to an increase of the expression of Nrf2 and HO-1. Using DHA before IFN-γ stimulation counteracted the elevation of the pro-inflammatory cytokines TNF-α, IL-1β, IL-6, CCL2, and C-C chemokine receptor type 2 (CCR2) [26]. With macrophages, Cai et al. have demonstrated that 24 h of DHA (20 µM) treatment increased the expression of arginase-1 and TGF-β and suppressed production of CCL2, C-X-C motif chemokine ligand 10 (CXCL10), IL-1α, and TNF-α in primary macrophage cultures [43]. Reduction of the production of cytokines TNF-α and IL-6 could be induced through toll-like receptor-3 (TLR-3) and TLR-4 activation in EOC20 microglia cells treated by polyinosinic–polycytidylic acid (synthetic double-stranded RNA consisting of one strand of poly(inosinic acid) and one strand of poly(cytidyl acid) paired by wobble pairing, structurally similar to the double-stranded RNA of certain viruses, triggering an immune response) or 10 µg/mL of imiquimod (immune response modifier) [27].

Other forms of DHA have been used to decrease LPS-induced inflammation in BV-2 cells. Triglycerides forms of DHA (20 μM), or endogenous derivatives, have the capability to significantly reduce the production of IL-6 and TNF-α [29]. N-docosahexaenoyl dopamine (DHDA, 2 μM) decreases production of IL-6 and CCL-20 (macrophage-inflammatory protein-3α). Authors have also demonstrated that the level of prostaglandine E2 is reduced by using DHDA [30]. Synaptamide, an endogenous metabolite of DHA, leads to similar effects on inflammation. Using primary cultures of rat microglia and BV-2 cells, Park et al. found that synaptamide suppressed LPS-induced TNF-α and iNOS mRNA expression in a dose dependent manner. Furthermore, synaptamide decreased expression of IL-1β, IL-6, and CCL2. The authors suggest that the anti-inflammatory effects of synaptamide could be due to its fixation on the GPR110 receptor, as the synaptamide effects were suppressed by blocking synaptamide binding to it. This interaction could lead to an upregulation of cyclic adenosine 3′,5′-monophosphate/protein kinase A (cAMP/PKA) signaling by inhibiting NF-κB p65 nuclear translocation [31,32]. Pro-inflammatory effects have also been demonstrated with resolvins (RvD) that are metabolites of DHA. RvD2 could counteract the mRNA pro-inflammatory upregulation induced by LPS (CD11b, ionized calcium binding adaptor molecule 1 (Iba-1), TNF-α, NF-κB p65, iNOS, IL-1, IL-18, IL-6, the nuclear factor of kappa light polypeptide gene enhancer in the B-cells inhibitor, alpha (IkBα), the the inhibitor of the nuclear factor kappa-B kinase subunit β (IKKβ), and IL-1β) and decrease the ROS production [44].

In their recent study, Chang et al. showed that the neuroprotective and anti-inflammatory properties of DHA could attenuate effects of Japanese encephalitis virus (JEV). This virus, when it invades the central nervous system, causes a robust inflammatory response that leads to neuronal cell death. When infected with the JEV, primary neuron/glia rat primary cultures (i.e., neurons, astrocytes, and microglial cells), the authors measured an increase of neurotoxin cytokines production (NO, TNF-α, IL-1 β, prostaglandine E2 (PGE2), and ROS) that is counteracted by a DHA treatment (25 or 50 μM, 12 h) [28].

Taken together, these studies demonstrate the ability of DHA or its derivatives to limit the inflammatory effects, to be neuroprotective, and even to promote anti-viral effects in different types of cell cultures and in different models of inflammation.

Some data obtained in vitro are also observed in in vivo models.

4.2. In Vivo Models

4.2.1. Neuroinflammation Mediated by LPS

A neuroinflammation model could be obtained by injection of LPS intraperitoneally (i.p. injection) or directly in the brain of C57BL/6J mice; i.p. injection of LPS works to increase the levels of TNF-α, IL-1β, IL-6, iNOS, and CCL2 in the brain. Expression of these mRNA and the number of Iba-1 positive cells significantly decrease when synaptamide is administered (5 mg/kg). These effects are not observed when synaptamide is injected in G-protein-coupled receptor 110 (GPR110) knock-out mice after LPS induction, suggesting that anti-inflammatory effects in vivo depend on GPR110 [32]. In another study, fish hydrolysate supplementation, containing low amounts of DHA (143 μg in 150 μL of fish hydrolysate supplement/day), reduced expression of TLR4, cytokines (IL-6, TNF-α, IL-1β,), CCL2, and the inhibitor of the nuclear factor κB (IκB) in mice with LPS induction. DHA alone reduced only IL-6 expression. Furthermore, the authors observed lower expressions of CD86, CD68, and CD11b M1-markers in DHA groups compared to LPS groups fed with fish hydrolysate, but no effect on CD206, CD36, and Arg1-M2 markers were observed in the hippocampus. The authors suggest that the DHA effect is potentialized in fish oil by other low molecular weight peptides [34]. As well as in in vitro experiment, RvD2 could inhibit microglia activation in PD model LPS-treated animals (RvD2 per 25, 50, and 100 ng/kg, 27 days post LPS-treatment). The activation of microglia is significantly lower after treating animals with RvD2 (25, 50, and 100 ng/kg, 27 days post LPS-treatment) than in the vehicle group (more ramified microglia and less CD11b content of the substantia nigra pars compacta in the treated group). Furthermore, IL-18, IL-6, NO, TNF-α, and IL-1β production induced by LPS were significantly inhibited by RvD2 in a dose dependent manner (2,5 μM to 20 μM), in vivo (serum) and in vitro, probably through inhibition of the TLR4/NF-kB pathway [44]. In their study, Fourrier et al. showed that 1-palmitoyl,2-docosahexaenoyl-PC (PC-DHA, 4.33 μg/g of mouse) attenuates the effect of LPS in mice brains when injected 24 h prior to LPS induction. Particularly, IL-6 production induced by LPS significantly decreased in the hippocampus. The in vitro results on BV-2 microglial cells suggest that these effects could be mediated by GP130 receptors [23]. Some authors did not observe significant effects of DHA dietary supplementation in vivo, even with a DHA increase in the brain. In [35], even though supplementing piglets with herring oil (DHA: 32.30% W/W total fatty acid), starting at postnatal day 2, led to a concentration increase of 27% at 14 postnatal days compared to piglets without supplementation, no attenuation of the LPS induced inflammation was observed. In a different model of LPS-induced inflammation (intracerebroventricular LPS injections in C57Bl/6 male mice), similar results were obtained. Indeed, DHA (fish oil, 1.4% of total fatty acids) increased more than 90% in fat-1 mice compared to their wildtype littermates, and a similar increase was observed in the brains of fish oil-fed mice compared to wildtype animals fed with a safflower diet. Despite this increase, no change in the expression of pro-inflammatory genes was found [36]. These contradictory results could be explained by the different modalities of LPS injection, the different animal models used, and the variations in analysis time.

4.2.2. Alzheimer Disease

As a model for Alzheimer disease, different transgenic kinds of mice have been studied to observe effects of DHA or its derivatives. In a triple-transgenic mouse model of Alzheimer disease, neuroprotection D1 (NPD1) (50 nM) downregulates Aβ42-induced expression of COX-2, TNF-α, and B-94, a TNF-α-inducible pro-inflammatory element [37]. Increasing brain DHA level after intracerebroventricular amyloid-β infusion led to a decrease in Iba-1 microglial cells counted in fat-1 mice and lower degenerative neurons in mice supplemented with fish oil containing 1.5% DHA (among other fatty acids). Authors also observed significantly higher branching complexity (CA1, CA3, and dentate gyrus) in fat-1 and wild type mice fed with fish oil than in wild type mice fed with safflower oil. These results suggest that dietary supplement in fatty acids, including DHA decreased microglial responses to amyloid-β infusion [38]. In another model of AD mice, induced by co-administration of D-galactose and aluminium chloride, supplementation with DHA (200 mg/kg) + EGb761 (*Ginkgo biloba* standardized extract) enhanced behavioral recovery and protein phosphatase 2A expression, a major protein phosphatase of tau in the brain, while it downregulated TNF-α expression (both in the hippocampus, CA3) [39]. Sharman et al. tried a different combination of supplements instead of using DHA, ALA (α-lipoic acid), epigallocatechin-3-gallate, or curcumin alone. Reductions in amyloid plaque load, microglial activation, Aβ40/Aβ42 levels, and memory deficits in male Tg2576 mice (a familial AD model) have been observed by using a combination of EGCG, DHA (50 mg/kg body weight), and ALA [41].

Surprisingly, TgCRND8 mice (transgenic mice model for AD) supplemented with 0.246% of DHA (wt/wt) in a whole-food diet, exhibited higher TNF-α expression compared with control groups, corresponding with observed behavioral impairment [40].

4.2.3. Multiple Sclerosis

Multiple sclerosis (MS) is an immune pathology leading to demyelination and dramatic alteration of central and peripheral nervous systems.

The cuprizone mouse presents a massive demyelination and is a validate model of MS. When cuprizone mice were fed with DHA + EPA (15 g/kg for 5 weeks), myelin integrity was improved, and behavioral deficits were reduced (better scores on the Morris water maze test and with the rotarod test). The authors measured iNOS and CD16 expression, as well as CD206, YM1/2, and Arg1. They found that supplementation with n-3 PUFA suppressed the increase of M1-associated genes but increased the expression of M2-related genes [24].

Experimental autoimmune encephalomyelitis (EAE) is another commonly used model for MS. Mancera et al. have shown that feeding EAE mice with DHA (250 mg/kg/day, 15 days before EAE induction and 41 days after EAE induction) with the triglyceride form of the omega-3 polyunsaturated fatty acid docosahexaenoic acid (TG-DHA) significantly improved the clinical score in a dose and time dependent manner, along with weight profile [29]. A recent study concluded that EAE onset and severity were reduced when mice were fed with a DHA diet (phospholipid-DHA, 0.3% or 1%, and triacylglycerol-DHA, 0.3%), compared to mice fed with low α-linolenic acid [42].

4.3. Some Other Neuropathologies

Beneficial effects of DHA or derivatives have been observed in other models of neuroinflammation. Even if these pathologies are not neurodegenerative diseases, they are of interest because of their neuroinflammation component accompanied by microglial activation and pro-inflammatory factor release.

For example, synaptamide could be used to treat chronic neuropathic pain. In vitro, the addition of synaptamide to the SIM-A9 murine microglia cell line prevented LPS-induced NO overproduction and ROS production. In vivo, rats with sciatic nerve chronic constriction injury (CCI) treated with synaptamide showed lower concentrations of hippocampal Iba-1, CD86, IL-6, and IL-1β than the CCI group without synaptamide treatment. Furthermore, more doublecortin-positive neurons and proliferating cell nuclear-positive cells have been counted in the dentate gyrus subgranular zone in the CCI synaptamide treated rats compared to the CCI rats. Behavioral improvements were also observed in the synaptamide-treated groups [45].

Others have shown anti-inflammatory effects of NPD-1 in old mice in a model of postoperative delirium, with reduction of IL-6, TNF-α, glial fibrillary acidic protein (GFAP), and Iba-1 compared with controls. IL-10 increase was also observed [46].

Taken together, all of these studies suggest strong benefits of DHA mediated by microglial cells on neuroinflammation in neurodegenerative disorders. However, the mechanisms of action of the anti-inflammatory in vivo are still not elucidated.

5. DHA, Clinical Trials, and Neurodegenerative Diseases

Various results of clinical trials have been published concerning the use of DHA in neurodegenerative diseases (summarized in Table 3). A majority of clinical trials focus on AD. In a first clinical trial on this pathology, a food supplementation composed of xanthophyll carotenoids and omega 3 fatty acids was tested [47]. Two conditions were tested: the first condition was lutein/meso-zeaxanthin/zeaxanthin at 10:10:2 mg/day and the second condition was the formulation used in the first condition plus 1 g/day of fish oil containing 430 mg DHA and 90 mg EPA. It turned out that the formulation containing DHA was the most effective in slowing down AD, showing that the consumption of xanthophyll carotenoids combined with DHA (fish oil) has a better protective effect than xanthophyll carotenoids used alone [47]. Disease progression is reduced with this formulation with improvement in memory, sight, and mood. An OmegAD clinical trial (NCT00211159), enrolling 204 participants, studied the effects of DHA-rich dietary supplementation on cognitive impairment in patients with AD. Preliminary results showed that DHA (capsule EPAX 1050TG; Pronova Biocare A/S, one capsule: 430 mg DHA and 150 mg EPA) supplementation for 6 months induces DNA hypomethylation in blood cells [48].These results provide a new possible mechanism of action for these compounds: they could modulate gene expression by hypomethylation. The authors postulate that it could then be interesting to treat AD with hypomethylating agents [48]. Another work from the OmegAD clinical trial (NCT00211159) studied the plasma levels of fatty acids following DHA intake. It was shown that the higher the plasma levels of omega-3 fatty acids, the better the cognitive functioning, regardless of gender [49]. However, body weight is important and DHA doses should be adjusted to it [49]. Another part of the clinical trial was to study immune function. Since it is difficult to work on microglia directly, the authors used peripheral blood mononuclear cells (PBMCs), which can infiltrate the brains of Alzheimer's patients like T lymphocytes and monocytes and participate in the development of inflammation. PBMCs were recovered before and after supplementation with DHA and EPA and then treated with the Aβ40 peptide. DHA/EPA supplementation prevented the reduction of specialized proresolving mediators (SPM, lipoxin A4, and RvD1 released from PBMCs) [50]. Furthermore, inflammation resolution is disrupted in patients with AD; DHA/EPA supplementation (EPAX1050TG; Pronova Biocare A/S, Lysaker, Norway) could improve it [50]. As the authors point out, it remains to be determined whether the same effects can be observed in microglia and whether the use of SPM or their precursors could be effective in the treatment of AD [50]. Another component of the study was to evaluate the effects of oral dietary omega-3 supplementation on inflammatory biomarkers and oxidative stress. Patients were supplemented for 6 months with a DHA/EPA complex (four capsules of EPAX 1050TG, i.e., 1.7 g of DHA and 0.6 g of EPA); urine samples were collected before and after supplementation [51]. In these samples, the levels of major F2-isoprostane, 8-iso-prostaglandinF2α (biomarker oxidative stress), and 15-keto-dihydro-prostaglandin F2α (biomarker of inflammation) were measured [51]. The results obtained indicate that DHA/EPA supplementation does not have a well-defined effect on oxidative stress as measured but may have a possible role in immunoregulation. Since AD affects the brain, it was interesting to know if fatty acids are able to cross the blood–brain barrier. Therefore, fatty acid profiling was performed in cerebrospinal fluid (CSF) to assess whether supplementation was able to alter this profile [52]. Patients received 6 months of DHA supplementation (four capsules of EPAX 1050TG, i.e., 1.7 g of DHA and 0.6 g of EPA). After 6 months, changes in the fatty acid profile were observed, with a significant increase in eicosapentaenoic acid (EPA), DHA, and total n-3 FA levels in CSF [52]. A correlation was also made with the markers of AD and, the more DHA levels increased in the CSF, the more changes there were in the biomarkers of the pathology (tau, phosphorylation of the tau protein, IL-1 receptor). Their results also showed that supplementation failed to stop disease progression and that DHA supplementation would likely need to be taken early to see an effect on disease progression [52]. In AD, deposits of the Aβ protein are present in the brain. Transthyretin (TTR) can bind to amyloid β and thus reduce its presence. Patients received DHA/EPA supplementation (four capsules of EPAX 1050TG, i.e., 1.7 g of DHA and 0.6 g of EPA) for 6 months; it was observed that this treatment could increase plasma levels of TTR, which could influence Aβ peptide deposits in the brain, results that need to be confirmed by further experiments [53]. DHA/EPA supplementation was also evaluated on gene expression in peripheral blood mononuclear cells [54]. Patients received a DHA/EPA complex (four capsules of EPAX 1050TG, i.e., 1.7 g of DHA and 0.6 g of EPA) for 6 months and the expression of 8000 genes was studied. Modulations of the expression of several genes (decrease (10) or increase (9)) were measured. The upregulated genes are *MS4A3* (role in signal transduction), *NAIP* (apoptosis inhibitory protein), *DRG1* (stress and hormone responses, cell growth, differentiation), *CD36* (cell adhesion, cell migration), *HSD17B11* (regulation of inflammation, modulation of intracellular glucocorticoid levels), *RAB27A* (signal transduction), *CASP4* (inflammatory caspase), *SUPT4H1* (RNA synthesis), and *UBE2V1* (ubiquitination). The negatively regulated genes are *RHOB* (inflammation), *VCP* (vesicle transport, fusion, and ubiquitin-dependent protein degradation), *LOC3999491*, *ZNF24* (transcription factor), *SORL1* (regulation of processing of amyloid precursor protein), *MAN2A1* (inflammation regulation), *PARP1* (differentiation, proliferation, tumor transformation, DNA damage reparation), *SSRP1* (action on transcription), *ARIH1* (ubiquitination process), and *ANAPC5* (cell cycle progression). Many of these genes are involved in inflammation regulation and neurodegeneration, and in ubiquitination processes [54]. The impact of DHA (four capsules of EPAX 1050TG, i.e., 1.7 g of DHA and 0.6 g of EPA) on inflammation was confirmed by another study performed in this clinical trial, which showed that DHA decreased the release of PGF2α from LPS-stimulated PBMCs and that it could be hypothesized that DHA could act via anti-inflammatory and neuroprotective lipid mediators on the resolution phase of inflammation [55]. Using the same protocol

(PBMCs treated with LPS following DHA/EPA supplementation, four capsules of EPAX 1050TG, i.e., 1.7 g of DHA and 0.6 g of EPA), the authors also showed that the increase in plasma DHA concentration was correlated with a reduction in the release of IL-1β, IL-6, and the granulocyte colony-stimulating factor of PBMCs [56].

Table 3. Clinical trials involving DHA.

Experimentation Type	DHA Forms	Concentration or Dose	Effects	References
AD Patients	Fish oil	1 g/day of fish oil (30 mg DHA, 90 mg EPA)	Slowing down AD	[47]
AD Patients	Capsule EPAX 1050TG	four capsules (One capsule: 430 mg DHA and 150 mg EPA)	Induction of DNA hypomethylation in blood cell, can be used as treatment in AD	[48]
AD patients		2.3 g of omega-3 fatty acid	Positive correlation between plasma levels of omega-3 fatty acids and cognitive functions	[49]
Peripheral blood mononuclear cells treated with the Aβ40 peptide	Capsule EPAX 1050TG		Prevention of the reduction of specialized proresolving mediators (lipoxin A4 and resolvin D1) released from PBMCs	[50]
Moderate AD patients	Capsule EPAX 1050TG	four capsules (One capsule: 430 mg DHA and 150 mg EPA	No clear effect on oxidative stress but potential role in immunoregulation	[51]
AD patients	Capsule EPAX 1050TG	four capsules (One capsule: 430 mg DHA and 150 mg EPA	Increase in eicosapentaenoic acid (EPA), DHA, and total n-3 FA levels in cerebrospinal fluid	[52]
			Increase plasma levels of transthyretin which could influence Aβ peptide deposits in the brain	[53]
Peripheral blood mononuclear cells of AD patients			Regulation of genes involved in inflammation regulation, neurodegeneration, and in ubiquitination processes	[54]
LPS-stimulated peripheral blood mononuclear cells			Anti-inflammatory effects, reduction in the release of IL-1β, IL-6, and granulocyte colony-stimulating factor	[55,56]
Cognitive impairment: no dementia and AD patients	Capsules	625 mg DHA and 600 mg EPA	No beneficial effect on cognition and mood	[57]
Mild or moderate AD patients	Algal origin	2 g of capsule containing 15% to 55% DHA	No slowdown in cognitive decline	[58]
Mild or moderate AD patients	Capsule EPAX 1050TG	four capsules (One capsule: 430 mg DHA and 150 mg EPA	No effect on neuropsychiatric symptoms, possible positive effects on depressive symptoms in non-ApoEω4 carriers and on agitation symptoms in ApoEω4 carriers	[59]
Patients with organic brain damage or mild cognitive impairment	Aravita capsules	40 mg/capsule of arachidonic acid and DHA	Significant improvements in memory	[60]
Spinocerebellar ataxia 38	Algal oil	600 mg/day	Improvement in clinical symptoms and no degradation of neurophysiological parameters	[61,62]

In an independent study of the OmegAD study, omega-3 fatty acid supplementation (capsules containing a total of 625 mg of DHA and 600 mg of EPA) increased plasma DHA and EPA concentrations in people with cognitive impairment no dementia, and AD. However, no beneficial effect on cognition and mood was observed in these populations. As emphasised by the authors, the sample used to conduct this study was quite small (76 participants), and the duration of the study was short (4 months). One hypothesis raised by the authors was to adapt the dose according to the pathology studied and the level of progression of the pathology [57]. These same observations were made in a study carried out by Paul S. Aisen's group, where DHA supplementation did not slow cognitive decline in patients with mild or moderate AD [58]. This study involved 402 patients and lasted for 18 months. DHA supplementation was performed with an algal-derived DHA (Martek Biosciences, Columbia, Maryland) in capsule form. Twice a day, patients took 1 g, for a total daily dose of 2 g, knowing that these capsules contained approximately 45% to 55% DHA by weight and did not contain eicosapentaenoic acid. The authors suggest that DHA may have an effect if the patients do not have overt dementia [58]. A third study shows that omega-3 supplementation (EPAX1050TG™ from Pro-nova Biocare A/S, Lysaker, Norway, four capsules) in patients with mild to moderate AD did not induce effects on neuropsychiatric symptoms, but had possible positive effects on depressive symptoms in non-ApoEω4 carriers and on agitation symptoms in ApoEω4 carriers [59]. In contrast to these three studies, a supplementation study using Aravita capsules (Suntory Ltd., Osaka, Japan), containing 40 mg/capsule of arachidonic acid (ARA) and DHA and 0.16 mg/capsule of astaxanthin (antioxidant of PUFA), showed significant improvements in the memory of patients with organic brain damage or mild cognitive impairment [60]. The authors hypothesize that these changes may be due to neural circuit remodeling (possible upregulation of synaptogenesis and/or neurogenesis with ARA), as well as improvement of membrane function and regional cerebral blood flow by DHA [60].

A few clinical trials have focused on spinocerebellar ataxias, which are very heterogeneous neurodegenerative diseases, clinically and genetically. The main characteristic of these pathologies is cerebellar syndrome, associated with walking and balance disorders. Spinocerebellar ataxia 38 (SCA38) is caused by a mutation in the elongation of the very long chain fatty acid protein 5 (ELOVL5) gene and is associated with reduced serum DHA levels. The team of Borroni et al. studied the effect of short-term (16–40 weeks) and long-term (2 years) DHA supplementation [61,62]. The DHA was derived from algal oil (Sofedus, Milan, Italy) and administered as sachets dosed at 600 mg/day. Improvement in clinical symptoms and no degradation of neurophysiological parameters were observed following fish oil-derived DHA intake.

The published results of clinical studies (mainly on AD) underline the interest in DHA, particularly through its ability to cross the blood–brain barrier and to influence inflammation. It is not known to what extent DHA crosses this barrier and whether microglia, the orchestral leader of inflammatory reactions in the brain, is affected by DHA. Many authors also mention the need to adapt the dose according to the severity of the disease or the weight of the patients. With DHA, the problem is not that it does not cross the blood–brain barrier, but that the quantity of DHA that crosses the barrier is not sufficient and does not only target the microglia. It is therefore important to look at the contribution of nanotechnologies in the targeting of therapeutic molecules of interest, such as DHA, to the microglia.

6. Brain Nanomedicine and Microglia

The DHA used in clinical studies comes from either oil, algae, or industry. In the OmegAD study, the DHA came from Pronova Biocare. This company uses a process to deodorize its EPAX triglyceride oils, inducing neither taste nor odor. They use enzymes to form the triglycerides. Their product can be in the form of chewable capsules or a liquid formula. In these different possibilities, DHA is not targeted to a cerebral distribution. DHA is delivered systemically, and authors have observed changes at the plasma, cerebrospinal fluid, and cognitive levels [49,52]. It might be interesting to look at targeting DHA to the brain by using nanotechnology-based drug delivery systems to counteract the low penetration efficiency of drugs to the central nervous system due to the presence of the BBB. Two possibilities exist to reach the brain: either a direct passage that will damage the BBB, or an indirect passage via nanomedicines targeting the brain either orally or by nasal delivery through a spray. The results presented in this section will focus on those obtained on microglia, whether from primary culture, lineages, or animal models, in the context of neurodegenerative diseases.

6.1. Nanoparticles

For this purpose, nanoparticles (size ranging from 1 to 100 nm) have been developed, which are 3D encapsulation systems that allow the transport of a molecule, and which can be functionalized by ligands, antibodies, or other molecules for targeting toward a target organ. These nanoparticles can be made from natural or synthetic polymers, metals, or from lipids.

6.1.1. Lipid-Based Nanoparticles

These lipid-based nanoparticles have the advantages of being non-toxic and having a high loading capacity for hydrophilic or non-hydrophilic molecules of interest, as well as a capacity to get through the BBB (Figure 3).

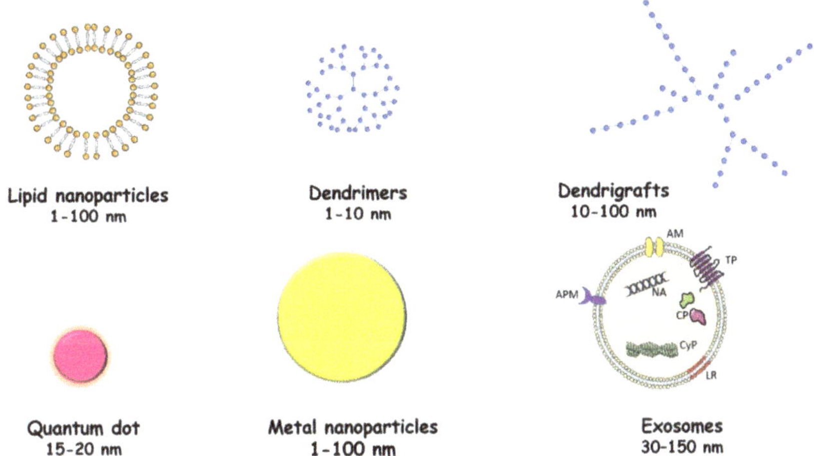

Figure 3. Representation of different delivery systems for therapeutic molecules. Different systems are used to deliver therapeutic molecules: lipidic nanoparticles, dendrimers and dendrigrafts, quantum dots and metallic nanoparticles, and exosomes. AM: adhesion molecules; TP: transmembrane proteins; APM: antigen presenting molecules; NA: nucleic acid; CP: cytosolic proteins; CyP: cytoskeletal proteins; LR: lipid rafts.

Nanoparticles have been formed from DHA and its hydroxylated derivative (DHAH); these are directly active without the need to load other molecules. These nanoparticles were tested on primary cultures of microglia obtained from rats. The viability of the cells is not affected by the use of these DHA/DHAH nanoparticles; however, an anti-inflammatory action of these nanoparticles is noted when the microglial cells are stimulated by LPS, with a decrease in the release of TNFα, IL-6, and IL-1β [63]. This work allowed us to show in vitro that these DHA/DHAH nanoparticles were not toxic and could reduce the release of proinflammatory cytokines by microglial cells [63]. This approach must still be validated in vivo in animals before testing in humans. This same team had already used lipid nanoparticles coated with chitosan, with the surface modified with a transactivator of transcription (TAT) peptide and loaded with GDNF (glial cell-derived neurotrophic factor)

(CS-NLC-TAT-GDNF) [64]. In a 1-methyl-4-phenyl-1,2,3,6-tetrahydropyridine (MPTP)-induced Parkinson's mouse model, intranasal administration of CS-NLC-TAT-GDNF led to modulation of microglial activation [64]. The use of lipid nanoparticles can induce toxicity and induce activation of brain microglia (involvement of the P2X7/caspase-1/IL-1β pathway). It has been shown that by modifying these lipid nanoparticles with PEGylation, this microglial activation could be reduced [65]. Nanoparticles can pass through the nasal epithelium and reach the brain by two different routes: (a) the extracellular route, the most common mechanism of delivery of therapeutics to the brain, via passive transport through the nasal epithelium, and (b) the intracellular route, involving endocytosis in the branches of the olfactory and trigeminal nerves followed by axonal transport in the brain [66].

Nanoparticles were synthesized from high-density lipoprotein (HDL) associated with apolipoprotein E [67,68]. These nanoparticles are able to enter cerebral vessels and accumulate around Aβ aggregates. First, ApoE-associated nanoparticles were shown to reduce Aβ deposition, attenuate microgliosis, improve neurological changes, and reduce memory deficits in an animal model of AD [67]. In a second step, α-Mangostin (α-M), a polyphenolic agent capable of inhibiting the formation of Aβ oligomers and fibrils and accelerating the cellular degradation of Aβ, was added to these nanoparticles [68]. These α-M-loaded nanoparticles are able to promote the uptake and degradation of Aβ1-42 by microglia more than unloaded nanoparticles [68]. This same system of reconstituted and modified HDL was used with monosialotetrahexosylganglioside (GM1), possessing high Aβ binding affinity [69]. This nanosystem promotes Aβ degradation by microglia following intranasal administration. This structure was used to load a neuroprotective peptide NAP, αNAP-GM1-rHDL, which was able to reduce Aβ deposition more efficiently than the nanostructure alone or α-NAP alone, in mouse models of AD after intranasal administration [69].

Oridonin, a natural diterpenoid compound isolated from the Chinese herb *Rabdosia rubescens*, was loaded into commercial lipid nanocarriers, Lipofundin® (MCT, 10% for infusion, B. Braun AG, Melsungen, Germany), and then given orally or injected into mice constituting an animal model of cerebral amyloidosis for AD, transgenic APP/PS1 mice. Regardless of the mode of injection, oridonin-loaded nanoparticles were able to attenuate microglia activation [66]. In an in vitro model of microglia (line N9) stimulated by LPS, these nanoparticles are able to inhibit the inflammatory response by reducing the NO concentration and decreasing mRNA expression of iNOS, IL-1β, and IL-6 [70].

Nanoparticles with a lipid core (capric/caprylic triglycerides) were loaded with indomethacin (a non-steroidal, anti-inflammatory drug) and their impacts on neuroinflammation were evaluated on organotypic rat hippocampal cultures after treatment with Aβ1-42 peptide, mimicking AD [71]. The use of these nanoparticles allows for the decrease of the TNF-α and the increase of IL-6 induced by Aβ1-42, but also the increase of the release of interleukin-10 [68]. The activation of microglia is reduced by the use of these nanoparticles [71].

The use of lipid nanoparticles seems to be a promising way to deliver molecules of interest to the brain by targeting microglia, while controlling their possible toxicity via surface modifications.

6.1.2. Metal Nanoparticles

Different types of metals can be used to create nanoparticles, such as titanium, gold, or silver (Figure 3).

Regarding titanium nanoparticles, in vitro experiments on BV-2 microglia cells showed that titanium dioxide nanotubes, which were functionalized with 3-aminopropyl triethoxysilane (APTES), allowing the addition of amine functions for drug molecule conjugation, do not induce toxicity or activation of these cells [72]. Titanium dioxide nanoparticles are able to move in the brain by decreasing the transendothelial electrical resistance and by disrupting the tight junctions between the endothelial cells of the brain capillaries [73]. Titanium dioxide (TiO_2) nanoparticles are able, without being functionalized by other molecules, to drive microglia toward the proinflammatory activation phenotype [74]. This change of phenotype would be specific to microglia, as astrocytes do not change their phenotype following the use of these nanoparticles [74].

Gold nanoparticles have also been developed. Some have been tested on primary cultures of microglia. They consist of 18 atoms of gold and are stabilized with glutathione ligands. The authors of this study showed that these Au18 gold nanoclusters (NCs) have, at low concentrations, anti-inflammatory signaling (reduction of (IL1-β) levels, unchanged levels of TNF-α or Ym1/2) but, at higher concentrations, they can have pro-inflammatory activity [75]. The authors suggested that the presence of glutathione could be the source of this anti-inflammatory activity. Gold nanoparticles (AuNCs) were functionalized with dihydrolipoic acid (DHLA-AuNCs), a neuroprotective antioxidant. The BV2 microglial line was used to evaluate their effects on microglial polarization. These nanoparticles induced a polarization toward the M2-like phenotype as well as a decrease of oxidative stress, a reduction of NF-kB signaling, and an increase of cell survival (increase of autophagy, inhibition, of apoptosis) [76]. Microglial changes also impacted neuronal cells by improving neurogenesis and reducing astrogliosis [76]. Another team used the root extract of *Paeonia moutan*, woody trees which are used in traditional Chinese medicine, to functionalize gold nanoparticles. Still, on a BV2 cell model, it was shown that these nanoparticles were able to decrease oxidative stress and inhibit the synthesis of pro-inflammatory cytokines following stimulation by LPS [73]. The effect of these nanoparticles on oxidative stress and inflammation was found in a model of parkinsonian mice, associated with an improvement in motor disorders [77]. An extract of *Ephedra sinica* Stapf was used to functionalize gold nanoparticles. These nanoparticles were able to decrease the production of pro-inflammatory mediators and cytokines (TNF-α, IL-1β, and IL-6) following primary microglia and BV-2 microglial cells induction by LPS (decrease of IκB kinase-α/β, NF-κB, Janus-activated kinase/signal transducers and activators of transcription, mitogen-activated protein kinase, and phospholipase D signaling pathways) [78]. Gold nanoparticles were synthesized with quercetin and then used on BV2 cells stimulated with LPS. The release of pro-inflammatory prostaglandin, prostaglandin E2, NO, upregulation of COX, inducible NO synthase mRNA, and protein levels were strongly inhibited by gold/quercetin nanoparticles [79]. The effects of these nanoparticles are superior to the use of quercetin alone. The use of these gold-quercetin nanoparticles could thus decrease the activation of microglia. A gold cluster with a positively charged tridecapeptide Sv (Au25Sv9) (peptide Sv: (H2N-CCYGGPKKKRKVG-COOH)) was produced, and it was shown that these particles were able to attenuate the cytotoxicity of stimulated microglia cells toward neuronal cells [80]. These nanoparticles inhibited IL-6, TNF-α, and NO secretions by suppressing the activation of NF-κB and p38 pathways. The action of these nanoparticles was also observed directly on neuronal cells, indicating that these nanoparticles could target microglia and neurons and could be an effective therapeutic approach [80].

Complex nanoparticles combining a metallic part and a polymer were produced in the following way: diblock polymer Man-PCB-PB was synthesized and then assembled to form nanoparticles to enclose the hydrophobic part fingolimod and zinc [81]. These nanoparticles were named Man-PCB-PB/ZnO/fingolimod NPs or MCPZF NPs. They were then linked to a signal transducer and activator of transcription 3 small interfering RNA (siSTAT3) to give Man-PCB-PB/ZnO/fingolimod/siSTAT3 NPs or MCPZFS NPs [81]. These nanoparticles promote phagocytosis of Aβ by microglia and decrease the release of pro-inflammatory cytokines [81].

Magnetic iron oxide (maghemite, γ-Fe2O3) nanoparticles (high surface-to-volume ratio, diameter 21 $\pm$ 3.5 nm, magnetic, biocompatible, relatively non-toxic, biodegradable) were used to deliver the following peptide: fibrin γ377–395 peptide [82]. The effect of these nanoparticles is different depending on the age and the state of AD. Indeed, in the early stages, the reduction of microglial cell activation following the action of these nanoparticles increases the number of neurons with hyperphosphorylated tau in transgenic mice [82]. Abnormal hyperphosphorylation of tau protein in sites that are not normally phosphorylated leads to the formation of neurofibrillary tangles (NFTs) in neuronal cell bodies and sometimes in glial cells. Hyperphosphorylation and NFT formation induce an inability for tau protein to bind to microtubules, resulting in alterations in axonal trafficking leading to changes in neuronal function and viability. These processes participate in synaptic dysfunction and neurodegeneration. On the other hand, in older mice, the reduction of microglial cell activation reduces the severity of tau pathology [82]. The number of neurons with hyperphosphorylated tau and the number of neurons with tangles are reduced in animals receiving the γ377–395 fibrin peptide-nanoparticle conjugate compared with control animals [82].

6.1.3. Polymer Nanoparticles and Mesoporous Silica Nanoparticles

Nanoparticles can be synthesized from synthetic or natural polymers. One team has developed poly(carboxybetaine) (PCB)-based zwitterionic nanoparticles (MCPZFS NP). These nanoparticles decrease microglia priming by lowering the levels of pro-inflammatory mediators and contributing to brain-derived neurotrophic factor (BDNF) secretion [81]. They also enhance Aβ recruitment to microglia, contributing to improved Aβ phagocytosis [81]. Beyond the action on microglia, these nanoparticles can also play on Aβ loading, neuronal damage, memory deficits, and neuroinflammation in APPswe/PS1dE9 mice [81]. Poly(lactic-co-glycolic acid) (PLGA), polyethylene glycol (PEG), and lipid chains as building blocks were used to synthesize 200 nm spherical polymeric nanoconstructs (SPNs) and 1000 nm discoidal polymeric nanoconstructs (DPNs). SPNs are more rapidly absorbed than DPNs and were used to encapsulate curcumin in the PLGA core. These curcumin-loaded nanoparticles decrease the production of proinflammatory cytokines-IL-1β, IL-6, and TNF-α in amyloid-β fiber-stimulated macrophages [83]. The α-M, previously used in a lipid nanoparticle, was encapsulated in the core of poly(ethylene glycol)-poly(l-lactide) (PEG-PLA) nanoparticles [NP(α-M)] [84]. This nanoformulation reduced Aβ deposition in AD and attenuated neuroinflammatory responses by microglia (using the BV-2 line) [84]. Apart from these effects on microglia, nanoencapsulation has improved the biodistribution and clearance of these molecules [84]. PLGA nanoparticles were used to deliver SurR9-C84A, a survivin mutant belonging to the inhibitors of the apoptosis protein family [85]. For this study, neuron monocultures and co-cultures of neurons and THP-1 (monocytes/macrophages) were used. These cultures were treated with LPS or β-amyloid to mimic the pathological inflammatory conditions of AD. Following this stimulation by LPS or β-amyloid, a decrease in THP-1 secretions was observed by the use of these nanoparticles [85]. This inhibition of secretions decreased the neuronal cell death induced by them.

Amphiphilic sugar-based molecules (AM) derived from mucic acid were synthesized to exhibit high affinity to scavenger receptors, allowing internalization of α-synuclein at the microglia [86]. Internalization of monomeric α-synuclein and formation of intracellular α-synuclein oligomers were decreased in microglial cells treated with these amphiphilic molecules. Following this observation, the antioxidant poly(ferric acid) was added to the core of these amphiphilic molecules by performing nanoprecipitation [86]. Microglial cells treated with these nanoparticles and stimulated by α-synuclein saw a decrease in their activation as well as in the neurotoxicity induced by α-synuclein aggregated at the level of microglia [86]. In vivo, the activation of microglia is also decreased after injection of these nanoparticles in the substantia nigra of mice stimulated by fibrillar α-synuclein. Targeting the receptors responsible for α-synuclein entry into microglia and adding an antioxidant may represent an interesting therapeutic approach via nanotechnology [86].

Polylactic acid (PLA)-coated mesoporous silica nanoparticles were loaded with resveratrol, which exhibits antioxidant activities among others. The PLA coating protects the resveratrol and prevents its systematic release. In the presence of oxidative stress, PLA is degraded and resveratrol can be released. Following this release, resveratrol was able to effectively reduce the activation of microglia cells stimulated by phorbol-myristate-acetate or lipopolysaccharide [87].

6.1.4. Cell-Derived Nanoparticles

Extracellular vesicles are membrane-containing vesicles from the endocytic pathway or plasma membrane, released into the extracellular space by virtually all cells. Three types of extracellular vesicles exist: (a) exosomes, the smallest vesicles (30–150 nm), derived from the inward budding of multivesicular bodies; (ii) microvesicles, or ectosomes (50 nm–1 μm), result from outward budding of the plasma membrane, released under physiological conditions or in response to specific stimuli; and (c) apoptotic bodies (50 nm–5 μm), which are produced by cells undergoing apoptosis.

Exosomes contain cellular proteins, lipids, nucleic acids, mRNAs, and microRNAs (miRNAs) from host cells (Figure 3). Exosomes can be internalized by cells and functionally modify the cells that internalize them. Exosomes can be derived from activated or non-activated cells or loaded with therapeutic molecules to be used as cargo.

The most commonly used cells for the use of exosomes as therapeutic cargo are mesenchymal stromal cells (MSCs). Exosomes derived from MSCs or MSCs preconditioned by hypoxia (increased miR-21 expression) were systemically administered to APP/PS1 transgenic mice mimicking AD [88]. In both cases, the use of MSC-derived exosomes or hypoxia-preconditioned MSCs had positive effects. Administration of exosomes from hypoxia-preconditioned MSCs improved memory and learning abilities; decreased plaque deposition and Aβ levels; increased expression of growth-associated protein 43, synapsin-1, and IL-10; and decreased levels of GFAP, Iba-1, TNF-α, IL-1β, and activation of STAT3 and NF-κB [85]. The use of these exosomes would correct synaptic dysfunction as well as inflammatory responses, which would lead to improvement of cognitive decline observed in AD via miR-21 signaling [88]. Human umbilical cord mesenchymal stem cells (hucMSC-exosomes) were injected into mouse models of AD and were found to improve cognitive decline and decrease Aβ deposition [89]. These hucMSC-exosomes also modulated microglial activation with a decrease in the number of activated microglia and a shift toward an M2 anti-inflammatory profile (increase in IL-10 and TGF-β cytokines) with an increase in Aβ-degrading enzymes [89]. Brain perfusion of neuroblastoma-derived exosomes may mediate Aβ clearance in an AD mouse model. Indeed, Aβ peptides can be taken up and transported by exosomes for presentation to microglia, resulting in their degradation [90,91].

Exosomes can also be loaded with a therapeutic molecule. The exosomes can be passively incubated with the therapeutic molecule, followed by purification. Curcumin was incubated at 22 °C for 5 min with exosomes, and the mixture was then effectively administered to the brain by intranasal route [92,93]. This decreased inflammation via microglia targeting [92,93]. Other exosome loading strategies exist, such as electroporation, incubation at room temperature, permeabilization with saponin, freeze/thaw cycles, sonication, and extrusion, and have been tested in neurodegenerative diseases but without microglia targeting or in other pathologies like cancer.

Exosomes are able to modulate the inflammatory response and phagocytosis activity of microglia and can be considered as a very interesting possibility in therapy.

6.1.5. Antioxidant Nanoparticles

Nanoparticles could also be synthesized directly from an antioxidant like quercetin. Authors have obtained quercetin nanoparticles with a very heterogeneous size ranging from 520 to 750 nm. In a model of AD, these quercetin nanoparticles were able to reduce neuronal damage, decrease the formation of amyloid plaques and neurofibrillary tangles, and modulate the activity of microglia [94].

In the context of PD, a nanoparticle (NP) formulation containing two polyphenol antioxidants, tannic acid (TA) and a ferulic acid diacid molecule, was proposed. These antioxidant nanoparticles inhibited α-synuclein fibrillation and lowered intracellular α-synuclein oligomerization in the BV-2 microglial line subjected to a high concentration of extracellular α-synuclein, thereby ameliorating microglial oxidative stress [95]. Microglial activation is also reduced with a modulation of the production of the pro-inflammatory cytokines TNF-α and IL-6 [95].

6.2. Dendrosomal Nanoparticles

An Iranian team used the dendrosome, a neutral, amphipathic, and biodegradable nanomaterial, to transport a molecule of interest such as curcumin. In a cuprizone-induced model of MS, these curcumin-loaded dendrosomes suppressed the accumulation of microglia and astrocytes, highlighting their possible use in therapy [96].

6.3. Dendrimers and Other Dendritic Polymers

Dendritic polymers belong to the synthetic polymers with linear, cross-linked, and branched polymers. Among these are the dendrimers which are obtained after generational synthesis (Figure 3), resulting in the formation of theoretically monodisperse structures with a narrow molecular weight distribution. In contrast to dendrimers, hyperbranched polymers are polydisperse, with a broad molecular weight distribution. The third members are dendrigrafts and have a configuration that shares commonalities between dendrimers and hyperbranched polymers (Figure 3).

Dendrimers, which are synthetic molecules with a tree-like structure, can be constructed by different methods: the hyperbranched structure can be built from the core, layer by layer, or by attaching dendrons to a central core [97,98]. Among the best-known dendrimers are poly(amidoamine) (PAMAM), poly(propylene imine) (PPI), phosphorus, and dimethylolpropionic acid-based dendrimers. One team has developed a dendrimer that targets the mitochondria (using triphenyl-phosphonium (TPP)) and delivers the antioxidant N-acetylcysteine (NAC):mitochondrial targeting hydroxyl PAMAM dendrimer-drug construct (TPP-D-NAC) [99]. The authors had already shown in previous studies that this PAMAM polymer was able to cross the BBB by selectively targeting activated microglia/macrophages and was able to deliver NAC to these cells (dendrimer D-NAC) [100]. These TPP-D-NAC dendrimers show preferential targeting to mitochondria (but not only) of activated microglial/macrophage cells. They showed superior efficacy in terms of oxidative stress inhibition to dendrimers developed from NAC but without organelle targeting (D-NAC) and to NAC used alone [99]. Concerning the dendrimers, the functionalization of the surface is an important factor to take into account, as it can play on the toxicity of the molecule [101]. Peroxisome proliferator-activated receptor (PPAR) α and PPARγ agonists can switch microglia from an M1-like to an M2-like phenotype. Because hydroxyl-terminated polyamidoamine dendrimers cross the altered BBB at the site of neuroinflammation and accumulate in activated microglia, they are conjugated with a PPARα/γ dual agonist [102,103]. The dual agonist dendrimer-PPARα/γ conjugate (D-tesaglitazar) induced the following: (a) an "M1 to M2" phenotype change, (b) a decrease in reactive oxygen species secretion, (c) an increase in the expression of phagocytosis and enzymatic degradation genes of pathogenic proteins, and (d) an increase in phagocytosis of β-amyloid [104].

Dendrigrafts have also been used to deliver treatment to the brain. Caspase-3 is involved in cell death and inhibiting it could help prevent the progression of neurodegenerative diseases and, in this study, PD. For this purpose, RNA interference was used as well as a vector constituted by dendrigraft poly-L-lysines, on which a peptide glycoprotein of the rabies virus with 29 amino acids was bound, allowing it to cross the BBB by transcytosis mediated by a specific receptor [105]. Plasmid DNA encoding the short hairpin RNA of caspase-3 was complexed with this vector to give nanoparticles. Injection by weekly intravenous administration of the nanoparticles reduced the levels of activated caspase-3, which decreased dopaminergic neuronal loss in the brains of rats with PD [105]. In addition, the rat model was obtained by treating them with rotenone, which increases TNF-α and NO levels in the brain. The use of these nanoparticles reduced the levels of TNF-α and NO in the brain [105]. These nanoparticles would thus have an anti-inflammatory effect on the microglia according to this study [105].

6.4. Quantum Dots

Quantum dots are fluorophores constituted by clusters of atoms at the nanometric scale containing a few hundred to a few thousand atoms of a semiconductor material wrapped by an additional semiconductor layer to improve the optical properties of the material (Figure 3). Quantum dots have been successfully used in immunocyto- and histochemistry, in flow cytometry, and in confocal microscopy [106,107].

Amino(polyethylene glycol)-2000 molybdenum disulfide quantum dots conjugated to (3-carboxypropyl)triphenyl-phosphonium bromide (TPP-MoS2 QDs) have been fabricated and tested in mouse models of AD [108]. These TPP-MoS2 QDs have the ability to cross the BBB and target mitochondria with TPP. These TPP-MoS2 QDs are able to induce a change in the phenotype of the microglia from pro-inflammatory M1 to anti-inflammatory M2 [108]. This contributes to the removal of Aβ aggregates [108].

7. Conclusions

DHA has shown cytoprotective effects on microglial cells but also on microglia activation in vitro and in vivo in animal models. In clinical studies, positive effects on the development of neurodegenerative diseases have been observed, but some conclusions are divergent and require further investigation, since DHA has a systemic and not a targeted release at the brain level. This may lead to the use of nanoparticles that could allow a release at the brain level of DHA and improve its effectiveness at the level of the microglia, thus leading to better management of neurodegenerative diseases associated with neuroinflammation.

Author Contributions: Conceptualization, A.V.; Supervision, A.V.; writing and editing, K.C., G.L., and A.V.; proofreading, I.G. All authors have read and agreed to the published version of the manuscript.

Funding: This research received no external funding.

Institutional Review Board Statement: Not applicable.

Informed Consent Statement: Not applicable.

Data Availability Statement: Not applicable.

Acknowledgments: This work was supported by the University of Bourgogne (Dijon, France).

Conflicts of Interest: The authors declare no conflict of interest.

References

1. Calcia, M.A.; Bonsall, D.R.; Bloomfield, P.S.; Selvaraj, S.; Barichello, T.; Howes, O.D. Stress and Neuroinflammation: A Systematic Review of the Effects of Stress on Microglia and the Implications for Mental Illness. *Psychopharmacology* **2016**, *233*, 1637–1650. [CrossRef]
2. Kwon, H.S.; Koh, S.-H. Neuroinflammation in Neurodegenerative Disorders: The Roles of Microglia and Astrocytes. *Transl. Neurodegener.* **2020**, *9*, 42. [CrossRef]
3. Glass, C.K.; Saijo, K.; Winner, B.; Marchetto, M.C.; Gage, F.H. Mechanisms Underlying Inflammation in Neurodegeneration. *Cell* **2010**, *140*, 918–934. [CrossRef]
4. Tang, Y.; Le, W. Differential Roles of M1 and M2 Microglia in Neurodegenerative Diseases. *Mol. Neurobiol.* **2016**, *53*, 1181–1194. [CrossRef]
5. Bachiller, S.; Jiménez-Ferrer, I.; Paulus, A.; Yang, Y.; Swanberg, M.; Deierborg, T.; Boza-Serrano, A. Microglia in Neurological Diseases: A Road Map to Brain-Disease Dependent-Inflammatory Response. *Front. Cell. Neurosci.* **2018**, *12*, 488. [CrossRef]
6. Hickman, S.; Izzy, S.; Sen, P.; Morsett, L.; Khoury, J.E. Microglia in Neurodegeneration. *Nat. Neurosci.* **2018**, *21*, 1359–1369. [CrossRef] [PubMed]
7. Franco, R.; Lillo, A.; Rivas-Santisteban, R.; Reyes-Resina, I.; Navarro, G. Microglial Adenosine Receptors: From Preconditioning to Modulating the M1/M2 Balance in Activated Cells. *Cells* **2021**, *10*, 1124. [CrossRef] [PubMed]
8. Layé, S.; Nadjar, A.; Joffre, C.; Bazinet, R.P. Anti-Inflammatory Effects of Omega-3 Fatty Acids in the Brain: Physiological Mechanisms and Relevance to Pharmacology. *Pharmacol. Rev.* **2018**, *70*, 12–38. [CrossRef] [PubMed]
9. Guesnet, P.; Alessandri, J.-M. Docosahexaenoic Acid (DHA) and the Developing Central Nervous System (CNS)—Implications for Dietary Recommendations. *Biochimie* **2011**, *93*, 7–12. [CrossRef]

10. Champeil-Potokar, G.; Denis, I.; Goustard-Langelier, B.; Alessandri, J.-M.; Guesnet, P.; Lavialle, M. Astrocytes in Culture Require Docosahexaenoic Acid to Restore the N-3/n-6 Polyunsaturated Fatty Acid Balance in Their Membrane Phospholipids. *J. Neurosci. Res.* **2004**, *75*, 96–106. [CrossRef] [PubMed]
11. Alessandri, J.-M.; Guesnet, P.; Vancassel, S.; Astorg, P.; Denis, I.; Langelier, B.; Aïd, S.; Poumès-Ballihaut, C.; Champeil-Potokar, G.; Lavialle, M. Polyunsaturated Fatty Acids in the Central Nervous System: Evolution of Concepts and Nutritional Implications throughout Life. *Reprod. Nutr. Dev.* **2004**, *44*, 509–538. [CrossRef]
12. Alessandri, J.-M.; Extier, A.; Astorg, P.; Lavialle, M.; Simon, N.; Guesnet, P. Métabolisme des acides gras oméga-3: Différences entre hommes et femmes. *Nutr. Clin. Métabolisme* **2009**, *23*, 55–66. [CrossRef]
13. Ferdinandusse, S.; Denis, S.; Mooijer, P.A.; Zhang, Z.; Reddy, J.K.; Spector, A.A.; Wanders, R.J. Identification of the Peroxisomal Beta-Oxidation Enzymes Involved in the Biosynthesis of Docosahexaenoic Acid. *J. Lipid Res.* **2001**, *42*, 1987–1995. [CrossRef]
14. Nury, T.; Zarrouk, A.; Yammine, A.; Mackrill, J.J.; Vejux, A.; Lizard, G. Oxiapoptophagy: A Type of Cell Death Induced by Some Oxysterols. *Br. J. Pharmacol.* **2021**, *178*, 3115–3123. [CrossRef] [PubMed]
15. Debbabi, M.; Zarrouk, A.; Bezine, M.; Meddeb, W.; Nury, T.; Badreddine, A.; Karym, E.M.; Sghaier, R.; Bretillon, L.; Guyot, S.; et al. Comparison of the Effects of Major Fatty Acids Present in the Mediterranean Diet (Oleic Acid, Docosahexaenoic Acid) and in Hydrogenated Oils (Elaidic Acid) on 7-Ketocholesterol-Induced Oxiapoptophagy in Microglial BV-2 Cells. *Chem. Phys. Lipids* **2017**, *207*, 151–170. [CrossRef] [PubMed]
16. Herr, D.R.; Yam, T.Y.A.; Tan, W.S.D.; Koh, S.S.; Wong, W.S.F.; Ong, W.-Y.; Chayaburakul, K. Ultrastructural Characteristics of DHA-Induced Pyroptosis. *Neuromol. Med.* **2020**, *22*, 293–303. [CrossRef]
17. Srikanth, M.; Chandrasaharan, K.; Zhao, X.; Chayaburakul, K.; Ong, W.-Y.; Herr, D.R. Metabolism of Docosahexaenoic Acid (DHA) Induces Pyroptosis in BV-2 Microglial Cells. *Neuromol. Med.* **2018**, *20*, 504–514. [CrossRef]
18. Geng, X.; Yang, B.; Li, R.; Teng, T.; Ladu, M.J.; Sun, G.Y.; Greenlief, C.M.; Lee, J.C. Effects of Docosahexaenoic Acid and Its Peroxidation Product on Amyloid-β Peptide-Stimulated Microglia. *Mol. Neurobiol.* **2020**, *57*, 1085–1098. [CrossRef] [PubMed]
19. Yang, B.; Li, R.; Michael Greenlief, C.; Fritsche, K.L.; Gu, Z.; Cui, J.; Lee, J.C.; Beversdorf, D.Q.; Sun, G.Y. Unveiling Anti-Oxidative and Anti-Inflammatory Effects of Docosahexaenoic Acid and Its Lipid Peroxidation Product on Lipopolysaccharide-Stimulated BV-2 Microglial Cells. *J. Neuroinflamm.* **2018**, *15*, 202. [CrossRef]
20. Sun, G.Y.; Li, R.; Yang, B.; Fritsche, K.L.; Beversdorf, D.Q.; Lubahn, D.B.; Geng, X.; Lee, J.C.; Greenlief, C.M. Quercetin Potentiates Docosahexaenoic Acid to Suppress Lipopolysaccharide-Induced Oxidative/Inflammatory Responses, Alter Lipid Peroxidation Products, and Enhance the Adaptive Stress Pathways in BV-2 Microglial Cells. *Int. J. Mol. Sci.* **2019**, *20*, 932. [CrossRef]
21. Lu, D.-Y.; Tsao, Y.-Y.; Leung, Y.-M.; Su, K.-P. Docosahexaenoic Acid Suppresses Neuroinflammatory Responses and Induces Heme Oxygenase-1 Expression in BV-2 Microglia: Implications of Antidepressant Effects for ω-3 Fatty Acids. *Neuropsychopharmacol. Off. Publ. Am. Coll. Neuropsychopharmacol.* **2010**, *35*, 2238–2248. [CrossRef]
22. Inoue, T.; Tanaka, M.; Masuda, S.; Ohue-Kitano, R.; Yamakage, H.; Muranaka, K.; Wada, H.; Kusakabe, T.; Shimatsu, A.; Hasegawa, K.; et al. Omega-3 Polyunsaturated Fatty Acids Suppress the Inflammatory Responses of Lipopolysaccharide-Stimulated Mouse Microglia by Activating SIRT1 Pathways. *Biochim. Biophys. Acta Mol. Cell Biol. Lipids* **2017**, *1862*, 552–560. [CrossRef]
23. Fourrier, C.; Remus-Borel, J.; Greenhalgh, A.D.; Guichardant, M.; Bernoud-Hubac, N.; Lagarde, M.; Joffre, C.; Layé, S. Docosahexaenoic Acid-Containing Choline Phospholipid Modulates LPS-Induced Neuroinflammation in Vivo and in Microglia in Vitro. *J. Neuroinflamm.* **2017**, *14*, 170. [CrossRef]
24. Chen, S.; Zhang, H.; Pu, H.; Wang, G.; Li, W.; Leak, R.K.; Chen, J.; Liou, A.K.; Hu, X. N-3 PUFA Supplementation Benefits Microglial Responses to Myelin Pathology. *Sci. Rep.* **2014**, *4*, 7458. [CrossRef] [PubMed]
25. Hjorth, E.; Zhu, M.; Toro, V.C.; Vedin, I.; Palmblad, J.; Cederholm, T.; Freund-Levi, Y.; Faxen-Irving, G.; Wahlund, L.-O.; Basun, H.; et al. Omega-3 Fatty Acids Enhance Phagocytosis of Alzheimer's Disease-Related Amyloid-β 42 by Human Microglia and Decrease Inflammatory Markers. *J. Alzheimers Dis.* **2013**, *35*, 697–713. [CrossRef] [PubMed]
26. Chang, C.-Y.; Kuan, Y.-H.; Li, J.-R.; Chen, W.-Y.; Ou, Y.-C.; Pan, H.-C.; Liao, S.-L.; Raung, S.-L.; Chang, C.-J.; Chen, C.-J. Docosahexaenoic Acid Reduces Cellular Inflammatory Response Following Permanent Focal Cerebral Ischemia in Rats. *J. Nutr. Biochem.* **2013**, *24*, 2127–2137. [CrossRef] [PubMed]
27. Pettit, L.K.; Varsanyi, C.; Tadros, J.; Vassiliou, E. Modulating the Inflammatory Properties of Activated Microglia with Docosahexaenoic Acid and Aspirin. *Lipids Health Dis.* **2013**, *12*, 16. [CrossRef] [PubMed]
28. Chang, C.-Y.; Wu, C.-C.; Wang, J.-D.; Li, J.-R.; Wang, Y.-Y.; Lin, S.-Y.; Chen, W.-Y.; Liao, S.-L.; Chen, C.-J. DHA Attenuated Japanese Encephalitis Virus Infection-Induced Neuroinflammation and Neuronal Cell Death in Cultured Rat Neuron/Glia. *Brain. Behav. Immun.* **2021**, *93*, 194–205. [CrossRef] [PubMed]
29. Mancera, P.; Wappenhans, B.; Cordobilla, B.; Virgili, N.; Pugliese, M.; Rueda, F.; Espinosa-Parrilla, J.F.; Domingo, J.C. Natural Docosahexaenoic Acid in the Triglyceride Form Attenuates In Vitro Microglial Activation and Ameliorates Autoimmune Encephalomyelitis in Mice. *Nutrients* **2017**, *9*, 681. [CrossRef] [PubMed]
30. Wang, Y.; Plastina, P.; Vincken, J.-P.; Jansen, R.; Balvers, M.; Ten Klooster, J.P.; Gruppen, H.; Witkamp, R.; Meijerink, J. N-Docosahexaenoyl Dopamine, an Endocannabinoid-like Conjugate of Dopamine and the n-3 Fatty Acid Docosahexaenoic Acid, Attenuates Lipopolysaccharide-Induced Activation of Microglia and Macrophages via COX-2. *ACS Chem. Neurosci.* **2017**, *8*, 548–557. [CrossRef]
31. Park, T.; Chen, H.; Kevala, K.; Lee, J.-W.; Kim, H.-Y. N-Docosahexaenoylethanolamine Ameliorates LPS-Induced Neuroinflammation via CAMP/PKA-Dependent Signaling. *J. Neuroinflamm.* **2016**, *13*, 284. [CrossRef]

32. Park, T.; Chen, H.; Kim, H.-Y. GPR110 (ADGRF1) Mediates Anti-Inflammatory Effects of N-Docosahexaenoylethanolamine. *J. Neuroinflamm.* **2019**, *16*, 225. [CrossRef] [PubMed]
33. Perry, V.H.; Teeling, J. Microglia and Macrophages of the Central Nervous System: The Contribution of Microglia Priming and Systemic Inflammation to Chronic Neurodegeneration. *Semin. Immunopathol.* **2013**, *35*, 601–612. [CrossRef] [PubMed]
34. Chataigner, M.; Martin, M.; Lucas, C.; Pallet, V.; Layé, S.; Mehaignerie, A.; Bouvret, E.; Dinel, A.-L.; Joffre, C. Fish Hydrolysate Supplementation Containing N-3 Long Chain Polyunsaturated Fatty Acids and Peptides Prevents LPS-Induced Neuroinflammation. *Nutrients* **2021**, *13*, 824. [CrossRef]
35. Caputo, M.P.; Radlowski, E.C.; Lawson, M.; Antonson, A.; Watson, J.E.; Matt, S.M.; Leyshon, B.J.; Das, A.; Johnson, R.W. Herring Roe Oil Supplementation Alters Microglial Cell Gene Expression and Reduces Peripheral Inflammation After Immune Activation in a Neonatal Piglet Model. *Brain. Behav. Immun.* **2019**, *81*, 455–469. [CrossRef] [PubMed]
36. Trépanier, M.-O.; Hopperton, K.E.; Giuliano, V.; Masoodi, M.; Bazinet, R.P. Increased Brain Docosahexaenoic Acid Has No Effect on the Resolution of Neuroinflammation Following Intracerebroventricular Lipopolysaccharide Injection. *Neurochem. Int.* **2018**, *118*, 115–126. [CrossRef]
37. Zhao, Y.; Calon, F.; Julien, C.; Winkler, J.W.; Petasis, N.A.; Lukiw, W.J.; Bazan, N.G. Docosahexaenoic Acid-Derived Neuroprotectin D1 Induces Neuronal Survival via Secretase- and PPARγ-Mediated Mechanisms in Alzheimer's Disease Models. *PLoS ONE* **2011**, *6*, e15816. [CrossRef]
38. Hopperton, K.E.; Trépanier, M.-O.; Giuliano, V.; Bazinet, R.P. Brain Omega-3 Polyunsaturated Fatty Acids Modulate Microglia Cell Number and Morphology in Response to Intracerebroventricular Amyloid-β 1-40 in Mice. *J. Neuroinflamm.* **2016**, *13*, 257. [CrossRef]
39. Abdelmeguid, N.E.; Khalil, M.I.M.; Elhabet, R.; Sultan, A.S.; Salam, S.A. Combination of Docosahexaenoic Acid and Ginko Biloba Extract Improves Cognitive Function and Hippocampal Tissue Damages in a Mouse Model of Alzheimer's Disease. *J. Chem. Neuroanat.* **2021**, *116*, 101995. [CrossRef]
40. Parrott, M.D.; Winocur, G.; Bazinet, R.P.; Ma, D.W.L.; Greenwood, C.E. Whole-Food Diet Worsened Cognitive Dysfunction in an Alzheimer's Disease Mouse Model. *Neurobiol. Aging* **2015**, *36*, 90–99. [CrossRef]
41. Sharman, M.J.; Gyengesi, E.; Liang, H.; Chatterjee, P.; Karl, T.; Li, Q.-X.; Wenk, M.R.; Halliwell, B.; Martins, R.N.; Münch, G. Assessment of Diets Containing Curcumin, Epigallocatechin-3-Gallate, Docosahexaenoic Acid and α-Lipoic Acid on Amyloid Load and Inflammation in a Male Transgenic Mouse Model of Alzheimer's Disease: Are Combinations More Effective? *Neurobiol. Dis.* **2019**, *124*, 505–519. [CrossRef]
42. Adkins, Y.; Soulika, A.M.; Mackey, B.; Kelley, D.S. Docosahexaenoic Acid (22:6n-3) Ameliorated the Onset and Severity of Experimental Autoimmune Encephalomyelitis in Mice. *Lipids* **2019**, *54*, 13–23. [CrossRef]
43. Cai, W.; Liu, S.; Hu, M.; Sun, X.; Qiu, W.; Zheng, S.; Hu, X.; Lu, Z. Post-Stroke DHA Treatment Protects Against Acute Ischemic Brain Injury by Skewing Macrophage Polarity Toward the M2 Phenotype. *Transl. Stroke Res.* **2018**, *9*, 669–680. [CrossRef]
44. Tian, Y.; Zhang, Y.; Zhang, R.; Qiao, S.; Fan, J. Resolvin D2 Recovers Neural Injury by Suppressing Inflammatory Mediators Expression in Lipopolysaccharide-Induced Parkinson's Disease Rat Model. *Biochem. Biophys. Res. Commun.* **2015**, *460*, 799–805. [CrossRef]
45. Tyrtyshnaia, A.A.; Egorova, E.L.; Starinets, A.A.; Ponomarenko, A.I.; Ermolenko, E.V.; Manzhulo, I.V. N-Docosahexaenoylethanolamine Attenuates Neuroinflammation and Improves Hippocampal Neurogenesis in Rats with Sciatic Nerve Chronic Constriction Injury. *Mar. Drugs* **2020**, *18*, 516. [CrossRef]
46. Zhou, Y.; Wang, J.; Li, X.; Li, K.; Chen, L.; Zhang, Z.; Peng, M. Neuroprotectin D1 Protects Against Postoperative Delirium-Like Behavior in Aged Mice. *Front. Aging Neurosci.* **2020**, *12*, 582674. [CrossRef]
47. Nolan, J.M.; Mulcahy, R.; Power, R.; Moran, R.; Howard, A.N. Nutritional Intervention to Prevent Alzheimer's Disease: Potential Benefits of Xanthophyll Carotenoids and Omega-3 Fatty Acids Combined. *J. Alzheimers Dis. JAD* **2018**, *64*, 367–378. [CrossRef]
48. Karimi, M.; Vedin, I.; Freund Levi, Y.; Basun, H.; Faxén Irving, G.; Eriksdotter, M.; Wahlund, L.-O.; Schultzberg, M.; Hjorth, E.; Cederholm, T.; et al. DHA-Rich n-3 Fatty Acid Supplementation Decreases DNA Methylation in Blood Leukocytes: The OmegAD Study. *Am. J. Clin. Nutr.* **2017**, *106*, 1157–1165. [CrossRef] [PubMed]
49. Eriksdotter, M.; Vedin, I.; Falahati, F.; Freund-Levi, Y.; Hjorth, E.; Faxen-Irving, G.; Wahlund, L.-O.; Schultzberg, M.; Basun, H.; Cederholm, T.; et al. Plasma Fatty Acid Profiles in Relation to Cognition and Gender in Alzheimer's Disease Patients During Oral Omega-3 Fatty Acid Supplementation: The OmegAD Study. *J. Alzheimers Dis. JAD* **2015**, *48*, 805–812. [CrossRef] [PubMed]
50. Wang, X.; Hjorth, E.; Vedin, I.; Eriksdotter, M.; Freund-Levi, Y.; Wahlund, L.-O.; Cederholm, T.; Palmblad, J.; Schultzberg, M. Effects of N-3 FA Supplementation on the Release of Proresolving Lipid Mediators by Blood Mononuclear Cells: The OmegAD Study. *J. Lipid Res.* **2015**, *56*, 674–681. [CrossRef] [PubMed]
51. Freund-Levi, Y.; Vedin, I.; Hjorth, E.; Basun, H.; Faxén Irving, G.; Schultzberg, M.; Eriksdotter, M.; Palmblad, J.; Vessby, B.; Wahlund, L.-O.; et al. Effects of Supplementation with Omega-3 Fatty Acids on Oxidative Stress and Inflammation in Patients with Alzheimer's Disease: The OmegAD Study. *J. Alzheimers Dis. JAD* **2014**, *42*, 823–831. [CrossRef]
52. Freund Levi, Y.; Vedin, I.; Cederholm, T.; Basun, H.; Faxén Irving, G.; Eriksdotter, M.; Hjorth, E.; Schultzberg, M.; Vessby, B.; Wahlund, L.-O.; et al. Transfer of Omega-3 Fatty Acids across the Blood-Brain Barrier after Dietary Supplementation with a Docosahexaenoic Acid-Rich Omega-3 Fatty Acid Preparation in Patients with Alzheimer's Disease: The OmegAD Study. *J. Intern. Med.* **2014**, *275*, 428–436. [CrossRef]

53. Faxén-Irving, G.; Freund-Levi, Y.; Eriksdotter-Jönhagen, M.; Basun, H.; Hjorth, E.; Palmblad, J.; Vedin, I.; Cederholm, T.; Wahlund, L.-O. Effects on Transthyretin in Plasma and Cerebrospinal Fluid by DHA-Rich n-3 Fatty Acid Supplementation in Patients with Alzheimer's Disease: The OmegAD Study. *J. Alzheimers Dis. JAD* **2013**, *36*, 1–6. [CrossRef] [PubMed]
54. Vedin, I.; Cederholm, T.; Freund-Levi, Y.; Basun, H.; Garlind, A.; Irving, G.F.; Eriksdotter-Jönhagen, M.; Wahlund, L.-O.; Dahlman, I.; Palmblad, J. Effects of DHA-Rich n-3 Fatty Acid Supplementation on Gene Expression in Blood Mononuclear Leukocytes: The OmegAD Study. *PLoS ONE* **2012**, *7*, e35425. [CrossRef]
55. Vedin, I.; Cederholm, T.; Freund-Levi, Y.; Basun, H.; Hjorth, E.; Irving, G.F.; Eriksdotter-Jönhagen, M.; Schultzberg, M.; Wahlund, L.-O.; Palmblad, J. Reduced Prostaglandin F2 Alpha Release from Blood Mononuclear Leukocytes after Oral Supplementation of Omega3 Fatty Acids: The OmegAD Study. *J. Lipid Res.* **2010**, *51*, 1179–1185. [CrossRef] [PubMed]
56. Vedin, I.; Cederholm, T.; Freund Levi, Y.; Basun, H.; Garlind, A.; Faxén Irving, G.; Jönhagen, M.E.; Vessby, B.; Wahlund, L.-O.; Palmblad, J. Effects of Docosahexaenoic Acid-Rich n-3 Fatty Acid Supplementation on Cytokine Release from Blood Mononuclear Leukocytes: The OmegAD Study. *Am. J. Clin. Nutr.* **2008**, *87*, 1616–1622. [CrossRef] [PubMed]
57. Phillips, M.A.; Childs, C.E.; Calder, P.C.; Rogers, P.J. No Effect of Omega-3 Fatty Acid Supplementation on Cognition and Mood in Individuals with Cognitive Impairment and Probable Alzheimer's Disease: A Randomised Controlled Trial. *Int. J. Mol. Sci.* **2015**, *16*, 24600–24613. [CrossRef] [PubMed]
58. Quinn, J.F.; Raman, R.; Thomas, R.G.; Yurko-Mauro, K.; Nelson, E.B.; Van Dyck, C.; Galvin, J.E.; Emond, J.; Jack, C.R.; Weiner, M.; et al. Docosahexaenoic Acid Supplementation and Cognitive Decline in Alzheimer Disease: A Randomized Trial. *JAMA* **2010**, *304*, 1903–1911. [CrossRef]
59. Freund-Levi, Y.; Basun, H.; Cederholm, T.; Faxén-Irving, G.; Garlind, A.; Grut, M.; Vedin, I.; Palmblad, J.; Wahlund, L.-O.; Eriksdotter-Jönhagen, M. Omega-3 Supplementation in Mild to Moderate Alzheimer's Disease: Effects on Neuropsychiatric Symptoms. *Int. J. Geriatr. Psychiatry* **2008**, *23*, 161–169. [CrossRef]
60. Kotani, S.; Sakaguchi, E.; Warashina, S.; Matsukawa, N.; Ishikura, Y.; Kiso, Y.; Sakakibara, M.; Yoshimoto, T.; Guo, J.; Yamashima, T. Dietary Supplementation of Arachidonic and Docosahexaenoic Acids Improves Cognitive Dysfunction. *Neurosci. Res.* **2006**, *56*, 159–164. [CrossRef]
61. Manes, M.; Alberici, A.; Di Gregorio, E.; Boccone, L.; Premi, E.; Mitro, N.; Pasolini, M.P.; Pani, C.; Paghera, B.; Perani, D.; et al. Docosahexaenoic Acid Is a Beneficial Replacement Treatment for Spinocerebellar Ataxia 38. *Ann. Neurol.* **2017**, *82*, 615–621. [CrossRef]
62. Manes, M.; Alberici, A.; Di Gregorio, E.; Boccone, L.; Premi, E.; Mitro, N.; Pasolini, M.P.; Pani, C.; Paghera, B.; Orsi, L.; et al. Long-Term Efficacy of Docosahexaenoic Acid (DHA) for Spinocerebellar Ataxia 38 (SCA38) Treatment: An Open Label Extension Study. *Parkinsonism Relat. Disord.* **2019**, *63*, 191–194. [CrossRef]
63. Hernando, S.; Herran, E.; Hernandez, R.M.; Igartua, M. Nanostructured Lipid Carriers Made of Ω-3 Polyunsaturated Fatty Acids: In Vitro Evaluation of Emerging Nanocarriers to Treat Neurodegenerative Diseases. *Pharmaceutics* **2020**, *12*, 928. [CrossRef]
64. Hernando, S.; Herran, E.; Figueiro-Silva, J.; Pedraz, J.L.; Igartua, M.; Carro, E.; Hernandez, R.M. Intranasal Administration of TAT-Conjugated Lipid Nanocarriers Loading GDNF for Parkinson's Disease. *Mol. Neurobiol.* **2018**, *55*, 145–155. [CrossRef]
65. Huang, J.; Lu, Y.; Wang, H.; Liu, J.; Liao, M.; Hong, L.; Tao, R.; Ahmed, M.M.; Liu, P.; Liu, S.; et al. The Effect of Lipid Nanoparticle PEGylation on Neuroinflammatory Response in Mouse Brain. *Biomaterials* **2013**, *34*, 7960–7970. [CrossRef]
66. Illum, L. Nasal Drug Delivery—Possibilities, Problems and Solutions. *J. Control. Release* **2003**, *87*, 187–198. [CrossRef]
67. Song, Q.; Huang, M.; Yao, L.; Wang, X.; Gu, X.; Chen, J.; Chen, J.; Huang, J.; Hu, Q.; Kang, T.; et al. Lipoprotein-Based Nanoparticles Rescue the Memory Loss of Mice with Alzheimer's Disease by Accelerating the Clearance of Amyloid-Beta. *ACS Nano* **2014**, *8*, 2345–2359. [CrossRef] [PubMed]
68. Song, Q.; Song, H.; Xu, J.; Huang, J.; Hu, M.; Gu, X.; Chen, J.; Zheng, G.; Chen, H.; Gao, X. Biomimetic ApoE-Reconstituted High Density Lipoprotein Nanocarrier for Blood-Brain Barrier Penetration and Amyloid Beta-Targeting Drug Delivery. *Mol. Pharm.* **2016**, *13*, 3976–3987. [CrossRef] [PubMed]
69. Huang, M.; Hu, M.; Song, Q.; Song, H.; Huang, J.; Gu, X.; Wang, X.; Chen, J.; Kang, T.; Feng, X.; et al. GM1-Modified Lipoprotein-like Nanoparticle: Multifunctional Nanoplatform for the Combination Therapy of Alzheimer's Disease. *ACS Nano* **2015**, *9*, 10801–10816. [CrossRef] [PubMed]
70. Zhang, Z.-Y.; Daniels, R.; Schluesener, H.J. Oridonin Ameliorates Neuropathological Changes and Behavioural Deficits in a Mouse Model of Cerebral Amyloidosis. *J. Cell. Mol. Med.* **2013**, *17*, 1566–1576. [CrossRef]
71. Bernardi, A.; Frozza, R.L.; Meneghetti, A.; Hoppe, J.B.; Battastini, A.M.O.; Pohlmann, A.R.; Guterres, S.S.; Salbego, C.G. Indomethacin-Loaded Lipid-Core Nanocapsules Reduce the Damage Triggered by Aβ1-42 in Alzheimer's Disease Models. *Int. J. Nanomed.* **2012**, *7*, 4927–4942. [CrossRef]
72. Sruthi, S.; Loiseau, A.; Boudon, J.; Sallem, F.; Maurizi, L.; Mohanan, P.V.; Lizard, G.; Millot, N. In Vitro Interaction and Biocompatibility of Titanate Nanotubes with Microglial Cells. *Toxicol. Appl. Pharmacol.* **2018**, *353*, 74–86. [CrossRef]
73. Sela, H.; Cohen, H.; Elia, P.; Zach, R.; Karpas, Z.; Zeiri, Y. Spontaneous Penetration of Gold Nanoparticles through the Blood Brain Barrier (BBB). *J. Nanobiotechnol.* **2015**, *13*, 71. [CrossRef]
74. De Astis, S.; Corradini, I.; Morini, R.; Rodighiero, S.; Tomasoni, R.; Lenardi, C.; Verderio, C.; Milani, P.; Matteoli, M. Nanostructured TiO2 Surfaces Promote Polarized Activation of Microglia, but Not Astrocytes, toward a Proinflammatory Profile. *Nanoscale* **2013**, *5*, 10963–10974. [CrossRef]

75. Sobska, J.; Waszkielewicz, M.; Podleśny-Drabiniok, A.; Olesiak-Banska, J.; Krężel, W.; Matczyszyn, K. Gold Nanoclusters Display Low Immunogenic Effect in Microglia Cells. *Nanomaterials* **2021**, *11*, 1066. [CrossRef]
76. Xiao, L.; Wei, F.; Zhou, Y.; Anderson, G.J.; Frazer, D.M.; Lim, Y.C.; Liu, T.; Xiao, Y. Dihydrolipoic Acid-Gold Nanoclusters Regulate Microglial Polarization and Have the Potential To Alter Neurogenesis. *Nano Lett.* **2020**, *20*, 478–495. [CrossRef] [PubMed]
77. Xue, J.; Liu, T.; Liu, Y.; Jiang, Y.; Seshadri, V.D.D.; Mohan, S.K.; Ling, L. Neuroprotective Effect of Biosynthesised Gold Nanoparticles Synthesised from Root Extract of Paeonia Moutan against Parkinson Disease – In Vitro & In Vivo Model. *J. Photochem. Photobiol. B* **2019**, *200*, 111635. [CrossRef] [PubMed]
78. Park, S.Y.; Yi, E.H.; Kim, Y.; Park, G. Anti-Neuroinflammatory Effects of Ephedra Sinica Stapf Extract-Capped Gold Nanoparticles in Microglia. *Int. J. Nanomed.* **2019**, *14*, 2861–2877. [CrossRef] [PubMed]
79. Ozdal, Z.D.; Sahmetlioglu, E.; Narin, I.; Cumaoglu, A. Synthesis of Gold and Silver Nanoparticles Using Flavonoid Quercetin and Their Effects on Lipopolysaccharide Induced Inflammatory Response in Microglial Cells. *3 Biotech* **2019**, *9*, 212. [CrossRef]
80. Yuan, Q.; Yao, Y.; Zhang, X.; Yuan, J.; Sun, B.; Gao, X. The Gold Nanocluster Protects Neurons Directly or via Inhibiting Cytotoxic Secretions of Microglia Cell. *J. Nanosci. Nanotechnol.* **2019**, *19*, 1986–1995. [CrossRef]
81. Liu, R.; Yang, J.; Liu, L.; Lu, Z.; Shi, Z.; Ji, W.; Shen, J.; Zhang, X. An "Amyloid-β Cleaner" for the Treatment of Alzheimer's Disease by Normalizing Microglial Dysfunction. *Adv. Sci. Weinh. Baden-Wurtt. Ger.* **2020**, *7*, 1901555. [CrossRef]
82. Glat, M.; Skaat, H.; Menkes-Caspi, N.; Margel, S.; Stern, E.A. Age-Dependent Effects of Microglial Inhibition in Vivo on Alzheimer's Disease Neuropathology Using Bioactive-Conjugated Iron Oxide Nanoparticles. *J. Nanobiotechnol.* **2013**, *11*, 32. [CrossRef]
83. Ameruoso, A.; Palomba, R.; Palange, A.L.; Cervadoro, A.; Lee, A.; Di Mascolo, D.; Decuzzi, P. Ameliorating Amyloid-β Fibrils Triggered Inflammation via Curcumin-Loaded Polymeric Nanoconstructs. *Front. Immunol.* **2017**, *8*, 1411. [CrossRef]
84. Yao, L.; Gu, X.; Song, Q.; Wang, X.; Huang, M.; Hu, M.; Hou, L.; Kang, T.; Chen, J.; Chen, H.; et al. Nanoformulated Alpha-Mangostin Ameliorates Alzheimer's Disease Neuropathology by Elevating LDLR Expression and Accelerating Amyloid-Beta Clearance. *J. Control. Release Off. J. Control. Release Soc.* **2016**, *226*, 1–14. [CrossRef] [PubMed]
85. Sriramoju, B.; Kanwar, R.K.; Kanwar, J.R. Nanoformulated Mutant SurR9-C84A: A Possible Key for Alzheimer's and Its Associated Inflammation. *Pharm. Res.* **2015**, *32*, 2787–2797. [CrossRef] [PubMed]
86. Bennett, N.K.; Chmielowski, R.; Abdelhamid, D.S.; Faig, J.J.; Francis, N.; Baum, J.; Pang, Z.P.; Uhrich, K.E.; Moghe, P.V. Polymer Brain-Nanotherapeutics for Multipronged Inhibition of Microglial α-Synuclein Aggregation, Activation, and Neurotoxicity. *Biomaterials* **2016**, *111*, 179–189. [CrossRef] [PubMed]
87. Shen, Y.; Cao, B.; Snyder, N.R.; Woeppel, K.M.; Eles, J.R.; Cui, X.T. ROS Responsive Resveratrol Delivery from LDLR Peptide Conjugated PLA-Coated Mesoporous Silica Nanoparticles across the Blood-Brain Barrier. *J. Nanobiotechnol.* **2018**, *16*, 13. [CrossRef] [PubMed]
88. Cui, G.-H.; Wu, J.; Mou, F.-F.; Xie, W.-H.; Wang, F.-B.; Wang, Q.-L.; Fang, J.; Xu, Y.-W.; Dong, Y.-R.; Liu, J.-R.; et al. Exosomes Derived from Hypoxia-Preconditioned Mesenchymal Stromal Cells Ameliorate Cognitive Decline by Rescuing Synaptic Dysfunction and Regulating Inflammatory Responses in APP/PS1 Mice. *FASEB J. Off. Publ. Fed. Am. Soc. Exp. Biol.* **2018**, *32*, 654–668. [CrossRef]
89. Ding, M.; Shen, Y.; Wang, P.; Xie, Z.; Xu, S.; Zhu, Z.; Wang, Y.; Lyu, Y.; Wang, D.; Xu, L.; et al. Exosomes Isolated From Human Umbilical Cord Mesenchymal Stem Cells Alleviate Neuroinflammation and Reduce Amyloid-Beta Deposition by Modulating Microglial Activation in Alzheimer's Disease. *Neurochem. Res.* **2018**, *43*, 2165–2177. [CrossRef] [PubMed]
90. Yuyama, K.; Sun, H.; Sakai, S.; Mitsutake, S.; Okada, M.; Tahara, H.; Furukawa, J.; Fujitani, N.; Shinohara, Y.; Igarashi, Y. Decreased Amyloid-β Pathologies by Intracerebral Loading of Glycosphingolipid-Enriched Exosomes in Alzheimer Model Mice. *J. Biol. Chem.* **2014**, *289*, 24488–24498. [CrossRef]
91. Yuyama, K.; Sun, H.; Usuki, S.; Sakai, S.; Hanamatsu, H.; Mioka, T.; Kimura, N.; Okada, M.; Tahara, H.; Furukawa, J.; et al. A Potential Function for Neuronal Exosomes: Sequestering Intracerebral Amyloid-β Peptide. *FEBS Lett.* **2015**, *589*, 84–88. [CrossRef]
92. Zhuang, X.; Xiang, X.; Grizzle, W.; Sun, D.; Zhang, S.; Axtell, R.C.; Ju, S.; Mu, J.; Zhang, L.; Steinman, L.; et al. Treatment of Brain Inflammatory Diseases by Delivering Exosome Encapsulated Anti-Inflammatory Drugs From the Nasal Region to the Brain. *Mol. Ther.* **2011**, *19*, 1769–1779. [CrossRef]
93. Sun, D.; Zhuang, X.; Xiang, X.; Liu, Y.; Zhang, S.; Liu, C.; Barnes, S.; Grizzle, W.; Miller, D.; Zhang, H.-G. A Novel Nanoparticle Drug Delivery System: The Anti-Inflammatory Activity of Curcumin Is Enhanced When Encapsulated in Exosomes. *Mol. Ther.* **2010**, *18*, 1606–1614. [CrossRef]
94. Rifaai, R.A.; Mokhemer, S.A.; Saber, E.A.; El-Aleem, S.A.A.; El-Tahawy, N.F.G. Neuroprotective Effect of Quercetin Nanoparticles: A Possible Prophylactic and Therapeutic Role in Alzheimer's Disease. *J. Chem. Neuroanat.* **2020**, *107*, 101795. [CrossRef] [PubMed]
95. Zhao, N.; Yang, X.; Calvelli, H.R.; Cao, Y.; Francis, N.L.; Chmielowski, R.A.; Joseph, L.B.; Pang, Z.P.; Uhrich, K.E.; Baum, J.; et al. Antioxidant Nanoparticles for Concerted Inhibition of α-Synuclein Fibrillization, and Attenuation of Microglial Intracellular Aggregation and Activation. *Front. Bioeng. Biotechnol.* **2020**, *8*, 112. [CrossRef] [PubMed]
96. Motavaf, M.; Sadeghizadeh, M.; Babashah, S.; Zare, L.; Javan, M. Protective Effects of a Nano-Formulation of Curcumin against Cuprizone-Induced Demyelination in the Mouse Corpus Callosum. *Iran. J. Pharm. Res. IJPR* **2020**, *19*, 310–320. [CrossRef] [PubMed]
97. Tomalia, D.A.; Baker, H.; Dewald, J.; Hall, M.; Kallos, G.; Martin, S.; Roeck, J.; Ryder, J.; Smith, P. A New Class of Polymers: Starburst-Dendritic Macromolecules. *Polym. J.* **1985**, *17*, 117–132. [CrossRef]

98. Grayson, S.M.; Fréchet, J.M. Convergent Dendrons and Dendrimers: From Synthesis to Applications. *Chem. Rev.* **2001**, *101*, 3819–3868. [CrossRef]
99. Sharma, A.; Liaw, K.; Sharma, R.; Zhang, Z.; Kannan, S.; Kannan, R.M. Targeting Mitochondrial Dysfunction and Oxidative Stress in Activated Microglia Using Dendrimer-Based Therapeutics. *Theranostics* **2018**, *8*, 5529–5547. [CrossRef] [PubMed]
100. Dai, H.; Navath, R.S.; Balakrishnan, B.; Guru, B.R.; Mishra, M.K.; Romero, R.; Kannan, R.M.; Kannan, S. Intrinsic Targeting of Inflammatory Cells in the Brain by Polyamidoamine Dendrimers upon Subarachnoid Administration. *Nanomedicine* **2010**, *5*, 1317–1329. [CrossRef]
101. Bertero, A.; Boni, A.; Gemmi, M.; Gagliardi, M.; Bifone, A.; Bardi, G. Surface Functionalisation Regulates Polyamidoamine Dendrimer Toxicity on Blood-Brain Barrier Cells and the Modulation of Key Inflammatory Receptors on Microglia. *Nanotoxicology* **2014**, *8*, 158–168. [CrossRef]
102. Kannan, S.; Dai, H.; Navath, R.S.; Balakrishnan, B.; Jyoti, A.; Janisse, J.; Romero, R.; Kannan, R.M. Dendrimer-Based Postnatal Therapy for Neuroinflammation and Cerebral Palsy in a Rabbit Model. *Sci. Transl. Med.* **2012**, *4*, 130ra46. [CrossRef]
103. Zhang, F.; Mastorakos, P.; Mishra, M.K.; Mangraviti, A.; Hwang, L.; Zhou, J.; Hanes, J.; Brem, H.; Olivi, A.; Tyler, B.; et al. Uniform Brain Tumor Distribution and Tumor Associated Macrophage Targeting of Systemically Administered Dendrimers. *Biomaterials* **2015**, *52*, 507–516. [CrossRef]
104. DeRidder, L.; Sharma, A.; Liaw, K.; Sharma, R.; John, J.; Kannan, S.; Kannan, R.M. Dendrimer-Tesaglitazar Conjugate Induces a Phenotype Shift of Microglia and Enhances β-Amyloid Phagocytosis. *Nanoscale* **2021**, *13*, 939–952. [CrossRef]
105. Liu, Y.; Guo, Y.; An, S.; Kuang, Y.; He, X.; Ma, H.; Li, J.; Lu, J.; Lv, J.; Zhang, N.; et al. Targeting Caspase-3 as Dual Therapeutic Benefits by RNAi Facilitating Brain-Targeted Nanoparticles in a Rat Model of Parkinson's Disease. *PLoS ONE* **2013**, *8*, e62905. [CrossRef]
106. Kahn, E.; Vejux, A.; Ménétrier, F.; Maiza, C.; Hammann, A.; Sequeira-Le Grand, A.; Frouin, F.; Tourneur, Y.; Brau, F.; Riedinger, J.-M.; et al. Analysis of CD36 Expression on Human Monocytic Cells and Atherosclerotic Tissue Sections with Quantum Dots: Investigation by Flow Cytometry and Spectral Imaging Microscopy. *Anal. Quant. Cytol. Histol.* **2006**, *28*, 14–26. [PubMed]
107. Kahn, E.; Lizard, G.; Monier, S.; Bessède, G.; Frouin, F.; Gambert, P.; Todd-Pokropek, A. Flow Cytometry and Factor Analysis Evaluation of Confocal Image Sequences of Morphologic and Functional Changes Occurring at the Mitochondrial Level during 7-Ketocholesterol-Induced Cell Death. *Anal. Quant. Cytol. Histol.* **2002**, *24*, 355–362. [PubMed]
108. Ren, C.; Li, D.; Zhou, Q.; Hu, X. Mitochondria-Targeted TPP-MoS2 with Dual Enzyme Activity Provides Efficient Neuroprotection through M1/M2 Microglial Polarization in an Alzheimer's Disease Model. *Biomaterials* **2020**, *232*, 119752. [CrossRef] [PubMed]

MDPI
St. Alban-Anlage 66
4052 Basel
Switzerland
Tel. +41 61 683 77 34
Fax +41 61 302 89 18
www.mdpi.com

International Journal of Molecular Sciences Editorial Office
E-mail: ijms@mdpi.com
www.mdpi.com/journal/ijms

www.ingramcontent.com/pod-product-compliance
Lightning Source LLC
LaVergne TN
LVHW071302170726
843515LV00006BA/1471

* 9 7 8 3 0 3 6 5 2 9 1 0 3 *